DISCRETE *q*-DISTRIBUTIONS

DISCRETE *q*-DISTRIBUTIONS

CHARALAMBOS A. CHARALAMBIDES
Department of Mathematics
University of Athens, Athens, Greece

Published by John Wiley & Sons, Inc., Hoboken, New Jersey

Published simultaneously in Canada

Library of Congress Cataloging-in-Publication Data:

Names: Charalambides, Ch. A., author.
Title: Discrete q-distributions / Charalambos A. Charalambides.
Description: Hoboken, New Jersey : John Wiley & Sons, 2016. | Includes bibliographical references and index.
Identifiers: LCCN 2015031840 (print) | LCCN 2015039486 (ebook) | ISBN 9781119119043 (cloth) | ISBN 9781119119050 (pdf) | ISBN 9781119119104 (epub)
Subjects: LCSH: Distribution (Probability theory) | Stochastic sequences. | Discrete geometry. | Combinatorial geometry.
Classification: LCC QA273.6 .C453 2016 (print) | LCC QA273.6 (ebook) | DDC 519.2/4–dc23
LC record available at http://lccn.loc.gov/2015031840

Set in 10/12pt TimesLTStd-Roman by SPi Global, Chennai, India

Printed in the United States of America

10 9 8 7 6 5 4 3 2 1

1 2016

To the memory of my parents
Angelos and Elpida

CONTENTS

PREFACE

The classical binomial and negative binomial (or Pascal) distributions are defined in the stochastic model of a sequence of independent and identically distributed Bernoulli trials. The Poisson distribution may be considered as a limiting case of the binomial distribution as the number of trials tends to infinity. Also, the logarithmic distribution may be considered as a limiting case of the zero-truncated negative binomial distribution as the number of successes tends to zero.

Poisson (1837) generalized the binomial distribution (and implicitly the negative binomial distribution) by assuming that the probability of success at a trial varies with the number of previous trials. The probability function of the number of successes up to a given number of trials was derived by Platonov (1976) in terms of the generalized signless Stirling numbers of the first kind. Balakrishnan and Nevzorov (1997) obtained this distribution as the distribution of the number of records up to a given time in a general record model.

The negative binomial distribution (and implicitly the binomial distribution) can be generalized to a different direction by assuming that the probability of success at a trial varies with the number of successes occurring in the previous trials. The probability function of the number of successes up to a given number of trials was derived by Woodbury (1949) essentially in terms of the generalized Stirling numbers of the second kind. Sen and Balakrishnan (1999) obtained the distribution of the number of trials up to a given number of successes in connection with a reliability model; their expression was also essentially in terms of the generalized Stirling numbers of the second kind.

It should be noticed that a stochastic model of a sequence of independent Bernoulli trials, in which the probability of success at a trial is assumed to vary with the number of trials and/or the number of successes, is advantageous in the sense that it permits incorporating the experience gained from previous trials and/or successes. If the probability of success at a trial is a very general function of the number of trials and/or the number successes, very little can be inferred from it about the distributions of the various random variables that may be defined in this model. The assumption that the probability of success (or failure) at a trial varies geometrically, with rate (proportion) q, leads to the introduction of discrete q-distributions. The study of these distributions is greatly facilitated by the wealth of existing q-sequences and q-functions, in q-combinatorics, and the theory of q-hypergeometric series.

This book is devoted to the study of discrete q-distributions. As to its contents, the following remarks and stressing may be useful. The mathematics prerequisites are modest. They are offered by a basic course in infinitesimal calculus. This should include some power series. The necessary basic q-combinatorics and q-hypergeometric series are included in an introductory chapter making the entire text self-contained.

In Chapter 1, after introducing the notions of a q-power, a q-factorial, and a q-binomial coefficient of a real number, two q-Vandermonde's (q-factorial convolution) formulae are derived. Furthermore, two q-Cauchy's (q-binomial convolution) formulae are presented as a corollary of the q-Vandermonde's formulae. Also, the q-binomial (q-Newton's binomial) and the negative q-binomial formulae are obtained. In addition, a general q-binomial formula is derived and, as limiting forms of it, q-exponential and q-logarithmic functions are deduced. The q-Stirling numbers of the first and second kind, which are the coefficients of the expansions of q-factorials into q-powers and of q-powers into q-factorials, respectively, are presented. Also, the generalized q-factorial coefficients are briefly discussed. Moreover, the q-factorial and q-binomial moments of a discrete q-distribution are briefly presented. In addition, two equivalent formulae connecting the usual factorial and binomial moments with the q-factorial and q-binomial moments, respectively, are deduced. Consequently, the q-factorial and q-binomial moments, apart from the interest in their own, can be used in the calculation of the usual factorial and binomial moments of a discrete q-distribution.

Chapter 2 deals with discrete q-distributions defined in the stochastic model of a sequence of independent Bernoulli trials, with success probability at a trial varying geometrically with the number of previous trials. Specifically, assuming that the odds of success at a trial is a geometrically varying (increasing or decreasing) sequence, a q-binomial distribution of the first kind and a negative q-binomial distribution of the first kind are introduced and studied. In addition, the Heine distribution, which is a q-Poisson distribution of the first kind, is obtained as a limiting distribution of the q-binomial distribution (or the negative q-binomial distribution) of the first kind, as the number of trials (or the number of successes) tends to infinity. Furthermore, considering a stochastic model that is developing in time or space, in which events (successes) may occur at continuous points, a Heine stochastic process, which is a q-analogue of a Poisson process, is presented. Also, assuming that the probability of success at a trial is a geometrically varying (increasing or decreasing) sequence, a q-Stirling distribution of the first kind is defined and discussed. Finally, supposing that the odds of failure at a trial is a geometrically increasing sequence, another q-Stirling distribution of the first kind is obtained.

Chapter 3 is devoted to the study of discrete q-distributions defined in the stochastic model of a sequence of independent Bernoulli trials with success probability varying geometrically with the number of previous successes. Introducing the notion of a geometric sequence of (Bernoulli) trials as a sequence of independent Bernoulli trials, with constant probability of success, which is terminated with the occurrence of the first success, an equivalent stochastic model is constructed as follows. A sequence of independent geometric sequences of trials with success probability at a geometric sequence of trials varying (increasing or decreasing) geometrically with the number of previous sequences (successes), is considered. In this model, a negative q-binomial distribution of the second kind and a q-binomial distribution of the second kind are introduced and examined. In addition, the Euler distribution, which is a q-Poisson distribution of the second kind, is obtained as a limiting distribution of the q-binomial distribution (or the negative q-binomial distribution) of the second kind, as the number of trials (or the number of successes) tends to infinity. Furthermore, considering a stochastic model that is developing in time or space, in which events (successes) may occur at continuous points, an Euler stochastic process, which is a q-analogue of a Poisson process, is presented. Also, the q-logarithmic distribution is deduced as an approximation of a zero-truncated negative q-binomial distribution of the second kind, as the number of successes increases indefinitely. Finally, assuming that the probability of success at a geometric sequence of trials is a geometrically varying (increasing or decreasing) sequence, a q-Stirling distribution of the second kind is introduced and discussed.

In Chapter 4, a stochastic model of a sequence of independent Bernoulli trials, with success probability varying geometrically both with the number of trials and the number of successes in a specific manner, is considered. In the first part of this chapter, after introducing a q-Pólya urn model, a q-Pólya distribution is defined and studied. An approximation of the q-Pólya distribution by a q-binomial distribution of the second kind is obtained. As a particular case, a q-hypergeometric distribution is presented. Also, an inverse q-Pólya distribution is introduced and discussed. An approximation of the inverse q-Pólya distribution by a negative q-binomial distribution of the second kind is obtained. As a particular case, an inverse q-hypergeometric distribution is examined. The second part of this chapter is concerned with the particular case in which the probability of success at a trial is a product of a function of the number of previous trials only and another function of the number previous successes and varies geometrically. In this stochastic model, generalized q-factorial coefficient distributions are defined and discussed.

In Chapter 5, after introducing the mode of stochastic convergence and the mode of convergence in distribution, the Chebyshev's law of large numbers is presented. In the particular case of a sequence of independent Bernoulli trials, with the probability of success varying with the number of trials, the Poisson's law of large numbers is concluded. In the other particular case of a sequence of independent geometric sequences of (Bernoulli) trials, with the probability of success varying with the number of geometric sequences of trials, another particular case of Chebyshev's law of large numbers is deduced. The central limit theorems for independent and not necessarily identically distributed random variables are presented next. Specifically, the Lyapunov and the Lindeberg–Feller central limit theorems are given and their use in investigating the limiting q-distributions are discussed. This chapter is concluded with some local limit theorems, which examine the converges of the probability (mass) functions of particular discrete q-deformed distributions to the density function of a Stieltjes–Wigert distribution.

A distinctive feature of the presentation of the material covered in this book is the comments (remarks) following most of the definitions and theorems. In these remarks, the particular concept or result presented is discussed, and extensions or generalizations of it are pointed out. Furthermore, several worked out examples illustrating the applications of the presented theory are included. In concluding each chapter, brief bibliographic notes, mainly of historical interest, are included.

At the end of each chapter, a collection of exercises is provided. Most of these exercises, which are of varying difficulty, aim to consolidate the concepts and results presented, while others complement, extend, or generalize some of the results. So, working these exercises must be considered an integral part of this text. Hints and answers to the exercises are included at the end of the book. Before trying to solve an exercise, the less experienced reader may first look up the hint to its solution.

The material of this book has been presented to graduate classes at the Department of Mathematics of the University of Athens, Greece.

Charalambos A. Charalambides

Athens, June 2015

ABOUT THE AUTHOR

Charalambos A. Charalambides is professor emeritus of mathematical statistics at the University of Athens, Greece. Dr. Charalambides received a diploma in mathematics (1969) and a Ph.D. in mathematical statistics (1972) from the University of Athens. He was visiting assistant professor at McGill University, Montreal, Canada (1973–1974), visiting associate professor at Temple University, Philadelphia, USA (1985–1986), and visiting professor at the University of Cyprus, Nicosia, Cyprus (1995–1996, 2003–2004, 2007–2008, 2010–2011). Since 1979, he has been an elected member of the International Statistical Institute (ISI). Professor Charalambides' research interests include enumerative combinatorics, combinatorial probability, and parametric inference/point estimation. He is the author of the textbooks *Enumerative Combinatorics*, Chapman & Hall/CRC Press, 2002, and *Combinatorial Methods in Discrete Distributions*, John Wiley & Sons, 2005, and co-editor of the volume *Probability and Statistical Models with Applications*, Chapman & Hall/CRC Press, 2001.

1

BASIC q-COMBINATORICS AND q-HYPERGEOMETRIC SERIES

1.1 INTRODUCTION

The basic q-sequences and q-functions of the calculus of q-hypergeometric series, which facilitate the study of discrete q-distributions, are thoroughly presented in this chapter. More precisely, after introducing the notions of a q-power, a q-factorial, and a q-binomial coefficient of a real number, two q-Vandermonde's (q-factorial convolution) formulae are derived. Also, two q-Cauchy's (q-binomial convolution) formulae are presented as a corollary of the two q-Vandermonde's formulae. Furthermore, the q-binomial and the negative q-binomial formulae are obtained. In addition, a general q-binomial formula is derived and, as limiting forms of it, q-exponential and q-logarithmic functions are deduced. The q-Stirling numbers of the first and second kind, which are the coefficients of the expansions of q-factorials into q-powers and of q-powers into q-factorials, respectively, are presented. Also, the generalized q-factorial coefficients are briefly discussed. Moreover, the q-factorial and q-binomial moments, which, apart from the interest in their own, are used as an intermediate step in the calculation of the usual factorial and binomial moments of a discrete q-distribution, are briefly presented. Finally, the probability function of a nonnegative integer-valued random variable is expressed in terms of its q-binomial (or q-factorial) moments.

Discrete q-Distributions, First Edition. Charalambos A. Charalambides.

1.2 q-FACTORIALS AND q-BINOMIAL COEFFICIENTS

Let x and q be real numbers, with $q \neq 1$, and k be an integer. The number

$$[x]_q = \frac{1-q^x}{1-q}$$

is called *q-number* and in particular $[k]_q$ is called *q-integer*. Note that

$$\lim_{q \to 1} [x]_q = x.$$

The base (parameter) q, in the theory of discrete q-distributions, varies in the interval $0 < q < 1$ or in the interval $1 < q < \infty$. In both these cases,

$$[x]_q \lesseqgtr [y]_q, \quad \text{if and only if } x \lesseqgtr y, \quad \text{respectively.}$$

In particular,

$$[x]_q \lesseqgtr 0, \quad \text{if and only if } x \lesseqgtr 0, \quad \text{respectively.}$$

In this book, unless stated otherwise, it is assumed that $0 < q < 1$ or $1 < q < \infty$.

The kth-order factorial of the q-number $[x]_q$, which is defined by

$$[x]_{k,q} = [x]_q[x-1]_q \cdots [x-k+1]_q, \quad k = 1, 2, \ldots,$$

is called *q-factorial* of x of order k. In particular,

$$[k]_q! = [1]_q[2]_q \cdots [k]_q, \quad k = 1, 2, \ldots,$$

is called q-factorial of k.

The q-factorial of x of negative order may be defined as follows. Clearly, the following *fundamental property* of the q-factorial

$$[x]_{r+k,q} = [x]_{r,q} \cdot [x-r]_{k,q}, \quad k = 1, 2, \ldots, r = 1, 2, \ldots,$$

is readily deduced from its definition. Requiring the validity of this fundamental property to be preserved, the definition of the factorial may be extended to zero or negative order. Specifically, it is required that the fundamental property is valid for any integer values of k and r. Then, substituting into it $r = 0$, it follows that

$$[x]_{k,q} = [x]_{0,q} \cdot [x]_{k,q},$$

for any integer k. This equation, if $x \neq 0, 1, \ldots, k-1$, whence $[x]_{k,q} \neq 0$, implies

$$[x]_{0,q} = 1,$$

while, if $x = 0, 1, \ldots, k-1$, reduces to an identity for any value $[x]_{0,q}$ is required to represent. Furthermore, from the fundamental property, with k a positive integer and $r = -k$, it follows that

$$[x]_{-k,q} \cdot [x+k]_{k,q} = 1$$

and, for $x \neq -1, -2, \ldots, -k$,

$$[x]_{-k,q} = \frac{1}{[x+k]_{k,q}} = \frac{1}{[x+k]_q[x+k-1]_q \cdots [x+1]_q}, \quad k = 1, 2, \ldots .$$

Notice that the last expression, for $x = 0$, yields

$$[0]_{-k,q} = \frac{1}{[k]_q!}, \quad k = 1, 2, \ldots .$$

The *q-binomial coefficient* (or *Gaussian polynomial*) is defined by

$$\begin{bmatrix} x \\ k \end{bmatrix}_q = \frac{[x]_{k,q}}{[k]_q!}, \quad k = 0, 1, \ldots,$$

and so

$$\lim_{q \to 1} \begin{bmatrix} x \\ k \end{bmatrix}_q = \binom{x}{k}, \quad k = 0, 1, \ldots .$$

Note that

$$[x]_{q^{-1}} = \frac{1 - q^{-x}}{1 - q^{-1}} = \frac{q^{-x}}{q^{-1}} \cdot \frac{1 - q^x}{1 - q} = q^{-x+1}[x]_q$$

and since

$$[x]_{k,q^{-1}} = \prod_{j=1}^{k} [x - j + 1]_{q^{-1}} = \prod_{j=1}^{k} q^{-x+j}[x - j + 1]_q = q^{-xk+1+2+\cdots+k}[x]_{k,q},$$

it follows that

$$[x]_{k,q^{-1}} = q^{-xk+\binom{k+1}{2}}[x]_{k,q}, \quad [k]_{q^{-1}}! = q^{-\binom{k}{2}}[k]_q!$$

and

$$\begin{bmatrix} x \\ k \end{bmatrix}_{q^{-1}} = q^{-k(x-k)} \begin{bmatrix} x \\ k \end{bmatrix}_q, \quad k = 0, 1, \ldots .$$

Using these expressions, a formula involving q-numbers, q-factorials, and q-binomial coefficients in a base q, with $1 < q < \infty$, can be converted, with respect to the base, into a similar formula in the base $p = q^{-1}$, with $0 < p < 1$.

Two useful versions of a triangular recurrence relation for the q-binomial coefficient, which constitutes a q-analogue of Pascal's triangle, are derived in the next theorem.

Theorem 1.1. *Let x and q be real numbers, with $q \neq 1$, and let k be a positive integer. Then, the q-binomial coefficient $\begin{bmatrix} x \\ k \end{bmatrix}_q$ satisfies the triangular recurrence relation*

$$\begin{bmatrix} x \\ k \end{bmatrix}_q = \begin{bmatrix} x-1 \\ k \end{bmatrix}_q + q^{x-k} \begin{bmatrix} x-1 \\ k-1 \end{bmatrix}_q, \quad k = 1, 2, \ldots, \tag{1.1}$$

with initial condition $\begin{bmatrix} x \\ 0 \end{bmatrix}_q = 1$. *Alternatively,*

$$\begin{bmatrix} x \\ k \end{bmatrix}_q = q^k \begin{bmatrix} x-1 \\ k \end{bmatrix}_q + \begin{bmatrix} x-1 \\ k-1 \end{bmatrix}_q, \quad k = 1, 2, \ldots . \tag{1.2}$$

Proof. The q-factorial of x of order k, since $[x]_{k,q} = [x]_q[x-1]_{k-1,q}$ and

$$[x]_q = [x-k+k]_q = [x-k]_q + q^{x-k}[k]_q, \quad [x-1]_{k-1,q}[x-k]_q = [x-1]_{k,q},$$

satisfies the triangular recurrence relation

$$[x]_{k,q} = [x-1]_{k,q} + q^{x-k}[k]_q[x-1]_{k-1,q},$$

with initial condition $[x]_{0,q} = 1$. Thus, dividing both members of it by $[k]_q!$ and using the expression

$$\begin{bmatrix} x \\ k \end{bmatrix}_q = \frac{[x]_{k,q}}{[k]_q!}, \quad k = 0, 1, \ldots,$$

the triangular recurrence relation (1.1) is readily deduced. Furthermore, replacing the base q by q^{-1} and using the relation

$$\begin{bmatrix} x \\ k \end{bmatrix}_{q^{-1}} = q^{-k(x-k)} \begin{bmatrix} x \\ k \end{bmatrix}_q, \quad k = 0, 1, \ldots,$$

(1.1) may be rewritten in the form (1.2).

Note that the triangular recurrence relation (1.2) may also be derived, independently of (1.1), by using the expression

$$[x]_q = [k+x-k]_q = [k]_q + q^k[x-k]_q,$$

which entails for the q-factorial of x of order k the triangular recurrence relation

$$[x]_{k,q} = [k]_q[x-1]_{k-1,q} + q^k[x-1]_{k,q}.$$

Hence, dividing both members of it by $[k]_q!$, (1.2) is obtained. □

Remark 1.1. *The lack of uniqueness of q-analogues of expressions and formulae.* The lack of uniqueness, due to the presence of powers of q in pseudo-isomorphisms as

$$[x+y]_q = [x]_q + q^x[y]_q \qquad \text{and} \qquad [x+y]_q = q^y[x] + [y]_q,$$

where $0 < q < 1$ or $1 < q < \infty$, should be remarked from the very beginning of the presentation of the basic q-sequences, q-functions and q-formulae. It should also be noticed that the two formulae may be considered as *equivalent* in the sense that any of these implies the other by replacing the base q by q^{-1}. In this framework, the existence of two versions of the q-analogue of Pascal's triangle, which may be considered as equivalent, is attributed to the lack of uniqueness.

The particular cases of the q-binomial coefficients $\begin{bmatrix} n \\ k \end{bmatrix}_q$ and $\begin{bmatrix} n+k-1 \\ k \end{bmatrix}_q$, with n and k positive integers, admit q-combinatorial interpretations, which are deduced in the following theorem, starting from a generating function of a number of partitions of an integer into parts of restricted size. Recall that a *partition* of a positive integer m into k parts is a nonordered collection of positive integers, $\{r_1, r_2, \ldots, r_k\}$, with $r_1 \geq r_2 \geq \cdots \geq r_k \geq 1$, for $k = 1, 2, \ldots, m$, whose sum equals m. In a partition of m into k parts, let $k_i \geq 0$ be the number of parts that are equal to i, for $i = 1, 2, \ldots, m$. Then,

$$k_1 + 2k_2 + \cdots + mk_m = m, \quad k_1 + k_2 + \cdots + k_m = k.$$

Theorem 1.2. *The q-binomial coefficient $\begin{bmatrix} n \\ k \end{bmatrix}_q$, for n and k positive integers, equals the k-combinations of the set $\{1, 2, \ldots, n\}$, $\{m_1, m_2, \ldots, m_k\}$, weighted by $q^{m_1+m_2+\cdots+m_k-\binom{k+1}{2}}$,*

$$\sum_{1 \leq m_1 < m_2 < \cdots < m_k \leq n} q^{m_1+m_2+\cdots+m_k-\binom{k+1}{2}} = \begin{bmatrix} n \\ k \end{bmatrix}_q. \tag{1.3}$$

Also, the q-binomial coefficient $\begin{bmatrix} n+k-1 \\ k \end{bmatrix}_q$, for n and k positive integers, equals the k-combinations of the set $\{1, 2, \ldots, n\}$ with repetition, $\{r_1, r_2, \ldots, r_k\}$, weighted by $q^{r_1+r_2+\cdots+r_k-k}$,

$$\sum_{1 \leq r_1 \leq r_2 \leq \cdots \leq r_k \leq n} q^{r_1+r_2+\cdots+r_k-k} = \begin{bmatrix} n+k-1 \\ k \end{bmatrix}_q. \tag{1.4}$$

Proof. Let $p(m, k; n)$ denotes the number of partitions of m into k parts, each of which is less than or equal to n, and consider its bivariate generating function

$$g_n(t, q) = \sum_{m=0}^{\infty} \sum_{k=0}^{m} p(m, k; n) t^k q^m,$$

for fixed n. Clearly, using the definition of a partition of m into k parts, each of which is less than or equal to n, it may be expressed as

$$g_n(t, q) = \sum_{m=0}^{\infty} \sum_{k=0}^{m} \left(\sum t^{k_1+k_2+\cdots+k_m} q^{k_1+2k_2+\cdots+mk_m} \right),$$

where in the inner sum the summation is extended over all integer solutions $k_i \geq 0$, for $i = 1, 2, \ldots, n$, and $k_i = 0$, for $i = n+1, n+2, \ldots, m$, of the equations $k_1 + 2k_2 + \cdots + mk_m = m$ and $k_1 + k_2 + \cdots + k_m = k$. Since this inner sum is summed over all $k = 0, 1, \ldots, m$ and $m = 0, 1, \ldots,$ it follows that

$$g_n(t, q) = \prod_{i=1}^{n} \left(\sum_{k_i=0}^{\infty} t^{k_i} q^{ik_i} \right) = \prod_{i=1}^{n} (1 - tq^i)^{-1}.$$

Furthermore, for the sequence of univariate generating functions

$$g_{n,k}(q) = \sum_{m=k}^{\infty} p(m, k; n)q^m, \quad k = 0, 1, \ldots,$$

on using the relation $g_n(t, q) = \sum_{k=0}^{\infty} g_{n,k}(q)t^k$ and since

$$(1 - tq^{n+1})g_n(tq, q) = (1 - tq)g_n(t, q),$$

we deduce the following first-order recurrence relation

$$g_{n,k}(q) = q\frac{1 - q^{n+k-1}}{1 - q^k} g_{n,k-1}(q), \quad k = 1, 2, \ldots, \; g_{n,0}(q) = 1.$$

Applying it repeatedly, we find the expression

$$g_{n,k}(q) = \sum_{m=k}^{\infty} p(m, k; n)q^m = q^k \begin{bmatrix} n + k - 1 \\ k \end{bmatrix}_q.$$

Since the number $p(m, k; n)$, of partitions of m into k parts, each of which is less than or equal to n, equals the number of solutions in positive integers of the equation $r_1 + r_2 + \cdots + r_k = m$, with $1 \le r_k \le r_{k-1} \le \cdots \le r_1 \le n$, or equivalently (by replacing r_{k-i+1} with r_i, for $i = 1, 2, \ldots, k$), with $1 \le r_1 \le r_2 \le \cdots \le r_k \le n$, the last expression may also be written in the form

$$\sum_{1 \le r_1 \le r_2 \le \cdots \le r_k \le n} q^{r_1 + r_2 + \cdots + r_k} = q^k \begin{bmatrix} n + k - 1 \\ k \end{bmatrix}_q,$$

which readily implies (1.4).

Furthermore, replacing $n + k - 1$ by n in (1.4) and then setting $m_i = r_i + i - 1$, for $i = 1, 2, \ldots, k$, the inequalities $1 \le r_1 \le r_2 \le \cdots \le r_k \le n - k + 1$ are transformed into $1 \le m_1 < m_2 < \cdots < m_k \le n$ and $r_1 + r_2 + \cdots + r_k = m_1 + m_2 + \cdots + m_k - \binom{k}{2}$. Consequently, expression (1.4) is transformed into

$$\sum_{1 \le m_1 < m_2 < \cdots < m_k \le n} q^{m_1 + m_2 + \cdots + m_k} = q^{\binom{k+1}{2}} \begin{bmatrix} n \\ k \end{bmatrix}_q,$$

from which (1.3) is deduced. □

Remark 1.2. *An alternative expression of a q-binomial coefficient.* Expression (1.4) of the q-binomial coefficient $\begin{bmatrix} n+k-1 \\ k \end{bmatrix}_q$, may be written alternatively as follows. Let $k_i \ge 0$ be the number of the (bound) variables $r_1, r_2, \ldots, r_k$, that are equal to i, for $i = 1, 2, \ldots, n$. Then, $r_1 + r_2 + \cdots + r_k = k_1 + 2k_2 + \cdots + nk_n$, with $k_1 + k_2 + \cdots + k_n = k$, and

$$\{(r_1, r_2, \ldots, r_k) : 1 \le r_1 \le r_2 \le \cdots \le r_k \le n\}$$
$$= \{(k_1, k_2, \ldots, k_n) : k_i \ge 0, i = 1, 2, \ldots, n, k_1 + k_2 + \cdots + k_n = k\}.$$

Consequently, expression (1.4) may be, alternatively, expressed as

$$\sum_{\substack{k_i \geq 0,\ i=1,2,\ldots,n \\ k_1+k_2+\cdots+k_n=k}} q^{k_1+2k_2+\cdots+nk_n-k} = \begin{bmatrix} n+k-1 \\ k \end{bmatrix}_q. \tag{1.5}$$

A few interesting combinatorial and probabilistic examples, in which q-numbers and q-binomial coefficients naturally emerged, are presented next. Specifically, in the following example, a number theoretic random variable is defined in a sequence of independent and identically distributed Bernoulli trials, and its probability function is expressed by a q-number.

Example 1.1. *Bernoulli trials and number theory.* Consider a sequence of independent Bernoulli trials, with constant failure probability q, and let X be the number of failures until the occurrence of the first success. Clearly, the random variable X follows a geometric distribution with probability function

$$P(X = x) = (1 - q)q^x, \quad x = 0, 1, \ldots,$$

where $0 < q < 1$. Also, consider a fixed positive integer n and let

$$C_x = \{x + kn : k = 0, 1, \ldots\}, \quad x = 0, 1, \ldots, n - 1.$$

Clearly, each of the possible values of the random variable X, $\{0, 1, \ldots\}$, belongs to one of these n congruence classes (pairwise disjoint sets), modulo n, $\{C_0, C_1, \ldots, C_{n-1}\}$. Furthermore, consider a sequence of independent Bernoulli trials, with constant failure probability q, and let X_n be the index of the congruence class in which the number of failures, until the occurrence of the first success, belongs. Since the random variable X_n assumes the value x if and only if X belongs to C_x, its probability function,

$$P(X_n = x) = P(X \in C_x) = \sum_{k=0}^{\infty} (1 - q)q^{x+kn} = \frac{(1 - q)q^x}{1 - q^n},$$

may be expressed as

$$P(X_n = x) = \frac{q^x}{[n]_q}, \quad x = 0, 1, \ldots, n - 1, \quad 0 < q < 1.$$

Note that the limiting probability function, as $q \to 1$,

$$\lim_{q \to 1} P(X_n = x) = \frac{1}{n}, \quad x = 0, 1, \ldots, n - 1,$$

is the discrete uniform probability function on the set $\{0, 1, \ldots, n - 1\}$. For this reason, the distribution of X_n is called *discrete q-uniform distribution.*

It is worth noting that the probability function $P(X = x|X \leq n-1)$, of a right truncated geometric distribution, since

$$P(X = x|X \leq n-1) = \frac{P(X = x, X \leq n-1)}{P(X \leq n-1)} = \frac{P(X = x)}{P(X \leq n-1)},$$

for $x = 0, 1, \ldots, n-1$, and

$$P(X \leq n-1) = \sum_{x=0}^{n-1} (1-q)q^x = 1 - q^n,$$

is readily deduced as

$$P(X = x|X \leq n-1) = \frac{q^x}{[n]_q}, \quad x = 0, 1, \ldots, n-1, \quad 0 < q < 1.$$

An interesting combinatorial interpretation of the q-binomial coefficient for $x = n$ a positive integer and q a power of a prime number, is given in the next example.

Example 1.2. *Number of subspaces of a vector space.* Let $V(n;q)$ be a vector space of dimension n over a finite field of $q = p^m$ elements, where p is a prime number and m is a positive integer. The vector space $V(n;q)$ contains q^n vectors. A subspace $S(k;q)$ of $V(n;q)$, of dimension $k \leq n$, contains k linearly independent vectors that are selected from $V(n;q)$. The number $G_{n,k}(q)$ of k-dimensional subspaces, $S(k;q)$, of an n-dimensional vector space $V(n;q)$ is of interest.

Let us first determine the number $H_{n,k}(q)$ of ordered k-tuples $(\mathbf{a}_1, \mathbf{a}_2, \ldots, \mathbf{a}_k)$ of linearly independent vectors in $V(n;q)$. The first nonzero vector $\mathbf{a}_1$, of an ordered k-tuple, can be selected from the $q^n - 1$ nonzero vectors of $V(n;q)$. Any nonzero vector $\mathbf{a}_1$ spans (generates) a one-dimensional subspace $S(1;q)$ of $V(n;q)$ containing q vectors. Therefore, excluding these q vectors, the second vector $\mathbf{a}_2$ can be selected from the $q^n - q$ vectors, which are linearly independent of $\mathbf{a}_1$. The ordered pair $(\mathbf{a}_1, \mathbf{a}_2)$ spans a two-dimensional subspace $S(2;q)$ of $V(n;q)$ containing q^2 vectors. Thus, excluding these q^2 vectors, the third vector $\mathbf{a}_3$ can be selected from the $q^n - q^2$ vectors, which are linearly independent of $\mathbf{a}_1, \mathbf{a}_2$. In general, the rth vector $\mathbf{a}_r$ can be selected from the $q^n - q^{r-1}$ vectors, which are linearly independent of $\mathbf{a}_1, \mathbf{a}_2, \ldots, \mathbf{a}_{r-1}$, for $r = 3, 4, \ldots, k$. Consequently, applying the multiplication principle, it follows that

$$H_{n,k}(q) = \prod_{r=1}^{k} (q^n - q^{r-1}).$$

Note that each ordered k-tuple $(\mathbf{a}_1, \mathbf{a}_2, \ldots, \mathbf{a}_k)$ of linearly independent vectors in $V(n;q)$ spans a k-dimensional subspace $S(k;q)$. Inversely, any k-dimensional subspace $S(k;q)$ is spanned by

$$H_{k,k}(q) = \prod_{r=1}^{k} (q^k - q^{r-1})$$

ordered k-tuples $(\mathbf{a}_1, \mathbf{a}_2, \ldots, \mathbf{a}_k)$ of linearly independent vectors in $V(n;q)$. The evaluation of this number is carried out the same way as above. Therefore, the number $G_{n,k}(q)$ of k-dimensional subspaces of an n-dimensional vector space $V(n;q)$ is given by the quotient

$$G_{n,k}(q) = \frac{H_{n,k}(q)}{H_{k,k}(q)} = \frac{\prod_{r=1}^{k}(q^n - q^{r-1})}{\prod_{r=1}^{k}(q^k - q^{r-1})} = \frac{(-1)^k q^{\binom{k}{2}} \prod_{r=1}^{k}(1 - q^{n-r+1})}{(-1)^k q^{\binom{k}{2}} \prod_{r=1}^{k}(1 - q^{k-r+1})}$$

and so

$$G_{n,k}(q) = \frac{[n]_{k,q}}{[k]_q!} = \begin{bmatrix} n \\ k \end{bmatrix}_q.$$

In the following example, the q-binomial coefficient $\begin{bmatrix} n+k-1 \\ k \end{bmatrix}_q$ is obtained as a generating function of a number of partitions of an integer into parts of restricted size, with variable (indeterminate) q, in addition to the generating function deduced in the proof of Theorem 1.2.

Example 1.3. *Partitions of integers into parts of restricted size.* Let us consider the number $P(m, k; n)$, of partitions of m into at most k parts, each of which is less than n. It is closely connected to the number $p(m, k; n)$ of partitions m into (exactly) k parts, each of which is less than or equal to n, which was discussed in the proof of Theorem 1.2. Specifically, since the size of a part in a partition is a positive integer, the restriction that a size is less than n is equivalent to the restriction that it is less than or equal to $n - 1$ and so $P(m, k; n) = \sum_{r=0}^{k} p(m, r; n - 1)$. The bivariate generating function

$$h_n(t, q) = \sum_{m=0}^{\infty} \sum_{k=0}^{m} P(m, k; n) t^k q^m,$$

for fixed n, on introducing the expression $P(m, k; n) = \sum_{r=0}^{k} p(m, r; n - 1)$ and subsequently interchanging the order of summation of the two inner sums,

$$\begin{aligned} h_n(t, q) &= \sum_{m=0}^{\infty} \sum_{k=0}^{m} \sum_{r=0}^{k} p(m, r; n - 1) t^k q^m \\ &= \sum_{m=0}^{\infty} \sum_{r=0}^{m} \left(\sum_{k=r}^{\infty} t^{k-r} \right) p(m, r; n - 1) t^r q^m, \end{aligned}$$

is expressed in terms of the bivariate generating function of $p(m, r; n - 1)$ as

$$h_n(t, q) = (1 - t)^{-1} \sum_{m=0}^{\infty} \sum_{r=0}^{m} p(m, r; n - 1) t^r q^m = (1 - t)^{-1} g_{n-1}(t, q)$$

and so

$$h_n(t, q) = \prod_{i=0}^{n-1} (1 - tq^i)^{-1}.$$

Also, for the sequence of univariate generating functions

$$h_{n,k}(q) = \sum_{m=k}^{\infty} P(m, k; n)q^m, \quad k = 0, 1, \dots,$$

on using the relation $h_n(t, q) = \sum_{k=0}^{\infty} h_{n,k}(q)t^k$ and since

$$(1 - tq^n)h_n(tq, q) = (1 - t)h_n(t, q),$$

we deduce the first-order recurrence relation

$$h_{n,k}(q) = \frac{1 - q^{n+k-1}}{1 - q^k} h_{n,k-1}(q), \quad k = 1, 2, \dots, h_{n,0}(q) = 1.$$

Applying it repeatedly, we get the expression

$$h_{n,k}(q) = \sum_{m=k}^{\infty} P(m, k; n)q^m = \begin{bmatrix} n+k-1 \\ k \end{bmatrix}_q.$$

Therefore, the number $P(m, k; n)$, of partitions of m into at most k parts, each of which is less than n, is the coefficient of q^m in the expansion of the q-binomial coefficient $\begin{bmatrix} n+k-1 \\ k \end{bmatrix}_q$ into powers of q.

1.3 q-VANDERMONDE'S AND q-CAUCHY'S FORMULAE

Two versions of a q-Vandermonde's (q-factorial convolution) formula are derived in the next theorem.

Theorem 1.3. *Let n be a positive integer and let x, y, and q be real numbers, with $q \neq 1$. Then,*

$$[x + y]_{n,q} = \sum_{k=0}^{n} \begin{bmatrix} n \\ k \end{bmatrix}_q q^{(n-k)(x-k)} [x]_{k,q} [y]_{n-k,q}. \tag{1.6}$$

Alternatively,

$$[x + y]_{n,q} = \sum_{k=0}^{n} \begin{bmatrix} n \\ k \end{bmatrix}_q q^{k(y-n+k)} [x]_{k,q} [y]_{n-k,q}. \tag{1.7}$$

Proof. Let us consider the sequence of sums

$$s_n(x, y; q) = \sum_{k=0}^{n} \begin{bmatrix} n \\ k \end{bmatrix}_q q^{(n-k)(x-k)} [x]_{k,q} [y]_{n-k,q}, \quad n = 1, 2, \dots,$$

with initial value

$$s_1(x, y; q) = q^x [y]_q + [x]_q = [x + y]_q.$$

Using the triangular recurrence relation (1.1), with $x = n$, the sequence $s_n(x, y; q)$, $n = 1, 2, \dots$, may be expressed as

$$s_n(x, y; q) = \sum_{k=0}^{n-1} \begin{bmatrix} n \\ k \end{bmatrix}_q q^{(n-k)(x-k)} [x]_{k,q} [y]_{n-k,q} + \sum_{k=1}^{n} \begin{bmatrix} n-1 \\ k-1 \end{bmatrix}_q q^{(n-k)(x-k+1)} [x]_{k,q} [y]_{n-k,q}.$$

Replacing $k - 1$ by k in the last sum and then using the relation

$$q^{x-k} [x]_{k,q} [y]_{n-k,q} + [x]_{k+1,q} [y]_{n-1-k,q} = [x + y - n + 1]_q [x]_{k,q} [y]_{n-1-k,q},$$

it follows that

$$\begin{aligned} s_n(x, y; q) &= \sum_{k=0}^{n-1} \begin{bmatrix} n-1 \\ k \end{bmatrix}_q q^{(n-1-k)(x-k)} q^{x-k} [x]_{k,q} [y]_{n-k,q} \\ &\quad + \sum_{k=0}^{n-1} \begin{bmatrix} n-1 \\ k \end{bmatrix}_q q^{(n-1-k)(x-k)} [x]_{k+1,q} [y]_{n-1-k,q} \\ &= [x + y - n + 1]_q \sum_{k=0}^{n-1} \begin{bmatrix} n-1 \\ k \end{bmatrix}_q q^{(n-1-k)(x-k)} [x]_{k,q} [y]_{n-1-k,q}. \end{aligned}$$

Hence, the sequence $s_n(x, y; q)$, $n = 1, 2, \dots$, satisfies the first-order recurrence relation

$$s_n(x, y; q) = [x + y - n + 1]_q s_{n-1}(x, y; q), \quad n = 1, 2, \dots,$$

with initial condition $s_1(x, y; q) = [x + y]_q$. Clearly, applying it successively, it follows that $s_n(x, y; q) = [x + y]_{n,q}$ and so (1.6) is shown. Furthermore, interchanging x by y and then replacing k by $n - k$, formula (1.6) is rewritten in the form (1.7).

Note that the alternative formula (1.7) may be derived, independently of (1.6), by following the steps of derivation of (1.6) and using the triangular recurrence relation (1.2) instead of (1.1) and

$$[x]_{k,q} [y]_{n-k,q} + q^{y-n+k+1} [x]_{k+1,q} [y]_{n-1-k,q} = [x + y - n + 1]_q [x]_{k,q} [y]_{n-1-k,q}$$

instead of

$$q^{x-k} [x]_{k,q} [y]_{n-k,q} + [x]_{k+1,q} [y]_{n-1-k,q} = [x + y - n + 1]_q [x]_{k,q} [y]_{n-1-k,q}.$$

The proof of the theorem is thus completed. □

Two versions of a q-Cauchy's (q-binomial convolution) formula, which by virtue of

$$\begin{bmatrix} x \\ k \end{bmatrix}_q = \frac{[x]_{k,q}}{[k]_q!}, \quad k = 0, 1, \dots,$$

may be considered as reformulations of the corresponding two versions of the q-Vandermonde's (q-factorial convolution) formula, are stated in the following corollary of Theorem 1.3.

Corollary 1.1. *Let n be a positive integer and let x, y, and q be real numbers, with $q \neq 1$. Then,*

$$\begin{bmatrix} x+y \\ n \end{bmatrix}_q = \sum_{k=0}^{n} q^{(n-k)(x-k)} \begin{bmatrix} x \\ k \end{bmatrix}_q \begin{bmatrix} y \\ n-k \end{bmatrix}_q. \tag{1.8}$$

Alternatively,

$$\begin{bmatrix} x+y \\ n \end{bmatrix}_q = \sum_{k=0}^{n} q^{k(y-n+k)} \begin{bmatrix} x \\ k \end{bmatrix}_q \begin{bmatrix} y \\ n-k \end{bmatrix}_q. \tag{1.9}$$

Two versions of a negative q-Vandermonde's formula are obtained in the next theorem.

Theorem 1.4. *Let n be a positive integer and let x, y, and q be real numbers, with $q \neq 1$. Then,*

$$[x+y]_{-n,q} = \sum_{k=0}^{\infty} \begin{bmatrix} -n \\ k \end{bmatrix}_q q^{-(n+k)(x-k)} [x]_{k,q} [y]_{-n-k,q}, \tag{1.10}$$

with $|q^{-(x+y+1)}| < 1$. Alternatively,

$$[x+y]_{-n,q} = \sum_{k=0}^{\infty} \begin{bmatrix} -n \\ k \end{bmatrix}_q q^{k(y+n+k)} [x]_{k,q} [y]_{-n-k,q}, \tag{1.11}$$

with $|q^{x+y+1}| < 1$.

Proof. Let us set

$$s_n(x,y;q) = \sum_{k=0}^{\infty} \begin{bmatrix} -n \\ k \end{bmatrix}_q q^{-(n+k)(x-k)} [x]_{k,q} [y]_{-n-k,q}, \quad |q^{-(x+y+1)}| < 1,$$

for $n = 1, 2, \ldots$. Note that, the first term of this sequence,

$$\begin{aligned} s_1(x,y;q) &= \sum_{k=0}^{\infty} \begin{bmatrix} -1 \\ k \end{bmatrix}_q q^{-(k+1)(x-k)} [x]_{k,q} [y]_{-k-1,q} \\ &= \sum_{k=0}^{\infty} (-1)^k q^{-(k+1)x + \binom{k+1}{2}} [x]_{k,q} [y]_{-k-1,q}, \end{aligned}$$

since

$$[x+y+1]_q [x]_{k,q} [y]_{-k-1,q} = q^{x-k} [x]_{k,q} [y]_{-k,q} + [x]_{k+1,q} [y]_{-k-1,q},$$

may be written as

$$\begin{aligned} s_1(x,y;q) = \frac{1}{[x+y+1]_q} \Bigg\{ &\sum_{k=0}^{\infty} (-1)^k q^{-kx + \binom{k}{2}} [x]_{k,q} [y]_{-k,q} \\ &- \sum_{k=0}^{\infty} (-1)^{k+1} q^{-(k+1)x + \binom{k+1}{2}} [x]_{k+1,q} [y]_{-k-1,q} \Bigg\} \end{aligned}$$

and so $s_1(x,y;q) = 1/[x+y+1]_q$. Using the triangular recurrence relation (1.1), with $x = -n$, the sequence $s_n(x,y;q)$, $n = 1,2,\ldots$, may be expressed as

$$s_n(x,y;q) = \sum_{k=0}^{\infty} \begin{bmatrix} -n-1 \\ k \end{bmatrix}_q q^{-(n+k)(x-k)} [x]_{k,q}[y]_{-n-k,q} + \sum_{k=1}^{\infty} \begin{bmatrix} -n-1 \\ k-1 \end{bmatrix}_q q^{-(n+k)(x-k+1)} [x]_{k,q}[y]_{-n-k,q}.$$

Replacing $k-1$ by k in the last sum and then using the relation

$$q^{x-k}[x]_{k,q}[y]_{-n-k,q} + [x]_{k+1,q}[y]_{-n-1-k,q} = [x+y+n+1]_q[x]_{k,q}[y]_{-n-1-k,q},$$

it follows that

$$\begin{aligned} s_n(x,y;q) &= \sum_{k=0}^{\infty} \begin{bmatrix} -n-1 \\ k \end{bmatrix}_q q^{-(n+1+k)(x-k)} q^{x-k} [x]_{k,q}[y]_{-n-k,q} \\ &\quad + \sum_{k=0}^{\infty} \begin{bmatrix} -n-1 \\ k \end{bmatrix}_q q^{-(n+1+k)(x-k)} [x]_{k+1,q}[y]_{-n-1-k,q} \\ &= [x+y+n+1]_q \sum_{k=0}^{\infty} \begin{bmatrix} -n-1 \\ k \end{bmatrix}_q q^{-(n+1+k)(x-k)} [x]_{k,q}[y]_{-n-1-k,q}. \end{aligned}$$

Hence, the sequence $s_n(x,y;q)$, $n = 1,2,\ldots$, satisfies the first-order recurrence relation

$$s_{n+1}(x,y;q) = \frac{s_n(x,y;q)}{[x+y+n+1]_q}, \qquad n = 1,2,\ldots,$$

with initial condition $s_1(x,y;q) = 1/[x+y+1]_q = [x+y]_{-1,q}$. Consequently, $s_n(x,y;q) = 1/[x+y+n]_{n,q} = [x+y]_{-n,q}$ and so (1.10) is shown. Furthermore, replacing the base q by q^{-1} and using the relations

$$[x+y]_{-n,q^{-1}} = q^{(x+y)n+\binom{n}{2}}[x+y]_{-n,q}, \qquad \begin{bmatrix} -n \\ k \end{bmatrix}_{q^{-1}} = q^{k(n+k)} \begin{bmatrix} -n \\ k \end{bmatrix}_q$$

and

$$[x]_{k,q^{-1}} = q^{-xk+\binom{k+1}{2}}[x]_{k,q}, \qquad [y]_{-n-k,q^{-1}} = q^{y(n+k)+\binom{n+k}{2}}[y]_{-n-k,q},$$

formula (1.10) may be rewritten in the form (1.11).

Note that the alternative expression (1.11) may be derived, independently of (1.10), by following the steps of derivation of (1.10) and using the triangular recurrence relation (1.2) instead of (1.1) and

$$[x]_{k,q}[y]_{-n-k,q} + q^{y+n+k+1}[x]_{k+1,q}[y]_{-n-1-k,q} = [x+y+n+1]_q[x]_{k,q}[y]_{-n-1-k,q}$$

instead of

$$q^{x-k}[x]_{k,q}[y]_{-n-k,q} + [x]_{k+1,q}[y]_{-n-1-k,q} = [x+y+n+1]_q[x]_{k,q}[y]_{-n-1-k,q}.$$

The proof of the theorem is thus completed. □

Remark 1.3. *Additional expressions of the negative q-Vandermonde's formula.* The two versions of the negative q-Vandermonde's formula (1.10) and (1.11) may be rewritten as

$$\frac{1}{[y]_{n,q}} = \sum_{k=0}^{\infty} \begin{bmatrix} n+k-1 \\ k \end{bmatrix}_q q^{k(y-n+1)} \frac{[x]_{k,q}}{[x+y]_{n+k,q}}, \quad |q^y| < 1 \tag{1.12}$$

and

$$\frac{1}{[y]_{n,q}} = \sum_{k=0}^{\infty} \begin{bmatrix} n+k-1 \\ k \end{bmatrix}_q q^{n(x-k)} \frac{[x]_{k,q}}{[x+y]_{n+k,q}}, \quad |q^{-y}| < 1, \tag{1.13}$$

respectively.

Indeed, the q-factorial of $x = -n$ of order k, with n a positive integer, may be expressed as

$$[-n]_{k,q} = (-1)^k q^{-nk-\binom{k}{2}} [n+k-1]_{k,q},$$

since

$$[-j]_q = -q^{-j}[j]_q, \quad j = n, n+1, \ldots, n+k-1,$$

and

$$n + (n+1) + \cdots + (n+k-1) = nk + (1+2+\cdots+(k-1)) = nk + \binom{k}{2}.$$

Equivalently, the q-binomial coefficient of $x = -n$, with n a positive integer, is expressed as

$$\begin{bmatrix} -n \\ k \end{bmatrix}_q = (-1)^k q^{-nk-\binom{k}{2}} \begin{bmatrix} n+k-1 \\ k \end{bmatrix}_q.$$

Expression (1.10), replacing $x+y+1$ by $-y$ and using the last relation together with the relations

$$[-y-1]_{-n,q} = \frac{1}{[-y+n-1]_{n,q}} = \frac{1}{(-1)^n q^{-ny+\binom{n}{2}} [y]_{n,q}},$$

and

$$\begin{aligned} [-x-y-1]_{-n-k,q} &= \frac{1}{[-x-y+n+k-1]_{n+k,q}} \\ &= \frac{1}{(-1)^{n+k} q^{-(n+k)(x+y)+\binom{n+k}{2}} [x+y]_{n+k,q}}, \end{aligned}$$

may be written as

$$\frac{1}{q^{-ny+\binom{n}{2}}[y]_{n,q}} = \sum_{k=0}^{\infty} \begin{bmatrix} n+k-1 \\ k \end{bmatrix}_q \frac{q^{-nk-\binom{k}{2}-(n+k)(x-k)}[x]_{k,q}}{q^{-(n+k)(x+y)+\binom{n+k}{2}}[x+y]_{n+k,q}}.$$

Finally, using the relation $\binom{n+k}{2} = \binom{n}{2} + \binom{k}{2} + nk$, the last expression, after the cancelations in the exponents of q, reduces to (1.12). Similarly, expression (1.11) can be rewritten as (1.13).

An interesting q-identity is deduced in the following example, using q-Vandermonde's formula (1.6).

Example 1.4. Let n be a positive integer and let x, y, and q be real numbers, with $q \neq 1$. Using q-Vandermonde's formula (1.6), show that

$$\sum_{k=0}^{n} \begin{bmatrix} n \\ k \end{bmatrix}_q q^{(n-k)(x-k)} \frac{[x]_{k,q}}{[y+k]_{k,q}} = \frac{[x+y+n]_{n,q}}{[y+n]_{n,q}}$$

and conclude that

$$\sum_{k=0}^{n} (-1)^{n-k} \begin{bmatrix} n \\ k \end{bmatrix}_q q^{\binom{n-k}{2}} \frac{[x]_q}{[x-k]_q} = 1 \Big/ \begin{bmatrix} x-1 \\ n \end{bmatrix}_q$$

and

$$\sum_{k=0}^{n} (-1)^{k} \begin{bmatrix} n \\ k \end{bmatrix}_q q^{\binom{k+1}{2}-n(y+k)} \frac{[y]_q}{[y+k]_q} = 1 \Big/ \begin{bmatrix} y+n \\ n \end{bmatrix}_q.$$

Replacing y by $y+n$, in q-Vandermonde's formula (1.6), we have

$$\sum_{k=0}^{n} \begin{bmatrix} n \\ k \end{bmatrix}_q q^{(n-k)(x-k)}[x]_{k,q}[y+n]_{n-k,q} = [x+y+n]_{n,q}.$$

Multiplying both sides by $[y]_{-n,q}$ and using the recurrence relation

$$[y]_{-k,q} = [y]_{-n,q}[y+n]_{n-k,q},$$

we get

$$\sum_{k=0}^{n} \begin{bmatrix} n \\ k \end{bmatrix}_q q^{(n-k)(x-k)}[x]_{k,q}[y]_{-k,q} = [x+y+n]_{n,q}[y]_{-n,q}$$

and since $[y]_{-k,q} = 1/[y+k]_{k,q}$, we deduce the required formula,

$$\sum_{k=0}^{n} \begin{bmatrix} n \\ k \end{bmatrix}_q q^{(n-k)(x-k)} \frac{[x]_{k,q}}{[y+k]_{k,q}} = \frac{[x+y+n]_{n,q}}{[y+n]_{n,q}}.$$

Setting $y = -x$ and since

$$[-x+k]_{k,q} = (-1)^k q^{\binom{k}{2}-kx}[x-1]_{k,q}, \quad [-x+n]_{n,q} = (-1)^n q^{\binom{n}{2}-nx}[x-1]_{n,q},$$

and $\binom{n-k}{2} = \binom{n}{2} - \binom{k}{2} - k(n-k)$, we get

$$\sum_{k=0}^{n} (-1)^{n-k} \begin{bmatrix} n \\ k \end{bmatrix}_q q^{\binom{n-k}{2}} \frac{[x]_q}{[x-k]_q} = 1 \Big/ \begin{bmatrix} x-1 \\ n \end{bmatrix}_q.$$

Similarly, substituting $x = -y$, we conclude that

$$\sum_{k=0}^{n} (-1)^{k} \begin{bmatrix} n \\ k \end{bmatrix}_q q^{\binom{k+1}{2}-n(y+k)} \frac{[y]_q}{[y+k]_q} = 1 \Big/ \begin{bmatrix} y+n \\ n \end{bmatrix}_q.$$

1.4 q-BINOMIAL AND NEGATIVE q-BINOMIAL FORMULAE

The number $\begin{bmatrix} n \\ k \end{bmatrix}_q$ is called *q-binomial coefficient*, since it is the coefficient of the general term of a q-binomial (q-Newton's binomial) formula. This formula is derived in the following theorem.

Theorem 1.5. *Let n be a positive integer and let t and q be real numbers, with $q \neq 1$. Then,*

$$\prod_{i=1}^{n}(1+tq^{i-1}) = \sum_{k=0}^{n} q^{\binom{k}{2}} \begin{bmatrix} n \\ k \end{bmatrix}_q t^k. \tag{1.14}$$

Proof. The multiplications and additions in the product $\prod_{i=1}^{n}(1+tq^{i-1})$ may be carried out by (a) selecting the term tq^{m_i-1} from the factor

$$a_{m_i}(t;q) = 1 + tq^{m_i-1},$$

for $i = 1, 2, \ldots, k$, with $\{m_1, m_2, \ldots, m_k\} \subseteq \{1, 2, \ldots, n\}$; (b) forming the products

$$q^{m_1+m_2+\cdots+m_k-k}t^k, \quad \{m_1, m_2, \ldots, m_k\} \subseteq \{1, 2, \ldots, n\};$$

and (c) adding together these products, a summand of the form $b_{n,k}(q)t^k$, for $k = 0, 1, \ldots, n$, is deduced, where, by (1.3),

$$b_{n,k}(q) = \sum_{1 \le m_1 < m_2 < \cdots < m_k \le n} q^{m_1+m_2+\cdots+m_k-k} = q^{\binom{k}{2}} \begin{bmatrix} n \\ k \end{bmatrix}_q.$$

Hence, according to the summation principle, expression (1.14) is established.

An alternative algebraic derivation of the q-binomial formula (1.14) may be carried out as follows. Consider the sequence of sums

$$s_n(t;q) = \sum_{k=0}^{n} q^{\binom{k}{2}} \begin{bmatrix} n \\ k \end{bmatrix}_q t^k, \quad n = 1, 2, \ldots.$$

Using the triangular recurrence relation (1.1), with $x = n$, together with the relation $\binom{k}{2} = \binom{k-1}{2} + \binom{k-1}{1}$, we deduce for the sequence $s_n(t;q)$, $n = 1, 2, \ldots$, the following first-order recurrence relation:

$$s_n(t;q) = (1 + tq^{n-1})s_{n-1}(t;q), \quad n = 2, 3, \ldots,$$

with initial condition $s_1(t;q) = 1 + t$. Therefore,

$$s_n(t;q) = \prod_{i=1}^{n}(1 + tq^{i-1})$$

and (1.14) is shown. □

An extension of the q-binomial formula (1.14) to a negative q-binomial formula is given in the next theorem.

Theorem 1.6. *Let n be a positive integer and let t and q be real numbers, with $|t| < 1$ and $|q| < 1$ or $|t| < |q|^{-(n-1)}$ and $|q| > 1$. Then,*

$$\prod_{j=1}^{n}(1 - tq^{j-1})^{-1} = \sum_{k=0}^{\infty} \begin{bmatrix} n+k-1 \\ k \end{bmatrix}_q t^k. \tag{1.15}$$

Proof. The general term of the product $\prod_{j=1}^{n}(1 - tq^{j-1})^{-1}$ may be expanded into a geometric series as follows:

$$a_j(t;q) = (1 - tq^{j-1})^{-1} = \sum_{k=0}^{\infty} t^k q^{(j-1)k}, \quad j = 1, 2, \ldots, n.$$

Thus, the multiplications and additions in the product $\prod_{j=1}^{n}(1 - tq^{j-1})^{-1}$ may be carried out by (a) selecting the term $t^{k_j}q^{(j-1)k_j}$, $k_j \geq 0$, from the factor

$$a_j(t;q) = 1 + tq^{j-1} + t^2q^{(j-1)2} + \cdots + t^kq^{(j-1)k} + \cdots,$$

for $j = 1, 2, \ldots, n$; (b) forming the products

$$q^{(k_1+2k_2+\cdots+nk_n)-(k_1+k_2+\cdots+k_n)}t^{k_1+k_2+\cdots+k_n};$$

and (c) adding together these products for $k_j \geq 0$, $j = 1, 2, \ldots, n$, with $k_1 + k_2 + \cdots + k_n = k$, a summand of the form $b_{n,k}(q)t^k$, for $k = 0, 1, \ldots$, is deduced, where, by (1.5),

$$b_{n,k}(q) = q^{-k} \sum_{\substack{k_j \geq 0,\, j=1,2,\ldots,n \\ k_1+k_2+\cdots+k_n=k}} q^{k_1+2k_2+\cdots+nk_n} = \begin{bmatrix} n+k-1 \\ k \end{bmatrix}_q.$$

Hence, according to the summation principle, expression (1.15) is established.

An alternative algebraic derivation of the negative q-binomial formula (1.15) may be carried out as follows. Consider the convergent sequence

$$s_n(t;q) = \sum_{k=0}^{\infty} \begin{bmatrix} n+k-1 \\ k \end{bmatrix}_q t^k, \quad n = 1, 2, \ldots, \quad |t| < 1, \quad |q| < 1.$$

Using the triangular recurrence relation

$$\begin{bmatrix} n+k-1 \\ k \end{bmatrix}_q = \begin{bmatrix} n+k-2 \\ k \end{bmatrix}_q + q^{n-1} \begin{bmatrix} n+k-2 \\ k-1 \end{bmatrix}_q,$$

it follows that the sequence $s_n(t;q)$, $n = 1, 2, \ldots$, satisfies the first-order recurrence relation

$$s_n(t;q) = (1 - tq^{n-1})^{-1} s_{n-1}(t;q), \quad n = 2, 3, \ldots,$$

with initial condition $s_1(t;q) = (1-t)^{-1}$. Therefore,

$$s_n(t;q) = \prod_{i=1}^{n} (1 - tq^{i-1})^{-1}, \quad n = 1, 2, \ldots, \quad |t| < 1, \quad |q| < 1,$$

and (1.15) is shown for $n = 1, 2, \ldots$, $|t| < 1$ and $|q| < 1$.

The case $s_n(t;q)$, $n = 1, 2, \ldots$, for $|t| < |q|^{-(n-1)}$ and $|q| > 1$, by setting $p = q^{-1}$ and $u = tq^{n-1}$ and using the relation

$$\begin{bmatrix} n+k-1 \\ k \end{bmatrix}_q = q^{(n-1)k} \begin{bmatrix} n+k-1 \\ k \end{bmatrix}_{q^{-1}} = p^{-(n-1)k} \begin{bmatrix} n+k-1 \\ k \end{bmatrix}_p,$$

reduces to

$$c_n(u;p) = \sum_{k=0}^{\infty} \begin{bmatrix} n+k-1 \\ k \end{bmatrix}_p u^k, \quad |u| < 1, |p| < 1, \quad n = 1, 2, \ldots,$$

which is exactly the previous case. □

The q-binomial coefficients satisfy an orthogonality relation. Specifically, the next theorem is shown.

Theorem 1.7. *Let n and k be positive integers and let q be a real number, with $q \neq 1$. Then, the following orthogonality relations hold true:*

$$\sum_{r=k}^{n} (-1)^{n-r} q^{\binom{n-r}{2}} \begin{bmatrix} n \\ r \end{bmatrix}_q \begin{bmatrix} r \\ k \end{bmatrix}_q = \delta_{n,k}, \quad \sum_{r=k}^{n} (-1)^{r-k} q^{\binom{r-k}{2}} \begin{bmatrix} n \\ r \end{bmatrix}_q \begin{bmatrix} r \\ k \end{bmatrix}_q = \delta_{n,k}, \tag{1.16}$$

where $\delta_{n,k} = 1$, if $k = n$ and $\delta_{n,k} = 0$, if $k \neq n$, is the Kronecker delta.

Proof. The first of the orthogonality relations (1.16), since

$$\begin{aligned} \begin{bmatrix} n \\ r \end{bmatrix}_q \begin{bmatrix} r \\ k \end{bmatrix}_q &= \frac{[n]_q!}{[r]_q![n-r]_q!} \cdot \frac{[r]_q!}{[k]_q![r-k]_q!} \\ &= \frac{[n]_q!}{[k]_q![n-k]_q!} \cdot \frac{[n-k]_q!}{[n-r]_q![r-k]_q!} = \begin{bmatrix} n \\ k \end{bmatrix}_q \begin{bmatrix} n-k \\ n-r \end{bmatrix}_q, \end{aligned}$$

may be expressed as

$$\sum_{r=k}^{n}(-1)^{n-r}q^{\binom{n-r}{2}}\begin{bmatrix}n\\r\end{bmatrix}_q\begin{bmatrix}r\\k\end{bmatrix}_q=\begin{bmatrix}n\\k\end{bmatrix}_q\sum_{r=k}^{n}(-1)^{n-r}q^{\binom{n-r}{2}}\begin{bmatrix}n-k\\n-r\end{bmatrix}_q$$
$$=\begin{bmatrix}n\\k\end{bmatrix}_q\sum_{j=0}^{n-k}(-1)^{j}q^{\binom{j}{2}}\begin{bmatrix}n-k\\j\end{bmatrix}_q.$$

The last sum, by the q-binomial formula (1.14), equals $\delta_{n,k}$ and so the first of (1.16) is proved.

The second of the orthogonality relations (1.16), using the identity

$$\begin{bmatrix}n\\r\end{bmatrix}_q\begin{bmatrix}r\\k\end{bmatrix}_q=\begin{bmatrix}n\\k\end{bmatrix}_q\begin{bmatrix}n-k\\r-k\end{bmatrix}_q,$$

can be similarly shown. □

A useful inversion of the q-binomial formula (1.14), is derived in the following corollary of Theorem 1.5. As a particular case of it another useful formula is deduced.

Corollary 1.2. *Let n be a positive integer and let t and q be real numbers, with $q \neq 1$. Then,*

$$\sum_{k=0}^{n}(-1)^{k}q^{\binom{k}{2}}\begin{bmatrix}n\\k\end{bmatrix}_q\prod_{i=1}^{k}(1-tq^{-(i-1)})=t^n. \tag{1.17}$$

In particular,

$$\sum_{k=0}^{n}(-1)^{k}(1-q)^{k}q^{\binom{k}{2}}\begin{bmatrix}n\\k\end{bmatrix}_q[t]_{k,q}=q^{nt}. \tag{1.18}$$

Proof. The q-binomial formula (1.14), by replacing q by q^{-1}, t by $-t$, n by k, the dummy variable k by r and using the relation

$$\begin{bmatrix}k\\r\end{bmatrix}_{q^{-1}}=q^{-r(k-r)}\begin{bmatrix}k\\r\end{bmatrix}_q,$$

is expressed as

$$\prod_{i=1}^{k}(1-tq^{-(i-1)})=\sum_{r=0}^{k}(-1)^{r}q^{-\binom{r}{2}-r(k-r)}\begin{bmatrix}k\\r\end{bmatrix}_q t^r.$$

Multiplying both members of the last expression by $(-1)^kq^{\binom{k}{2}}\begin{bmatrix}n\\k\end{bmatrix}_q$, reducing the exponent of q in the general term of the sum, using the relation

$$\binom{k-r}{2}=\binom{k}{2}-\binom{r}{2}-r(k-r),$$

and summing the resulting expression for $k = 0, 1, \ldots, n$, we get

$$\sum_{k=0}^{n} (-1)^k q^{\binom{k}{2}} \begin{bmatrix} n \\ k \end{bmatrix}_q \prod_{i=1}^{k} (1 - tq^{-(i-1)}) = \sum_{k=0}^{n} \begin{bmatrix} n \\ k \end{bmatrix}_q \sum_{r=0}^{k} (-1)^{k-r} q^{\binom{k-r}{2}} \begin{bmatrix} k \\ r \end{bmatrix}_q t^r$$
$$= \sum_{r=0}^{n} \left(\sum_{k=r}^{n} (-1)^{k-r} q^{\binom{k-r}{2}} \begin{bmatrix} n \\ k \end{bmatrix}_q \begin{bmatrix} k \\ r \end{bmatrix}_q \right) t^r.$$

Since, by (1.16),

$$\sum_{k=r}^{n} (-1)^{k-r} q^{\binom{k-r}{2}} \begin{bmatrix} n \\ k \end{bmatrix}_q \begin{bmatrix} k \\ r \end{bmatrix}_q = \delta_{n,r},$$

expression (1.17) is readily deduced. The last expression, by replacing t by q^t and using the relation

$$\prod_{i=1}^{k} (1 - q^{t-i+1}) = (1 - q)^k [t]_{k,q},$$

reduces to (1.18). □

The inverse of the q-binomial formula (1.17) and its particular case (1.18) are rewritten in the following remark in a form useful in the theory of discrete q-distributions.

Remark 1.4. *Another q-binomial formula.* Expressions of (1.17) and (1.18) may be rewritten as

$$\sum_{k=0}^{n} \begin{bmatrix} n \\ k \end{bmatrix}_q t^{n-k} \prod_{i=1}^{k} (1 - tq^{i-1}) = \sum_{k=0}^{n} \begin{bmatrix} n \\ k \end{bmatrix}_q t^k \prod_{i=1}^{n-k} (1 - tq^{i-1}) = 1 \tag{1.19}$$

and

$$\sum_{k=0}^{n} \begin{bmatrix} n \\ k \end{bmatrix}_q q^{(n-k)(t-k)} (1 - q)^k [t]_{k,q} = 1, \tag{1.20}$$

respectively.

Indeed, expression (1.17), by replacing t by t^{-1} and using the relation

$$(-1)^k q^{\binom{k}{2}} \prod_{i=1}^{k} (1 - t^{-1} q^{-(i-1)}) = t^{-k} \prod_{i=1}^{k} (1 - tq^{i-1}),$$

is rewritten as (1.19). Also, expression (1.18), by replacing q by q^{-1} and using the relation

$$(-1)^k (1 - q^{-1})^k q^{-\binom{k}{2}} \begin{bmatrix} n \\ k \end{bmatrix}_{q^{-1}} [t]_{k,q^{-1}} = \begin{bmatrix} n \\ k \end{bmatrix}_q q^{-k(t+n-k)} (1 - q)^k [t]_{k,q},$$

reduces to (1.20).

Some interesting applications of the q-binomial and the negative q-binomial formulae are worked out in the following examples.

Example 1.5. For r, s, and n positive integers and $q \neq 1$ a real number, show that

$$\sum_{k=0}^{n} q^{(n-k)(r-k)} \begin{bmatrix} r \\ k \end{bmatrix}_q \begin{bmatrix} s \\ n-k \end{bmatrix}_q = \begin{bmatrix} r+s \\ n \end{bmatrix}_q$$

and conclude that

$$\sum_{k=0}^{r-m} q^{(r-k)(r-k-m)} \begin{bmatrix} r \\ k \end{bmatrix}_q \begin{bmatrix} r \\ k+m \end{bmatrix}_q = \begin{bmatrix} 2r \\ r+m \end{bmatrix}_q, \quad m = 0, 1, \ldots, r,$$

and

$$\sum_{k=0}^{r} q^{(r-k)^2} \begin{bmatrix} r \\ k \end{bmatrix}_q^2 = \begin{bmatrix} 2r \\ r \end{bmatrix}_q.$$

The expansion of the identity

$$\prod_{i=1}^{r}(1+tq^{i-1}) \prod_{i=1}^{s}(1+(tq^r)q^{i-1}) = \prod_{i=1}^{r+s}(1+tq^{i-1})$$

into powers of t, using the q-binomial formula (1.14), yields

$$\sum_{k=0}^{r} q^{\binom{k}{2}} \begin{bmatrix} r \\ k \end{bmatrix}_q t^k \sum_{j=0}^{s} q^{\binom{j}{2}+rj} \begin{bmatrix} s \\ j \end{bmatrix}_q t^j = \sum_{n=0}^{r+s} q^{\binom{n}{2}} \begin{bmatrix} r+s \\ n \end{bmatrix}_q t^n.$$

Executing the multiplication in the left-hand side, the general term $a_n t^n$, for all $n = 0, 1, \ldots, r+s$, is formed by multiplying the term $q^{\binom{k}{2}} \begin{bmatrix} r \\ k \end{bmatrix}_q t^k$ of the first factor by the term $q^{\binom{n-k}{2}+r(n-k)} \begin{bmatrix} s \\ n-k \end{bmatrix}_q t^{n-k}$ of the second factor, for all values $k = 0, 1, \ldots, n$. Thus, by the addition principle and using for the exponent of q the relation $\binom{k}{2} + \binom{n-k}{2} = \binom{n}{2} - k(n-k)$, it follows that

$$a_n = q^{\binom{n}{2}} \sum_{k=0}^{n} q^{(n-k)(r-k)} \begin{bmatrix} r \\ k \end{bmatrix}_q \begin{bmatrix} s \\ n-k \end{bmatrix}_q$$

and

$$\sum_{n=0}^{r+s} q^{\binom{n}{2}} \left(\sum_{k=0}^{n} q^{(n-k)(r-k)} \begin{bmatrix} r \\ k \end{bmatrix}_q \begin{bmatrix} s \\ n-k \end{bmatrix}_q \right) t^n = \sum_{n=0}^{r+s} q^{\binom{n}{2}} \begin{bmatrix} r+s \\ n \end{bmatrix}_q t^n.$$

Equating the coefficients of t^n of both sides, we find

$$\sum_{k=0}^{n} q^{(n-k)(r-k)} \begin{bmatrix} r \\ k \end{bmatrix}_q \begin{bmatrix} s \\ n-k \end{bmatrix}_q = \begin{bmatrix} r+s \\ n \end{bmatrix}_q.$$

Note that this formula constitutes a particular case, with $x = r$ and $y = s$ positive integers, of q-Cauchy's (q-binomial convolution) formula (1.8). Substituting in this formula $s = r$, $n = r - m$ and since

$$\begin{bmatrix} r \\ r-m-k \end{bmatrix}_q = \begin{bmatrix} r \\ k+m \end{bmatrix}_q, \quad \begin{bmatrix} 2r \\ r-m \end{bmatrix}_q = \begin{bmatrix} 2r \\ r+m \end{bmatrix}_q,$$

we conclude that

$$\sum_{k=0}^{r-m} q^{(r-k)(r-k-m)} \begin{bmatrix} r \\ k \end{bmatrix}_q \begin{bmatrix} r \\ k+m \end{bmatrix}_q = \begin{bmatrix} 2r \\ r+m \end{bmatrix}_q.$$

Also, setting $r = s = n$, and since $\left[\begin{smallmatrix} r \\ r-k \end{smallmatrix}\right]_q = \left[\begin{smallmatrix} r \\ k \end{smallmatrix}\right]_q$,

$$\sum_{k=0}^{r} q^{(r-k)^2} \begin{bmatrix} r \\ k \end{bmatrix}_q^2 = \begin{bmatrix} 2r \\ r \end{bmatrix}_q.$$

Example 1.6. For r, s, and n positive integers and $q \neq 1$ a real number, show that

$$\sum_{k=0}^{n} q^{r(n-k)} \begin{bmatrix} r+k-1 \\ k \end{bmatrix}_q \begin{bmatrix} s+n-k-1 \\ n-k \end{bmatrix}_q = \begin{bmatrix} r+s+n-1 \\ n \end{bmatrix}_q.$$

The expansion of the identity

$$\prod_{i=1}^{r} (1 - tq^{i-1})^{-1} \prod_{i=1}^{s} (1 - (tq^r)q^{i-1})^{-1} = \prod_{i=1}^{r+s} (1 - tq^{i-1})^{-1}, |t| < 1,$$

using the negative q-binomial formula (1.15), yields

$$\sum_{k=0}^{\infty} \begin{bmatrix} r+k-1 \\ k \end{bmatrix}_q t^k \sum_{j=0}^{\infty} \begin{bmatrix} s+j-1 \\ j \end{bmatrix}_q q^{rj} t^j = \sum_{n=0}^{\infty} \begin{bmatrix} r+s+n-1 \\ n \end{bmatrix}_q t^n$$

and

$$\sum_{n=0}^{\infty} \sum_{k=0}^{n} q^{r(n-k)} \begin{bmatrix} r+k-1 \\ k \end{bmatrix}_q \begin{bmatrix} s+n-k-1 \\ n-k \end{bmatrix}_q t^n = \sum_{n=0}^{\infty} \begin{bmatrix} r+s+n-1 \\ n \end{bmatrix}_q t^n,$$

for $|t| < 1$. Therefore,

$$\sum_{k=0}^{n} q^{r(n-k)} \begin{bmatrix} r+k-1 \\ k \end{bmatrix}_q \begin{bmatrix} s+n-k-1 \\ n-k \end{bmatrix}_q = \begin{bmatrix} r+s+n-1 \\ n \end{bmatrix}_q.$$

Note that this formula constitutes a particular case, with $x = -r$ and $y = -s$ negative integers, of q-Cauchy's formula (1.8).

Example 1.7. For n, m, and k positive integers, with $k \leq m \leq n$, show that

$$\sum_{r=k}^{k+n-m} q^{k(n-r)} \begin{bmatrix} r-1 \\ k-1 \end{bmatrix}_q \begin{bmatrix} n-r \\ m-k \end{bmatrix}_q = \begin{bmatrix} n \\ m \end{bmatrix}_q.$$

Multiplying both members of the negative q-binomial formula

$$\sum_{j=0}^{\infty} \begin{bmatrix} m+j \\ m \end{bmatrix}_q t^j = \prod_{i=1}^{m+1} (1 - tq^{i-1})^{-1},$$

by t^m and substituting $n = m + j$, we get the expansion

$$\sum_{n=m}^{\infty} \begin{bmatrix} n \\ m \end{bmatrix}_q t^n = t^m \prod_{i=1}^{m+1} (1 - tq^{i-1})^{-1}.$$

In the same way, we deduce that

$$\sum_{r=k}^{\infty} \begin{bmatrix} r-1 \\ k-1 \end{bmatrix}_q t^r = t^k \prod_{i=1}^{k} (1 - tq^{i-1})^{-1}$$

and

$$\sum_{j=m-k}^{\infty} \begin{bmatrix} j \\ m-k \end{bmatrix}_q q^{kj} t^j = t^{m-k} \prod_{i=1}^{m-k+1} (1 - (tq^k)q^{i-1})^{-1}.$$

Expanding the identity

$$t^k \prod_{i=1}^{k} (1 - tq^{i-1})^{-1} \cdot t^{m-k} \prod_{i=1}^{m-k+1} (1 - (tq^k)q^{i-1})^{-1} = t^m \prod_{i=1}^{m+1} (1 - tq^{i-1})^{-1},$$

using the preceding three expansions, we get

$$\left(\sum_{r=k}^{\infty} \begin{bmatrix} r-1 \\ k-1 \end{bmatrix}_q t^r \right) \cdot \left(\sum_{j=m-k}^{\infty} \begin{bmatrix} j \\ m-k \end{bmatrix}_q q^{kj} t^j \right) = \sum_{n=m}^{\infty} \begin{bmatrix} n \\ m \end{bmatrix}_q t^n$$

and

$$\sum_{n=m}^{\infty} \left(\sum_{r=k}^{k+n-m} q^{k(n-r)} \begin{bmatrix} r-1 \\ k-1 \end{bmatrix}_q \begin{bmatrix} n-r \\ m-k \end{bmatrix}_q \right) t^n = \sum_{n=m}^{\infty} \begin{bmatrix} n \\ m \end{bmatrix}_q t^n,$$

for $|t| < 1$. Therefore,

$$\sum_{r=k}^{k+n-m} q^{k(n-r)} \begin{bmatrix} r-1 \\ k-1 \end{bmatrix}_q \begin{bmatrix} n-r \\ m-k \end{bmatrix}_q = \begin{bmatrix} n \\ m \end{bmatrix}_q.$$

1.5 GENERAL q-BINOMIAL FORMULA AND q-EXPONENTIAL FUNCTIONS

A general q-binomial formula and, alternatively, a general negative q-binomial formula are given in the following theorem. The adopted proof is due to Heine (1847, 1878).

Theorem 1.8. *Let x, t, and q be real numbers, with $|t| < 1$ and $|q| < 1$. Then,*

$$\prod_{i=1}^{\infty} \frac{1 + tq^{i-1}}{1 + tq^{x+i-1}} = \sum_{k=0}^{\infty} q^{\binom{k}{2}} \begin{bmatrix} x \\ k \end{bmatrix}_q t^k. \tag{1.21}$$

Alternatively,

$$\prod_{i=1}^{\infty} \frac{1 - tq^{x+i-1}}{1 - tq^{i-1}} = \sum_{k=0}^{\infty} \begin{bmatrix} x + k - 1 \\ k \end{bmatrix}_q t^k. \tag{1.22}$$

Proof. Let us consider the convergent series

$$h(x; t) = \sum_{k=0}^{\infty} q^{\binom{k}{2}} \begin{bmatrix} x \\ k \end{bmatrix}_q t^k, \quad |t| < 1, \quad |q| < 1$$

and let us compute the difference

$$h(x; t) - h(x - 1; t) = \sum_{k=0}^{\infty} q^{\binom{k}{2}} \left\{ \begin{bmatrix} x \\ k \end{bmatrix}_q - \begin{bmatrix} x - 1 \\ k \end{bmatrix}_q \right\} t^k, \quad |t| < 1, \quad |q| < 1.$$

Using the recurrence relation (1.1), it follows that

$$\begin{aligned} h(x; t) - h(x - 1; t) &= \sum_{k=1}^{\infty} q^{\binom{k}{2}+x-k} \begin{bmatrix} x - 1 \\ k - 1 \end{bmatrix}_q t^k \\ &= tq^{x-1} \sum_{k=1}^{\infty} q^{\binom{k-1}{2}} \begin{bmatrix} x - 1 \\ k - 1 \end{bmatrix}_q t^{k-1} \\ &= tq^{x-1} h(x - 1; t). \end{aligned}$$

Also

$$\begin{aligned} h(x; t) - h(x; tq) &= \sum_{k=0}^{\infty} q^{\binom{k}{2}} \begin{bmatrix} x \\ k \end{bmatrix}_q (1 - q^k) t^k \\ &= t(1 - q^x) \sum_{k=1}^{\infty} q^{\binom{k-1}{2}} \begin{bmatrix} x - 1 \\ k - 1 \end{bmatrix}_q (tq)^{k-1} \\ &= t(1 - q^x) h(x - 1; tq). \end{aligned}$$

Eliminating $h(x-1;tq)$ from the expressions

$$h(x;tq) = (1+tq^x)h(x-1;tq)$$

and

$$h(x;t) - h(x;tq) = t(1-q^x)h(x-1;tq),$$

we get the relation

$$h(x;t) = \frac{1+t}{1+tq^x}h(x;tq).$$

Iterating this relation $n-1$ times, we find

$$h(x;t) = \prod_{i=1}^{n} \frac{1+tq^{i-1}}{1+tq^{x+i-1}}h(x;tq^n).$$

Letting $n \to \infty$, and since $\lim_{n\to\infty} h(x;tq^n) = h(x;0) = 1$, for $|q| < 1$, we deduce the expression

$$h(x;t) = \prod_{i=1}^{\infty} \frac{1+tq^{i-1}}{1+tq^{x+i-1}},$$

which shows (1.21). The alternative expression (1.22) is deduced from (1.21), by replacing successively x by $-x$ and t by $-tq^x$ and finally using the relation

$$\begin{bmatrix} -x \\ k \end{bmatrix}_q = (-1)^k q^{-xk-\binom{k}{2}} \begin{bmatrix} x+k-1 \\ k \end{bmatrix}_q.$$

Hence, the proof of the theorem is completed. □

A q-analogue of the exponential function can be obtained from (1.21) by replacing t by $(1-q)t$ and then taking the limit as $x \to \infty$. Since, for $|q| < 1$,

$$\lim_{x\to\infty}(1-q)[x-j]_q = \lim_{x\to\infty}(1-q^{x-j}) = 1, \quad \lim_{x\to\infty}(1-q)^k[x]_{k,q} = 1,$$

a *q-exponential function* is deduced as

$$E_q(t) = \prod_{i=1}^{\infty}(1+t(1-q)q^{i-1}) = \sum_{k=0}^{\infty} q^{\binom{k}{2}} \frac{t^k}{[k]_q!}, \quad -\infty < t < \infty. \tag{1.23}$$

Another *q-exponential function* can be similarly obtained from (1.22) as

$$e_q(t) = \prod_{i=1}^{\infty}(1-t(1-q)q^{i-1})^{-1} = \sum_{k=0}^{\infty} \frac{t^k}{[k]_q!}, \quad |t| < 1/(1-q). \tag{1.24}$$

The limiting expressions of the q-exponential functions, as $q \to 1$,

$$\lim_{q\to1} E_q(t) = e^t, \quad \lim_{q\to1} e_q(t) = e^t,$$

justify their name. Clearly, these functions satisfy the relation

$$E_q(t)e_q(-t) = 1, \quad |t| < 1/(1-q), \quad |q| < 1.$$

Also, since $[k]_{q^{-1}}! = q^{-\binom{k}{2}}[k]_q!$, the following relation is readily obtained:

$$E_{q^{-1}}(t) = e_q(t), \quad |t| < 1/(1-q), \quad |q| < 1.$$

A q-analogue of the logarithmic function can be obtained from (1.22) by subtracting 1 from both sides, dividing the resulting expression by $[x]_q$ and taking the limit as $x \to 0$. Since

$$\lim_{x\to 0} \frac{1}{[x]_q} \begin{bmatrix} x+k-1 \\ k \end{bmatrix}_q = \frac{1}{[k]_q} \cdot \frac{\lim_{x\to 0}[x+1]_q[x+2]_q \cdots [x+k-1]_q}{[k-1]_q!} = \frac{1}{[k]_q},$$

a *q-logarithmic function* is deduced as

$$-l_q(1-t) = \lim_{x\to 0} \frac{1}{[x]_q} \left(\prod_{i=1}^{\infty} \frac{1-tq^{x+i-1}}{1-tq^{i-1}} - 1 \right) = \sum_{k=1}^{\infty} \frac{t^k}{[k]_q}, \quad |t| < 1. \tag{1.25}$$

The limiting expression of the q-logarithmic function, as $q \to 1$,

$$\lim_{q\to 1} l_q(1-t) = \log(1-t), \quad |t| < 1,$$

justifies its name.

1.6 q-STIRLING NUMBERS

Consider the noncentral q-factorial of t of order n and noncentrality parameter r,

$$[t-r]_{n,q} = [t-r]_q[t-r-1]_q \cdots [t-r-n+1]_q,$$

with t, q, and r real numbers, $q \neq 1$, and n a positive integer. Using the relation

$$[t-r-j]_q = q^{-r-j}([t]_q - [r+j]_q), \quad j = 0, 1, \ldots,$$

it may be expressed as

$$[t-r]_{n,q} = q^{-\binom{n}{2}-rn}([t]_q - [r]_q)([t]_q - [r+1]_q) \cdots ([t]_q - [r+n-1]_q).$$

This is a polynomial of the q-number $[t]_q$ of degree n. Executing the multiplications and arranging the terms in ascending order of powers of $[t]_q$, we get

$$[t-r]_{n,q} = q^{-\binom{n}{2}-rn} \sum_{k=0}^{n} s_q(n,k;r)[t]_q^k, \quad n = 0, 1, \ldots . \tag{1.26}$$

Inversely, the nth power of the q-number $[t]_q$ may be expressed in the form of a polynomial of noncentral q-factorials of t. Specifically,

$$[t]_q^n = \sum_{k=0}^{n} q^{\binom{k}{2}+rk} S_q(n,k;r)[t-r]_{k,q}, \quad n = 0,1,\ldots, \tag{1.27}$$

or equivalently

$$[t+r]_q^n = \sum_{k=0}^{n} q^{\binom{k}{2}+rk} S_q(n,k;r)[t]_{k,q}, \quad n = 0,1,\ldots . \tag{1.28}$$

The coefficients $s_q(n,k;r)$ and $S_q(n,k;r)$ are called *noncentral q-Stirling numbers of the first and second kind*, respectively. Note that for $r = 0$ the noncentral q-Stirling numbers of the first and second kind reduce to $s_q(n,k;0) = s_q(n,k)$ and $S_q(n,k;0) = S_q(n,k)$, the usual (central) q-Stirling numbers of the first and second kind, respectively.

Furthermore, consider the expansion of the noncentral ascending q-factorial of t of order n and noncentrality parameter r,

$$\begin{aligned}[t+r+n-1]_{n,q} &= [t+r]_q[t+r+1]_q \cdots [t+r+n-1]_q \\ &= q^{\binom{n}{2}+rn}([t]_q + q^{-1}[r]_{q^{-1}}) \cdots ([t]_q + q^{-1}[r+n-1]_{q^{-1}}),\end{aligned}$$

into a polynomial of $[t]_q$,

$$[t+r+n-1]_{n,q} = q^{\binom{n}{2}+rn} \sum_{k=0}^{n} |s_{q^{-1}}(n,k;r)|[t]_q^k, \quad n = 0,1,\ldots, \tag{1.29}$$

for $q \neq 0$. Since

$$[t+r+n-1]_{n,q} = [-1]_q^n[-t-r]_{n,q^{-1}}, \quad [-t]_q = [-1]_q[t]_{q^{-1}},$$

where $[-1]_q = -q^{-1}$, on using (1.26), with $-t$ instead of t and q^{-1} instead of q, it follows that

$$|s_{q^{-1}}(n,k;r)| = [-1]_q^{n-k} s_{q^{-1}}(n,k;r),$$

or, equivalently, that

$$|s_q(n,k;r)| = [-1]_q^{-(n-k)} s_q(n,k;r). \tag{1.30}$$

The coefficient $|s_q(n,k;r)|$, which for $0 < q < 1$ or $1 < q < \infty$ and $r \geq 0$ is positive, is called *signless (or absolute) noncentral q-Stirling number of the first kind.*

The noncentral q-Stirling numbers of the first and second kind constitute a pair of orthogonal bivariate sequences. This is shown in the next theorem.

Theorem 1.9. *The q-Stirling numbers of the first and second kind satisfy the orthogonality relations*

$$\sum_{m=k}^{n} s_q(n,m;r)S_q(m,k;r) = \delta_{n,k}, \quad \sum_{m=k}^{n} S_q(n,m;r)s_q(m,k;r) = \delta_{n,k}, \tag{1.31}$$

where $\delta_{n,k} = 1$*, if* $k = n$ *and* $\delta_{n,k} = 0$*, if* $k \neq n$*, is the Kronecker delta.*

Proof. Expanding the noncentral q-factorial of t of order n and noncentrality parameter r, $[t-r]_{n,q}$, into powers of $[t]_q$, by using (1.26), and in the resulting expression expanding the powers of $[t]_q$ into noncentral q-factorials $[t-r]_{k,q}$, for $k = 0, 1, \ldots, n$, by using (1.27), we get

$$\begin{aligned}[t-r]_{n,q} &= q^{-\binom{n}{2}-rn} \sum_{m=0}^{n} s_q(n,m;r)[t]_q^m \\ &= q^{-\binom{n}{2}-rn} \sum_{m=0}^{n} s_q(n,m;r) \sum_{k=0}^{m} q^{\binom{k}{2}+rk} S_q(m,k;r)[t-r]_{k,q}.\end{aligned}$$

Furthermore, interchanging the order of summation, we deduce the relation

$$[t-r]_{n,q} = \sum_{k=0}^{n} q^{-\binom{n}{2}+\binom{k}{2}-r(n-k)} \left\{ \sum_{m=k}^{n} s_q(n,m;r)S_q(m,k;r) \right\} [t-r]_{k,q},$$

which implies the first of (1.31). Similarly, expanding the nth power of the q-number $[t]_q$, $[t]_q^n$, into noncentral q-factorials $[t-r]_{m,q}$, $m = 0, 1, \ldots, n$, and in the resulting expression expanding the noncentral q-factorials $[t-r]_{m,q}$, $m = 0, 1, \ldots, n$, into powers of $[t]_q$, we deduce the second of (1.31). □

Triangular recurrence relations for the noncentral q-Stirling numbers of the first and second kind are derived in the following theorem.

Theorem 1.10. *The noncentral q-Stirling numbers of the first kind* $s_q(n,k;r)$*,* $k = 0, 1, \ldots, n$*,* $n = 0, 1, \ldots$*, satisfy the triangular recurrence relation*

$$s_q(n+1,k;r) = s_q(n,k-1;r) - [n+r]_q s_q(n,k;r), \tag{1.32}$$

for $k = 1, 2, \ldots, n+1$ *and* $n = 0, 1, \ldots$*, with initial conditions*

$$s_q(0,0;r) = 1, s_q(0,k;r) = 0, k > 0, s_q(n,0;r) = q^{\binom{n}{2}+rn}[-r]_{n,q}, n > 0.$$

Also, the noncentral q-Stirling numbers of the second kind $S_q(n,k;r)$*,* $k = 0, 1, \ldots, n$*,* $n = 0, 1, \ldots$*, satisfy the triangular recurrence relation*

$$S_q(n+1,k;r) = S_q(n,k-1;r) + [k+r]_q S_q(n,k;r), \tag{1.33}$$

for $k = 1, 2, \ldots, n+1$ *and* $n = 0, 1, \ldots,$ *with initial conditions*

$$S_q(0,0;r) = 1, \quad S_q(0,k;r) = 0, k > 0, \quad S_q(n,0;r) = [r]_q^n, n > 0.$$

Proof. Expanding both members of the recurrence relation

$$[t-r]_{n+1,q} = [t-(n+r)]_q [t-r]_{n,q} = q^{-n-r}([t]_q - [n+r]_q)[t-r]_{n,q}$$

into powers of $[t]_q$, according to (1.26), we get the relation

$$q^{-\binom{n+1}{2}-r(n+1)} \sum_{k=0}^{n+1} s_q(n+1,k;r)[t]_q^k = q^{-\binom{n}{2}-rn-n-r} \left\{ \sum_{j=0}^{n} s_q(n,j;r)[t]_q^{j+1} - \sum_{k=0}^{n} [n+r]_q s_q(n,k;r)[t]_q^k \right\}$$

or equivalently, the relation

$$\sum_{k=0}^{n+1} s_q(n+1,k;r)[t]_q^k = \sum_{k=1}^{n+1} s_q(n,k-1;r)[t]_q^k - \sum_{k=0}^{n} [n+r]_q s_q(n,k;r)[t]_q^k,$$

which implies (1.32). The initial conditions follow directly from (1.26).

Also, expanding both members of the recurrence relation

$$[t+r]_q^{n+1} = [t+r]_q [t+r]_q^n$$

into factorials of the q-number $[t]_q$, according to (1.28), we have

$$\sum_{k=0}^{n+1} q^{\binom{k}{2}+rk} S_q(n+1,k;r)[t]_{k,q} = \sum_{j=0}^{n} q^{\binom{j}{2}+rj} S_q(n,j;r)[t+r]_q [t]_{j,q}.$$

Since

$$[t+r]_q [t]_{j,q} = [(t-j)+(j+r)]_q [t]_{j,q} = (q^{j+r}[t-j]_q + [j+r]_q)[t]_{j,q},$$

whence

$$[t+r]_q [t]_{j,q} = q^{j+r}[t]_{j+1,q} + [j+r]_q [t]_{j,q},$$

we deduce the relation

$$\begin{aligned} \sum_{k=0}^{n+1} & q^{\binom{k}{2}+rk} S_q(n+1,k;r)[t]_{k,q} \\ &= \sum_{j=0}^{n} q^{\binom{j+1}{2}+r(j+1)} S_q(n,j;r)[t]_{j+1,q} + \sum_{j=0}^{n} q^{\binom{j}{2}+rj} [j+r]_q S_q(n,j;r)[t]_{j,q} \\ &= \sum_{k=1}^{n+1} q^{\binom{k}{2}+rk} S_q(n,k-1;r)[t]_{k,q} + \sum_{k=0}^{n} q^{\binom{k}{2}+rk} [k+r]_q S_q(n,k;r)[t]_{k,q}, \end{aligned}$$

which implies (1.33). The initial conditions follow directly from (1.28). □

The triangular recurrence relation (1.33) can be used for the determination of the (power) generating function of the sequence of the noncentral q-Stirling numbers of the second kind $S_q(n,k;r)$, $n = k, k+1, \ldots$, for fixed k. This generating function transformed yields the expansion of the reciprocal noncentral q-factorials into reciprocal q-powers, which inverted provides the expansion of the reciprocal q-powers into reciprocal noncentral q-factorials. Specifically, we have the following theorem.

Theorem 1.11. *The generating function of the sequence of the noncentral q-Stirling numbers of the second kind $S_q(n,k;r)$, $n = k, k+1, \ldots$, for fixed k, is given by*

$$\varphi_k(u;q,r) = \sum_{n=k}^{\infty} S_q(n,k;r)u^n = u^k \prod_{j=0}^{k} (1 - [r+j]_q u)^{-1}, \tag{1.34}$$

for $|u| < 1/[r+k]_q$ and $k = 0, 1, \ldots$.

Proof. Note first that the series in (1.34) is convergent, since

$$\lim_{n\to\infty} \frac{S_q(n,k;r)}{[r+k]_q^n}([r+k]_q u)^n = \frac{1}{[k]_q!} \lim_{n\to\infty} ([r+k]_q u)^n = 0,$$

for $|u| < 1/[r+k]_q$. Furthermore, multiplying the triangular recurrence relation (1.33) by u^{n+1} and summing the resulting expression for $n = k-1, k, \ldots$, and since $S_q(k-1,k;r) = 0$, we get

$$\sum_{n=k-1}^{\infty} S_q(n+1,k;r)u^{n+1} = u \sum_{n=k-1}^{\infty} S_q(n,k-1;r)u^n + [k+r]_q u \sum_{n=k}^{\infty} S_q(n,k;r)u^n,$$

for $k = 1, 2, \ldots$. Consequently

$$\varphi_k(u;q,r) = u\varphi_{k-1}(u;q,r) + [k+r]_q u\varphi_k(u;q,r), \quad k = 1, 2, \ldots$$

and so

$$\varphi_k(u;q,r) = u(1 - [k+r]_q u)^{-1}\varphi_{k-1}(u;q,r), \quad k = 1, 2, \ldots .$$

Applying this recurrence relation repeatedly and since

$$\varphi_0(u;q,r) = \sum_{n=0}^{\infty} S_q(n,0;r)u^n = \sum_{n=0}^{\infty} ([r]_q u)^n = (1 - [r]_q u)^{-1},$$

we deduce (1.34). □

Corollary 1.3. *The reciprocal noncentral q-factorial $1/[t-r]_{k+1,q}$ is expanded into reciprocal q-powers $1/[t]_q^{n+1}$, $n = k, k+1, \ldots$, $k = 0, 1, \ldots$, as*

$$\frac{1}{[t-r]_{k+1,q}} = q^{\binom{k+1}{2}+r(k+1)} \sum_{n=k}^{\infty} S_q(n,k;r)\frac{1}{[t]_q^{n+1}}, \quad t > k+r. \tag{1.35}$$

Inversely, the reciprocal q-power $1/[t]_q^{k+1}$ *is expanded into reciprocal noncentral q-factorials* $1/[t-r]_{n+1,q}$, $n = k, k+1, \ldots, k = 0, 1, \ldots,$ *as*

$$\frac{1}{[t]_q^{k+1}} = \sum_{n=k}^{\infty} q^{-\binom{n+1}{2}-r(n+1)} s_q(n,k;r) \frac{1}{[t-r]_{n+1,q}}, \quad t > k+r. \tag{1.36}$$

Proof. Setting in (1.34) $u = 1/[t]_q$ and since

$$([t]_q - [r]_q)([t]_q - [r+1]_q) \cdots ([t]_q - [r+k]_q) = q^{\binom{k+1}{2}+r(k+1)} [t-r]_{k+1,q},$$

we conclude (1.35). In the expansion (1.35), replacing the bound variable n by m and the fixed number k by n and then multiplying the resulting expression by

$$q^{-\binom{n+1}{2}-r(n+1)} s_q(n,k;r)$$

and summing for $n = k, k+1, \ldots,$ we find

$$\begin{aligned}\sum_{n=k}^{\infty} q^{-\binom{n+1}{2}-r(n+1)} s_q(n,k;r) \frac{1}{[t-r]_{n+1,q}} &= \sum_{n=k}^{\infty} \sum_{m=n}^{\infty} S_q(m,n;r) s_q(n,k;r) \frac{1}{[t]_q^{m+1}} \\ &= \sum_{m=k}^{\infty} \left\{ \sum_{n=k}^{m} S_q(m,n;r) s_q(n,k;r) \right\} \frac{1}{[t]_q^{m+1}}.\end{aligned}$$

Therefore, by the second of the orthogonality relations (1.31), we deduce the relation

$$\sum_{n=k}^{\infty} q^{-\binom{n+1}{2}-r(n+1)} s_q(n,k;r) \frac{1}{[t]_{n+1,q}} = \sum_{m=k}^{\infty} \delta_{m,k} \frac{1}{[t]_q^{m+1}},$$

which implies (1.36). □

Expressions of the noncentral q-Stirling numbers in terms of multiple sums of products of q-numbers, over combinations of positive integers are provided in the next theorem.

Theorem 1.12. *The noncentral q-Stirling number of the first kind* $s_q(n,k;r)$, $k = 1, 2, \ldots, n$, $n = 1, 2, \ldots,$ *is given by the multiple sum*

$$s_q(n,k;r) = (-1)^{n-k} \sum [r+i_1]_q [r+i_2]_q \cdots [r+i_{n-k}]_q, \tag{1.37}$$

where the summation is extended over all $(n-k)$*-combinations* $\{i_1, i_2, \ldots, i_{n-k}\}$ *of the n nonnegative integers* $\{0, 1, \ldots, n-1\}$.

Also, the noncentral q-Stirling number of the second kind $S_q(n,k;r)$, $k = 1, 2, \ldots, n$, $n = 1, 2, \ldots$, *is given by the multiple sum*

$$S_q(n,k;r) = \sum [r+i_1]_q[r+i_2]_q \cdots [r+i_{n-k}]_q, \tag{1.38}$$

where the summation is extended over all $(n-k)$*-combinations* $\{i_1, i_2, \ldots, i_{n-k}\}$*, with repetition, of the* $k+1$ *nonnegative integers* $\{0, 1, \ldots, k\}$.

Proof. According to the definition (1.26), and since

$$[t-r]_{n,q} = q^{-\binom{n}{2}-rn}([t]_q - [r]_q)([t]_q - [r+1]_q) \cdots ([t]_q - [r+n-1]_q),$$

we have

$$([t]_q - [r]_q)([t]_q - [r+1]_q) \cdots ([t]_q - [r+n-1]_q) = \sum_{k=1}^{n} s_q(n,k;r)[t]_q^k,$$

for $n = 2, 3, \ldots$ The ith factor of the product of the left-hand side, $f_i(t;q) = [t]_q - [r+i]_q$, $i = 0, 1, \ldots, n-1$, is a monomial with constant term $-[r+i]_q$. Executing the multiplications, the kth-order power of $[t]_q$ is formed by multiplying the constant terms of any $n-k$ factors $\{i_1, i_2, \ldots, i_{n-k}\}$, out of the n factors $\{0, 1, \ldots, n-1\}$, together with the first-order terms $[t]_q$ of the remaining k factors. Since the coefficient of the first-order term $[t]_q$, in any factor, equals one, by the multiplication principle, (1.37) is deduced.

Also, expanding each factor in (1.34) by using the geometric series, we get

$$\begin{aligned}\varphi_k(u;q,r) = \sum_{n=k}^{\infty} S_q(n,k;r)u^n &= u^k \prod_{j=0}^{k}\left(\sum_{m_j=0}^{\infty} [r+j]_q^{m_j} u^{m_j}\right)\\ &= \sum_{n=k}^{\infty}\left\{\sum [r]_q^{m_0}[r+1]_q^{m_1} \cdots [r+k]_q^{m_k}\right\} u^n\end{aligned}$$

and so

$$S_q(n,k;r) = \sum [r]_q^{m_0}[r+1]_q^{m_1} \cdots [r+k]_q^{m_k},$$

where the summation is extended over all integers $m_j \geq 0$, $j = 0, 1, \ldots, k$, with $m_0 + m_1 + \cdots + m_k = n-k$. This expression is equivalent to (1.38). □

Remark 1.5. *The sign of the q-Stirling numbers*. The sign of the noncentral q-Stirling numbers of the first and second kind in a base q, with $0 < q < 1$ or $1 < q < \infty$, according to expressions (1.37) and (1.38), depends on the noncentrality parameter r as follows:

The noncentral q-Stirling number of the first kind $s_q(n,k;r)$ has the sign of $(-1)^{n-k}$ for $r \geq 0$, whereas it is a nonnegative q-number for $r+n-1 < 0$. Also, the noncentral q-Stirling number of the second kind $S_q(n, k; r)$ is a nonnegative q-number for $r \geq 0$, whereas it has sign of $(-1)^{n-k}$ for $r + k < 0$.

In particular, for $r = 0$, the q-Stirling number of the first kind $s_q(n, k)$ has the sign of $(-1)^{n-k}$. Also, the q-Stirling number of the second kind $S_q(n, k)$ is a nonnegative q-number.

Expressions of the noncentral q-Stirling numbers in the form of single summations of elementary terms are given in the following theorem.

Theorem 1.13. *The noncentral q-Stirling number of the first kind $s_q(n, k; r)$, $k = 1, 2, \ldots, n$, $n = 1, 2, \ldots$, is given by*

$$s_q(n, k; r) = \frac{1}{(1-q)^{n-k}} \sum_{j=k}^{n} (-1)^{j-k} q^{\binom{n-j}{2} + r(n-j)} \begin{bmatrix} n \\ j \end{bmatrix}_q \binom{j}{k}, \tag{1.39}$$

Also, the noncentral q-Stirling number of the second kind $S_q(n, k)$, $k = 1, 2, \ldots, n$, $n = 1, 2, \ldots$, is given by

$$S_q(n, k; r) = \frac{1}{(1-q)^{n-k}} \sum_{j=k}^{n} (-1)^{j-k} q^{r(j-k)} \binom{n}{j} \begin{bmatrix} j \\ k \end{bmatrix}_q. \tag{1.40}$$

In addition,

$$S_q(n, k; r) = \frac{1}{[k]_q!} \sum_{j=0}^{k} (-1)^{k-j} q^{\binom{j+1}{2} - (r+j)k} \begin{bmatrix} k \\ j \end{bmatrix}_q [r + j]_q^n. \tag{1.41}$$

Proof. The q-binomial formula (1.14), by replacing successively q by q^{-1}, t by $-q^{t-r}$, the dummy variable k by j and n by k, it becomes

$$\prod_{i=1}^{k} (1 - q^{t-r} q^{-(i-1)}) = \sum_{j=0}^{k} (-1)^j q^{-\binom{j}{2}} \begin{bmatrix} k \\ j \end{bmatrix}_{q^{-1}} q^{tj-rj}.$$

Furthermore, using the relations

$$\prod_{i=1}^{n} (1 - q^{t-r} q^{-(i-1)}) = (1-q)^n [t-r]_{n,q}, \quad \begin{bmatrix} n \\ j \end{bmatrix}_{q^{-1}} = q^{-j(n-j)} \begin{bmatrix} n \\ j \end{bmatrix}_q,$$

it is expressed as

$$[t-r]_{n,q} = \frac{1}{(1-q)^n} \sum_{j=0}^{n} (-1)^j q^{-\binom{j}{2} - j(n-j+r)} \begin{bmatrix} n \\ j \end{bmatrix}_q q^{jt}. \tag{1.42}$$

Multiplying both members of (1.42) by $q^{\binom{n}{2}+rn}$, reducing the exponent of q by using the relation $\binom{n-j}{2} = \binom{n}{2} - \binom{j}{2} - j(n-j)$, and expanding the q-function $q^{jt} = (1-(1-q)[t]_q)^j$ into powers of the q-number $[t]_q$, we get

$$q^{\binom{n}{2}+rn}[t-r]_{n,q} = \sum_{j=0}^{n} (-1)^j q^{\binom{n-j}{2}+r(n-j)} \begin{bmatrix} n \\ j \end{bmatrix}_q \sum_{k=0}^{j} (-1)^k \binom{j}{k} \frac{(1-q)^k}{(1-q)^n} [t]_q^k.$$

Furthermore, interchanging the order of summation, we find the expansion

$$q^{\binom{n}{2}+rn}[t-r]_{n,q} = \sum_{k=0}^{n} \left\{ \frac{1}{(1-q)^{n-k}} \sum_{j=k}^{n} (-1)^{j-k} q^{\binom{n-j}{2}+r(n-j)} \begin{bmatrix} n \\ j \end{bmatrix}_q \binom{j}{k} \right\} [t]_q^k.$$

Comparing this expansion to (1.26), we readily deduce (1.39).

Also, expanding the nth power of the q-number of $t+r$, into powers of q^t, we have

$$[t+r]_q^n = \frac{(1-q^{t+r})^n}{(1-q)^n} = \frac{1}{(1-q)^n} \sum_{j=0}^{n} (-1)^j \binom{n}{j} q^{jr} q^{jt}.$$

Then, expanding the powers of q^t into factorials, by using (1.18), we get

$$\begin{aligned} [t+r]_q^n &= \frac{1}{(1-q)^n} \sum_{j=0}^{n} (-1)^j q^{jr} \binom{n}{j} \sum_{k=0}^{j} (-1)^k (1-q)^k q^{\binom{k}{2}} \begin{bmatrix} j \\ k \end{bmatrix}_q [t]_{k,q} \\ &= \sum_{k=0}^{n} q^{\binom{k}{2}} \left\{ \frac{1}{(1-q)^{n-k}} \sum_{j=k}^{n} (-1)^{j-k} q^{jr} \binom{n}{j} \begin{bmatrix} j \\ k \end{bmatrix}_q \right\} [t]_{k,q}. \end{aligned}$$

Comparing this expansion to (1.28), we readily deduce (1.40). The last expression can be transformed into (1.41) by using the expression

$$\begin{bmatrix} j \\ k \end{bmatrix}_q = \frac{1}{[k]_q!(1-q)^k} \sum_{i=0}^{k} (-1)^i q^{\binom{i+1}{2}-ik} \begin{bmatrix} k \\ i \end{bmatrix}_q q^{ij},$$

which is deduced from (1.42) by replacing n by k, the bound variable j by i, setting $t = j$, $r = 0$ and dividing both members of the resulting expression by $[k]_q!$. Specifically, we get

$$\begin{aligned} S_q(n,k;r) &= \frac{1}{[k]_q!} \sum_{j=0}^{n} (-1)^{j-k} \binom{n}{j} \sum_{i=0}^{k} (-1)^i q^{\binom{i+1}{2}-k(r+i)} \begin{bmatrix} k \\ i \end{bmatrix}_q \frac{q^{(r+i)j}}{(1-q)^n} \\ &= \frac{1}{[k]_q!} \sum_{i=0}^{k} (-1)^{k-i} q^{\binom{i+1}{2}-k(r+i)} \begin{bmatrix} k \\ i \end{bmatrix}_q \sum_{j=0}^{n} (-1)^j \binom{n}{j} \frac{(q^{r+i})^j}{(1-q)^n}. \end{aligned}$$

Since, by the classical binomial formula,

$$\sum_{j=0}^{n}(-1)^j\binom{n}{j}\frac{(q^{r+i})^j}{(1-q)^n}=\frac{(1-q^{r+i})^n}{(1-q)^n}=[r+i]_q^n,$$

the last expression implies (1.41). □

The explicit expressions of the q-Stirling numbers of the first and second kind, (1.39) and (1.40), with $r=0$, can be inverted to express the binomial coefficients in terms of the q-binomial coefficients and vice versa. These inversions are obtained in the following corollary of Theorem 1.13.

Corollary 1.4. *Let k and j be positive integers and let q be a real number, with $q\neq 1$. Then,*

$$\binom{k}{j}=\sum_{m=j}^{k}(-1)^{m-j}(1-q)^{m-j}s_q(m,j)\begin{bmatrix}k\\m\end{bmatrix}_q \tag{1.43}$$

and

$$\begin{bmatrix}k\\j\end{bmatrix}_q=\sum_{m=j}^{k}(-1)^{m-j}(1-q)^{m-j}S_q(m,j)\binom{k}{m}. \tag{1.44}$$

Proof. Expression (1.39), with $r=0$, may be inverted to obtain the binomial coefficient $\binom{k}{j}$ in terms of q-binomial coefficients as follows. Specifically, in (1.39), with $r=0$, we replace n by m, the bound (dummy) variable j by i and multiply the resulting expression by

$$(-1)^{m-j}(1-q)^{m-j}\begin{bmatrix}k\\m\end{bmatrix}_q.$$

Then, summing for $m=j,j+1,\dots,k$, we get

$$\sum_{m=j}^{k}(-1)^{m-j}(1-q)^{m-j}s_q(m,j)\begin{bmatrix}k\\m\end{bmatrix}_q=\sum_{m=j}^{k}\begin{bmatrix}k\\m\end{bmatrix}_q\sum_{i=j}^{m}(-1)^{m-i}q^{\binom{m-i}{2}}\begin{bmatrix}m\\i\end{bmatrix}_q\binom{i}{j}$$

$$=\sum_{i=j}^{k}\binom{i}{j}\left\{\sum_{m=i}^{k}(-1)^{m-i}q^{\binom{m-i}{2}}\begin{bmatrix}m\\i\end{bmatrix}_q\begin{bmatrix}k\\m\end{bmatrix}_q\right\}$$

and using the orthogonality relation (1.16), we deduce expression (1.43).

Similarly, in (1.40), with $r=0$, we replace n by m, the bound variable j by i and multiply the resulting expression by

$$(-1)^{m-j}(1-q)^{m-j}\binom{k}{m}.$$

Then, summing for $m = j, j+1, \ldots, k$, we get

$$\sum_{m=j}^{k} (-1)^{m-j}(1-q)^{m-j} S_q(m,j) \binom{k}{m} = \sum_{m=j}^{k} \binom{k}{m} \sum_{i=j}^{m} (-1)^{m-i} \binom{m}{i} \begin{bmatrix} i \\ j \end{bmatrix}_q$$

$$= \sum_{i=j}^{k} \begin{bmatrix} i \\ j \end{bmatrix}_q \sum_{m=i}^{k} (-1)^{m-i} \binom{m}{i} \binom{k}{m}$$

and using the orthogonality relation

$$\sum_{m=i}^{k} (-1)^{m-i} \binom{m}{i} \binom{k}{m} = \delta_{i,k},$$

we deduce relation (1.44). □

1.7 GENERALIZED q-FACTORIAL COEFFICIENTS

Let t, q, s, and r be real numbers, with $q \neq 1$. The noncentral generalized q-factorial of t of order n, scale parameter s and noncentrality parameter r,

$$[st+r]_{n,q} = [st+r]_q[st+r-1]_q \cdots [st+r-n+1]_q$$

$$= q^{-\binom{n}{2}+rn}([s]_q[t]_{q^s} - [-r]_q) \cdots ([s]_q[t]_{q^s} - [n-r-1]_q),$$

may be expressed as a polynomial of q^s-factorials of t as

$$[st+r]_{n,q} = q^{-\binom{n}{2}+rn} \sum_{k=0}^{n} q^{s\binom{k}{2}} C_q(n,k;s,r)[t]_{k,q^s}, \quad n = 0, 1, \ldots . \tag{1.45}$$

The coefficient $C_q(n,k;s,r)$ is called *noncentral generalized q-factorial coefficient.*

Furthermore, the expansion of the generalized ascending q-factorial of t of order n, scale parameter s and noncentrality parameter r,

$$[st+r+n-1]_{n,q} = [st+r]_q[st+r+1]_q \cdots [st+r+n-1]_q$$

$$= [-1]_q^n[-st-r]_{n,q^{-1}},$$

into a polynomial of the q^s-factorials of t may be deduced from (1.45) as

$$[st+r+n-1]_{n,q} = q^{\binom{n}{2}+rn} \sum_{k=0}^{n} q^{s\binom{k}{2}} |C_{q^{-1}}(n,k;-s,-r)|[t]_{k,q^s}, \tag{1.46}$$

for $n = 0, 1, \ldots$, where

$$|C_{q^{-1}}(n,k;-s,-r)| = [-1]_q^n C_{q^{-1}}(n,k;-s,-r).$$

The coefficient $|C_{q^{-1}}(n,k;-s,-r)|$, which for $0 < q < 1$ or $1 < q < \infty$ and s and r positive numbers, is nonnegative, is called *absolute noncentral generalized q-factorial coefficient.*

The noncentral generalized q-factorial coefficients as the scale parameter tends to zero or infinity converge to the noncentral q-Stirling numbers of the first and second kind, respectively. Specifically, expression (1.45) may be written as

$$[t+r]_{n,q} = q^{-\binom{n}{2}+rn} \sum_{k=0}^{n} q^{s\binom{k}{2}} \{[s]_q^{-k} C_q(n,k;s,r)\}([s]_q^k [t/s]_{k,q^s}).$$

Since $\lim_{s\to 0}[s]_q^k [t/s]_{k,q^s} = [t]_q^k$, it follows that

$$\lim_{s\to 0}[s]_q^{-k} C_q(n,k;s,r) = s_q(n,k;-r). \tag{1.47}$$

Similarly, writing expression (1.45) in the form

$$[s]_{q^{1/s}}^{-n}[s(t+r)]_{n,q^{1/s}} = q^{-\frac{1}{s}\binom{n}{2}} \sum_{k=0}^{n} q^{\binom{k}{2}+rk} \{q^{r(n-k)}[s]_{q^{1/s}}^{-n} C_{q^{1/s}}(n,k;s,rs)\}[t]_{k,q}$$

and since

$$\lim_{s\to\infty}[s]_{q^{1/s}}^{-n}[s(t+r)]_{n,q^{1/s}} = [t+r]_q^n,$$

it follows, by virtue of (1.28), that

$$\lim_{s\to\infty} q^{r(n-k)}[s]_{q^{1/s}}^{-n} C_{q^{1/s}}(n,k;s,rs) = S_q(n,k;r). \tag{1.48}$$

A triangular recurrence relation for the generalized q-factorial coefficients are derived in the following theorem.

Theorem 1.14. *The noncentral generalized q-factorial coefficients $C_q(n,k;s,r)$, $k=0,1,\ldots,n$, $n=0,1,\ldots$, satisfy the triangular recurrence relation*

$$C_q(n+1,k;s,r) = [s]_q C_q(n,k-1;s,r) + ([sk]_q - [n-r]_q)C_q(n,k;s,r), \tag{1.49}$$

for $k=1,2,\ldots,n+1$ and $n=0,1,\ldots$, with initial conditions

$$C_q(0,0;s,r)=1, \quad C_q(0,k;s,r)=0, \quad k>0,$$

and

$$C_q(n,0;s,r) = q^{\binom{n}{2}-rn}[r]_{n,q}, \quad n>0.$$

Proof. Expanding both members of the recurrence relation

$$[st+r]_{n+1,q} = [st+r-n]_q[st+r]_{n,q} = q^{-n+r}([s]_q[t]_{q^s} - [n-r]_q)[st]_{n,q}$$

into q^s-factorials of t, using (1.45), we find

$$\sum_{k=0}^{n+1} q^{s\binom{k}{2}} C_q(n+1,k;s,r)[t]_{k,q^s} = \sum_{k=0}^{n} q^{s\binom{k}{2}} C_q(n,k;s,r)[s]_q[t]_{q^s}[t]_{k,q^s}$$
$$+ \sum_{k=0}^{n} q^{s\binom{k}{2}} [n-r]_q C_q(n,k;s,r)[t]_{k,q^s}.$$

Furthermore, using the expressions

$$[t]_{q^s}[t]_{k,q^s} = q^{sk}[t]_{k+1,q^s} + [k]_{q^s}[t]_{k,q^s}, \quad [s]_q[k]_{q^s} = [sk]_q,$$

we get the relation

$$\sum_{k=0}^{n+1} q^{s\binom{k}{2}} C_q(n+1,k;s,r)[t]_{k,q^s} = \sum_{k=0}^{n} q^{s\binom{k+1}{2}} [s]_q C_q(n,k;s,r)[t]_{k+1,q^s}$$
$$+\sum_{k=0}^{n} q^{s\binom{k}{2}} ([sk]_q - [n-r]_q) C_q(n,k;s,r)[t]_{k,q^s}.$$

Equating the coefficients of $[t]_{k,q^s}$ in both sides of the last relation, we get (1.49). The initial conditions follow directly from (1.45). □

Remark 1.6. The absolute noncentral generalized q-factorial coefficients

$$|C_{q^{-1}}(n,k;-s,-r)| = [-1]_q^n C_{q^{-1}}(n,k;-s,-r),$$

for $k = 0,1,\ldots,n$ and $n = 0,1,\ldots$, with s and r positive numbers, according to (1.49), satisfy the triangular recurrence relation

$$|C_{q^{-1}}(n+1,k;\ -s,-r)| = [s]_q|C_{q^{-1}}(n,k-1;-s,-r)|$$
$$+ ([sk]_q + q^{-(n+r)}[n+r]_q)|C_{q^{-1}}(n,k;\ -s,-r)|,$$

for $k = 1,2,\ldots,n+1$ and $n = 0,1,\ldots$, with initial conditions

$$|C_{q^{-1}}(0,0;-s,-r)| = 1, \quad |C_{q^{-1}}(0,k;-s,-r)| = 0, k > 0,$$

and

$$|C_{q^{-1}}(n,0;-s,-r)| = q^{-\binom{n}{2}+rn}[r+n-1]_{n,q}, n > 0.$$

The noncentral generalized q-factorial coefficient $C_q(n,k;s,r)$ is a polynomial in $[s]_q$ of degree n. Specifically, they have the following theorem.

Theorem 1.15. *The noncentral generalized q-factorial coefficients are connected with the noncentral q-Stirling numbers of the first and second kind by*

$$C_q(n,k;s,s\rho-r) = q^{-s\rho(n-k)} \sum_{m=k}^{n} s_q(n,m;r) S_{q^s}(m,k;\rho)[s]_q^m. \tag{1.50}$$

Proof. Expanding the noncentral generalized q-factorial

$$[s(t+\rho)-r]_{n,q} = [st+(s\rho-r)]_{n,q}$$

into powers of $[s(t+\rho)]_q = [s]_q[t+\rho]_{q^s}$ by using (1.26) and, in the resulting expression, expanding the powers of $[t+\rho]_{q^s}$ into q^s-factorials of t, by using (1.28), we deduce the expression

$$[s(t+\rho)-r]_{n,q} = q^{-\binom{n}{2}-rn}\sum_{m=0}^{n} s_q(n,m;r)[s]_q^m[t+\rho]_{q^s}^m$$
$$= q^{-\binom{n}{2}-rn}\sum_{m=0}^{n} s(n,m;r)[s]_q^m\sum_{k=0}^{m} q^{s\binom{k}{2}+s\rho k}S_{q^s}(m,k;\rho)[t]_{k,q^s}$$
$$= q^{-\binom{n}{2}-rn}\sum_{k=0}^{n} q^{s\binom{k}{2}+s\rho k}\left\{\sum_{m=k}^{n} s_q(n,m;r)S_{q^s}(m,k;\rho)[s]_q^m)\right\}[t]_{k,q^s},$$

and since, by (1.45),

$$[st+(s\rho-r)]_{n,q} = q^{-\binom{n}{2}+(s\rho-r)n}\sum_{k=0}^{n} q^{s\binom{k}{2}}C_q(n,k;s,s\rho-r)[t]_{k,q^s},$$

we deduce (1.50). □

Remark 1.7. *The sign of the generalized q-factorial coefficients.* Expression (1.50) in the particular case $\rho = 0$ may be written as

$$C_q(n,k;s,r) = \sum_{m=k}^{n} s_q(n,m;-r)S_{q^s}(m,k)[s]_q^m.$$

Thus, according to this expression and Remark 1.4, the noncentral generalized q-factorial coefficients $C_q(n,k;s,r)$, $k=0,1,\ldots,n$, $n=0,1,\ldots$, in a base q, with $0<q<1$ or $1<q<\infty$, for s and r positive numbers and $n<r+1$, are nonnegative q-numbers.

Also, expression (1.50) for the numbers $|C_q(n,k;-s,-r)|$, $k=0,1,\ldots,n$, $n=0,1,\ldots$, may be written as

$$|C_q(n,k;-s,-r)| = \sum_{m=k}^{n} |s_q(n,m;r)|S_{q^s}(m,k)[s]_{q^{-1}}^m.$$

Thus, according to this expression and Remark 1.4, the absolute noncentral generalized q-factorial coefficients $|C_q(n,k;-s,-r)|$, $k=0,1,\ldots,n$, $n=0,1,\ldots$, in a base q, with $0<q<1$ or $1<q<\infty$, for s and r positive numbers, are nonnegative q-numbers.

Theorem 1.16. *The reciprocal q-factorial* $1/[t]_{k+1,q^s}$ *is expanded into reciprocal noncentral generalized q-factorials* $1/[st+r]_{n+1,q}$, $n=k,k+1,\ldots$, $k=0,1,\ldots$, *as*

$$\frac{1}{[t]_{k+1,q^s}} = q^{s\binom{k+1}{2}}\sum_{n=k}^{\infty} q^{-\binom{n+1}{2}+r(n+1)}[s]_qC_q(n,k;s,r)\frac{1}{[st+r]_{n+1,q}}, \tag{1.51}$$

for $t>k$.

Proof. Let us consider the series

$$C_k(t;q) = \sum_{n=k}^{\infty} q^{-\binom{n+1}{2}+r(n+1)} C_q(n,k;s,r) \frac{1}{[st+r]_{n+1,q}}, \quad t > k,$$

for $k = 0, 1, \ldots$. Multiplying both sides of the triangular recurrence relation (1.49) by

$$\frac{q^{-\binom{n+1}{2}+r(n+1)}}{[st+r]_{n+1,q}} = \frac{q^{-\binom{n+2}{2}+r(n+2)}([st]_q - [n-r+1]_q)}{[st+r]_{n+2,q}},$$

we find the relation

$$\begin{aligned}
&([st]_q C_q(n+1,k;s,r) - [n-r+1]_q C_q(n+1,k;s,r)) \frac{q^{-\binom{n+2}{2}+r(n+2)}}{[st+r]_{n+2,q}} \\
&= ([sk]_q C_q(n,k;s,r) - [n-r]_q C_q(n,k;s,r) \\
&\quad + [s]_q C_q(n,k-1;s,r)) \frac{q^{-\binom{n+1}{2}+r(n+1)}}{[st+r]_{n+1,q}}.
\end{aligned}$$

Summing it for $n = k-1, k, \ldots$ and since $C_q(k-1,k;s) = 0$, we obtain for $C_k(t;q)$ the relation

$$[st]_q C_k(t;q) = [sk]_q C_k(t;q) + [s]_q C_{k-1}(t;q), \quad k = 1, 2, \ldots .$$

Using the expressions

$$[st]_q = [s]_q [t]_{q^s}, \quad [sk]_q = [s]_q [k]_{q^s}, \quad [t]_{q^s} - [k]_{q^s} = q^{sk}[t-k]_{q^s},$$

we deduce the recurrence relation

$$C_k(t;q) = \frac{q^{-sk}}{[t-k]_{q^s}} C_{k-1}(t;q), \quad k = 1, 2, \ldots,$$

and applying it repeatedly, we find

$$C_k(t;q) = \frac{q^{-s\binom{k+1}{2}}}{[t-1]_{k,q^s}} C_0(t;q).$$

Since $C_q(n,0;s,r) = q^{\binom{n}{2}-rn}[r]_{n,q}$, for $n > 0$, we get the initial value

$$\begin{aligned}
[st]_q C_0(t;q) &= \sum_{n=0}^{\infty} q^{-n+r} \frac{[st]_q [r]_{n,q}}{[st+r]_{n+1,q}} \\
&= \sum_{n=0}^{\infty} \left(\frac{[r]_{n,q}}{[st+r]_{n,q}} - \frac{[r]_{n+1,q}}{[st+r]_{n+1,q}} \right) = 1,
\end{aligned}$$

and so $C_0(t;q) = 1/[st]_q = 1/([s]_q[t]_{q^s})$. Therefore,

$$C_k(t;q) = \frac{q^{-s\binom{k+1}{2}}}{[s]_q[t]_{k+1,q^s}}$$

and (1.51) is established. □

Explicit expressions for the noncentral generalized q-factorial coefficients are derived in the following theorem.

Theorem 1.17. *The noncentral generalized q-factorial coefficient $C_q(n,k;s,r)$, for $k = 1,2,\ldots,n$, $n = 1,2,\ldots$, is given by*

$$C_q(n,k;s,r) = \frac{[s]_q^k}{(1-q)^{n-k}} \sum_{j=k}^{n} (-1)^{j-k} q^{\binom{n-j}{2}-r(n-j)} \begin{bmatrix} n \\ j \end{bmatrix}_q \begin{bmatrix} j \\ k \end{bmatrix}_{q^s}. \tag{1.52}$$

Also

$$C_q(n,k;s,r) = \frac{q^{\binom{n}{2}-rn}}{[k]_q!} \sum_{j=0}^{k} (-1)^{k-j} q^{s\binom{j+1}{2}-skj} \begin{bmatrix} k \\ j \end{bmatrix}_{q^s} [sj+r]_{n,q}. \tag{1.53}$$

Proof. Let us consider (1.42), with t replaced by st, and multiply both its sides by $q^{\binom{n}{2}}$. Then, using the relation $\binom{n-j}{2} = \binom{n}{2} - \binom{j}{2} - j(n-j)$, we deduce the expression

$$q^{\binom{n}{2}}[st]_{n,q} = \frac{1}{(1-q)^n} \sum_{j=0}^{n} (-1)^j q^{\binom{n-j}{2}} \begin{bmatrix} n \\ j \end{bmatrix}_q (q^s)^{jt}.$$

Expanding the q^s-function $q^{s(jt)}$ into q^s-factorials of t, using (1.18), we get

$$\begin{aligned} q^{\binom{n}{2}}[st]_{n,q} &= \frac{1}{(1-q)^n} \sum_{j=0}^{n} (-1)^j q^{\binom{n-j}{2}} \begin{bmatrix} n \\ j \end{bmatrix}_q \sum_{k=0}^{j} (-1)^k (1-q^s)^k q^{s\binom{k}{2}} \begin{bmatrix} j \\ k \end{bmatrix}_{q^s} [t]_{q^s}^k \\ &= \sum_{k=0}^{n} q^{s\binom{k}{2}} \left\{ \frac{[s]_q^k}{(1-q)^{n-k}} \sum_{j=k}^{n} (-1)^{j-k} q^{\binom{n-j}{2}} \begin{bmatrix} n \\ j \end{bmatrix}_q \begin{bmatrix} j \\ k \end{bmatrix}_{q^s} \right\} [t]_q^k. \end{aligned}$$

Comparing this expansion to (1.45), we readily deduce (1.52). The last expression can be transformed into (1.53) by using the expression

$$\begin{bmatrix} j \\ k \end{bmatrix}_{q^s} = \frac{1}{[k]_q!(1-q)^k[s]_q^k} \sum_{i=0}^{k} (-1)^i q^{s\binom{i+1}{2}-sik} \begin{bmatrix} k \\ i \end{bmatrix}_{q^s} q^{sij},$$

which is deduced from (1.42) by replacing n by k, q by q^s, the bound variable j by i, setting $t = j$ and $r = 0$, dividing both members of the resulting expression by $[k]_{q^s}!$

and using the relation $(1-q^s)=(1-q)[s]_q$. Specifically, we get

$$C_q(n,k;s,r)=\frac{q^{-rn}}{[k]_{q^s}!}\sum_{j=0}^{n}(-1)^{j-k}q^{\binom{n-j}{2}}\begin{bmatrix}n\\j\end{bmatrix}_q\sum_{i=0}^{k}(-1)^i q^{s\binom{i+1}{2}-sik}\begin{bmatrix}k\\i\end{bmatrix}_{q^s}\frac{q^{(si+r)j}}{(1-q)^n}$$

$$=\frac{q^{-rn}}{[k]_{q^s}!}\sum_{i=0}^{k}(-1)^{k-i}q^{s\binom{i+1}{2}-sik}\begin{bmatrix}k\\i\end{bmatrix}_{q^s}\sum_{j=0}^{n}(-1)^j q^{\binom{n-j}{2}}\begin{bmatrix}n\\j\end{bmatrix}_q\frac{(q^{si+r})^j}{(1-q)^n}.$$

Since, by the q-binomial formula (1.14),

$$\sum_{j=0}^{n}(-1)^j q^{\binom{n-j}{2}}\begin{bmatrix}n\\j\end{bmatrix}_q\frac{(q^{si+r})^j}{(1-q)^n}=q^{\binom{n}{2}}\sum_{j=0}^{n}(-1)^j q^{-\binom{j}{2}}\begin{bmatrix}n\\j\end{bmatrix}_{q^{-1}}\frac{(q^{si+r})^j}{(1-q)^n}$$

$$=q^{\binom{n}{2}}\frac{\prod_{m=1}^{n}(1-q^{si+r-m+1})}{(1-q)^n}$$

$$=q^{\binom{n}{2}}[si+r]_{n,q},$$

the last expression implies (1.53). □

1.8 q-FACTORIAL AND q-BINOMIAL MOMENTS

The calculation of the mean and variance and generally the calculation of the moments of a discrete q-distribution is quite difficult. Several techniques have been used for the calculation of the mean and variance of particular q-distributions. In this section, we introduce the q-factorial and q-binomial moments of a discrete q-distribution, the calculation of which is as easy as that of the usual factorial and binomial moments of the classical discrete distributions. These moments, apart from the interest in their own, are used as an intermediate step in the calculation of the usual factorial and binomial moments of the q-distributions.

Definition 1.1. *Let X be a nonnegative integer-valued random variable, with probability function $f(x)=P(X=x)$, $x=0,1,\ldots$. The expected values*

$$E([X]_{m,q})=\sum_{x=m}^{\infty}[x]_{m,q}f(x),\quad m=1,2,\ldots \tag{1.54}$$

and

$$E\left(\begin{bmatrix}X\\m\end{bmatrix}_q\right)=\sum_{x=m}^{\infty}\begin{bmatrix}x\\m\end{bmatrix}_q f(x),\quad m=1,2,\ldots, \tag{1.55}$$

provided they exist, are called the mth-order q-factorial and the mth-order q-binomial moments, respectively, of the random variable X.

Note that, the q-factorial and the q-binomial moments are closely connected by

$$E\left(\begin{bmatrix} X \\ m \end{bmatrix}_q\right) = \frac{E([X]_{m,q})}{[m]_q!}, \quad E([X]_{m,q}) = [m]_q! E\left(\begin{bmatrix} X \\ m \end{bmatrix}_q\right).$$

Also in particular, for $m = 1$, the q-expected value or the q-mean of X, denoted by $E([X]_q)$ or by μ_q, is defined by

$$\mu_q = E([X]_q) = \sum_{x=1}^{\infty} [x]_q f(x), \tag{1.56}$$

provided the series is convergent. Furthermore, the q-variance of X, denoted by $V([X]_q)$ or by σ_q^2, is defined by

$$\sigma_q^2 = V([X]_q) = E[([X]_q - \mu_q)^2] = \sum_{x=1}^{\infty} ([x]_q - \mu_q)^2 f(x), \tag{1.57}$$

provided the series is convergent. Clearly

$$V([X]_q) = E([X]_q^2) - [E([X]_q)]^2. \tag{1.58}$$

Also, since $q[X-1]_q = [X]_q - 1$, it follows that

$$q[X]_{2,q} = q[X]_q[X-1]_q = [X]_q([X]_q - 1) = [X]_q^2 - [X]_q$$

and so the q-variance may be expressed in terms of the q-factorial moments as

$$V([X]_q) = qE([X]_{2,q}) + E([X]_q) - [E([X]_q)]^2. \tag{1.59}$$

Remark 1.8. *q-Deformed distributions in Quantum Physics.* Consider a non-negative integer-valued random variable X with probability mass function $f_X(x) = P(X = x)$, $x = 0, 1, \ldots$. Furthermore, consider the q-number transformation $Y = [X]_q$, which in the language of quantum physics is known as a q-deformation. The distribution of the random variable Y, with probability function

$$f_Y([x]_q) = P(Y = [x]_q) = P(X = x) = f_X(x), \quad x = 0, 1, \ldots,$$

is called *q-deformed distribution*. The mean and the variance of the q-deformed distribution of Y are the q-mean and the q-variance of the distribution of X.

The usual binomial and factorial moments are expressed in terms of the q-binomial and the q-factorial moments, respectively, through the q-Stirling numbers of the first kind, in the following theorem.

Theorem 1.18. *The usual binomial moments are expressed in terms of the q-binomial moments by*

$$E\left[\binom{X}{j}\right] = \sum_{m=j}^{\infty} (-1)^{m-j}(1-q)^{m-j} s_q(m,j) E\left(\begin{bmatrix} X \\ m \end{bmatrix}_q\right), \tag{1.60}$$

for $j = 1, 2, \ldots,$ *and equivalently, the usual factorial moments are expressed in terms of the q-factorial moments by*

$$E[(X)_j] = j! \sum_{m=j}^{\infty} (-1)^{m-j}(1-q)^{m-j} s_q(m,j) \frac{E([X]_{m,q})}{[m]_q!}, \tag{1.61}$$

for $j = 1, 2, \ldots,$ *where* $s_q(m,j)$ *is the q-Stirling number of the first kind.*

Proof. Multiplying both sides of expression (1.43),

$$\binom{x}{j} = j! \sum_{m=j}^{x} (-1)^{m-j}(1-q)^{m-j} s_q(m,j) \begin{bmatrix} x \\ m \end{bmatrix}_q, \quad j = 1, 2, \ldots,$$

by the probability function $f(x)$ of the random variable X and summing for all $x = 0, 1, \ldots,$ we deduce, according to (1.55), expression (1.60). Furthermore, since

$$E\left[\binom{X}{j}\right] = \frac{E[(X)_j]}{j!}, \quad E\left(\begin{bmatrix} X \\ m \end{bmatrix}_q\right) = \frac{E([X]_{m,q})}{[m]_q!},$$

expression (1.61) is readily deduced from (1.60). □

The probability function of a nonnegative integer-valued random variable X may be expressed in terms of its q-binomial (or q-factorial) moments, by inverting expression (1.55) (or (1.54)). Such an expression is derived in the following theorem.

Theorem 1.19. *The probability function* $f(x) = P(X = x)$, $x = 0, 1, \ldots,$ *of a nonnegative integer-valued random variable X, is expressed in terms of its q-binomial moments* $E\left(\begin{bmatrix} X \\ m \end{bmatrix}_q\right)$, $m = 0, 1, \ldots,$ *by*

$$f(x) = \sum_{m=x}^{\infty} (-1)^{m-x} q^{\binom{m-x}{2}} \begin{bmatrix} m \\ x \end{bmatrix}_q E\left(\begin{bmatrix} X \\ m \end{bmatrix}_q\right), \quad x = 0, 1, \ldots, \tag{1.62}$$

provided that the series is absolutely convergent.

Proof. Writing expression (1.55) with x replaced by k and then multiplying it by

$$(-1)^{m-x} q^{\binom{m-x}{2}} \begin{bmatrix} m \\ x \end{bmatrix}_q,$$

and summing the resulting expression for $m = x, x+1, \ldots,$ we get

$$\sum_{m=x}^{\infty} (-1)^{m-x} q^{\binom{m-x}{2}} \begin{bmatrix} m \\ x \end{bmatrix}_q E\left(\begin{bmatrix} X \\ m \end{bmatrix}_q\right) = \sum_{m=x}^{\infty} (-1)^{m-x} q^{\binom{m-x}{2}} \begin{bmatrix} m \\ x \end{bmatrix}_q \sum_{k=m}^{\infty} \begin{bmatrix} k \\ m \end{bmatrix}_q f(k)$$
$$= \sum_{k=x}^{\infty} \left\{ \sum_{m=x}^{k} (-1)^{m-x} q^{\binom{m-x}{2}} \begin{bmatrix} m \\ x \end{bmatrix}_q \begin{bmatrix} k \\ m \end{bmatrix}_q \right\} f(k).$$

Therefore, using the second of the orthogonality relations (1.16),

$$\sum_{m=x}^{k} (-1)^{m-x} q^{\binom{m-x}{2}} \begin{bmatrix} m \\ x \end{bmatrix}_q \begin{bmatrix} k \\ m \end{bmatrix}_q = \delta_{k,x},$$

we deduce (1.62). □

1.9 REFERENCE NOTES

The introduction of the q-number and its notation stems from Jackson (1910a), who published important and influential papers on the subject. A list of his publications is included in the obituary note by Chaudry (1962).

Gauss (1863) introduced the q-binomial coefficients (or Gaussian polynomials) and presented their triangular recurrence relations, derived in Theorem 1.1, and the vertical recurrence relations, which are given as Exercise 1.2. Also, the summation formula given as Exercise 1.4 is from Gauss (1863). The distribution of the number theoretic random variable examined in Example 1.1 was discussed by Rawlings (1994a). The combinatorial interpretation of the q-binomial coefficient as the number of subspaces of a vector space, presented in Example 1.2, was given in Goldman and Rota (1970). Also, its appearance as generating function of the number of partitions of an integer into parts of restricted size, with variable (indeterminate) q, presented in Example 1.3, was noted by Sylvester (1882).

The q-Vandermonde's (q-factorial convolution) formulae, and equivalently the q-Cauchy's (q-binomial convolution) formulae, together with the general q-binomial formulae were derived by Cauchy (1843), Jacobi (1846), and Heine (1847, 1878). It is worth noticing that the origin of the general q-binomial formulae is quite uncertain; Hardy (1940) attributed these formulae to Euler. The derivation of the power series expressions of the two q-exponential functions (1.23) and (1.24) are, indeed, from Euler (1748). The limit formulas for the q-exponential functions, which are given in Exercise 1.11, are from Rawlings (1994b). Several other interesting q-series expansions are presented in the classical book of Andrews (1976); Exercises 1.12 and 1.13, on the univariate and multivariate Rogers–Szegö polynomials, are taken from this book. A motivated introduction and a clear presentation of the q-gamma function and the q-beta integral can be found in the excellent book of Andrews et al. (1999). An authoritative and comprehensive account of the basic q-hypergeometric series is given by Gasper and Rahman (2004).

The q-Stirling numbers of the second kind were introduced by Carlitz (1933) in connection with an enumeration problem in abelian groups. In a second paper, Carlitz (1948) found it convenient to generalize these numbers to what are nowadays called noncentral q-Stirling numbers of the second kind. Furthermore, Gould (1961) studied the q-Stirling numbers of the first and second kind, which were defined as sums of all k-factor products that are formed from the first n q-natural numbers, without and with repeated factors, respectively. The q-Lah numbers appeared in Hahn (1949). Also, Garsia and Remmel (1980) discussed these numbers, as q-Laguerre numbers.

The central and noncentral generalized q-factorial coefficients were discussed in Charalambides (1996, 2002, 2004, 2005b). The q-factorial moments and their connection to the usual factorial moments were discussed in Charalambides and Papadatos (2005) and Charalambides (2005a). Jackson (1910a, 1910b, 1951) extensively studied q-derivatives and q-integrals. The generalized Stirling and Lah numbers were introduced by Tauber (1962, 1965) and further studied by Comtet (1972) and Platonov (1976).

1.10 EXERCISES

1.1 Let m and k be positive integers, with $k \leq m$.

(a) Show that

$$\begin{bmatrix} m \\ k \end{bmatrix}_q = \begin{bmatrix} m \\ m-k \end{bmatrix}_q.$$

(b) Furthermore, let x be a real number. Show that

$$\begin{bmatrix} x \\ m \end{bmatrix}_q \begin{bmatrix} m \\ k \end{bmatrix}_q = \begin{bmatrix} x \\ k \end{bmatrix}_q \begin{bmatrix} x-k \\ m-k \end{bmatrix}_q = \begin{bmatrix} x \\ m-k \end{bmatrix}_q \begin{bmatrix} x-m+k \\ k \end{bmatrix}_q$$

and

$$\begin{bmatrix} x \\ m \end{bmatrix}_q \begin{bmatrix} x-m \\ k \end{bmatrix}_q = \begin{bmatrix} x \\ k \end{bmatrix}_q \begin{bmatrix} x-k \\ m \end{bmatrix}_q = \begin{bmatrix} x \\ m+k \end{bmatrix}_q \begin{bmatrix} m+k \\ m \end{bmatrix}_q.$$

1.2 *Vertical recurrence relations for the q-binomial coefficients.* For n, k, and m positive integers, show that

$$\sum_{r=m}^{n} q^{r-k} \begin{bmatrix} r-1 \\ k-1 \end{bmatrix}_q = \begin{bmatrix} n \\ k \end{bmatrix}_q - \begin{bmatrix} m-1 \\ k \end{bmatrix}_q$$

and, alternatively, that

$$\sum_{r=m}^{n} q^{(n-r)k} \begin{bmatrix} r-1 \\ k-1 \end{bmatrix}_q = \begin{bmatrix} n \\ k \end{bmatrix}_q - q^{(n-m+1)k} \begin{bmatrix} m-1 \\ k \end{bmatrix}_q.$$

In particular, conclude that

$$\sum_{r=k}^{n} q^{r-k} \begin{bmatrix} r-1 \\ k-1 \end{bmatrix}_q = \begin{bmatrix} n \\ k \end{bmatrix}_q$$

and, alternatively, that

$$\sum_{r=k}^{n} q^{(n-r)k} \begin{bmatrix} r-1 \\ k-1 \end{bmatrix}_q = \begin{bmatrix} n \\ k \end{bmatrix}_q.$$

1.3 *A horizontal recurrence relation for the q-binomial coefficients.* For n, k, and m positive integers, show that

$$\sum_{r=k}^{m} (-1)^{r-k} q^{\binom{r+1}{2}} \begin{bmatrix} n+1 \\ r+1 \end{bmatrix}_q = q^{\binom{k+1}{2}} \begin{bmatrix} n \\ k \end{bmatrix}_q + (-1)^{m-k} q^{\binom{m+2}{2}} \begin{bmatrix} n \\ m+1 \end{bmatrix}_q$$

and conclude that

$$\sum_{r=k}^{n}(-1)^{r-k}q^{\binom{r+1}{2}}\begin{bmatrix} n+1 \\ r+1 \end{bmatrix}_q = q^{\binom{k+1}{2}}\begin{bmatrix} n \\ k \end{bmatrix}_q.$$

1.4 *A Gauss summation formula*. Show that

$$\sum_{k=0}^{n}(-1)^k\begin{bmatrix} n \\ k \end{bmatrix}_q = \begin{cases} \frac{[2m]_q!}{[m]_{q^2}!}\left(\frac{1-q}{1+q}\right)^m, & n=2m, \\ 0, & n=2m+1, \end{cases}$$

for m a nonnegative integer.

1.5 Let x, y, and q be real numbers, with $q \neq 1$, and let n be a positive integer. Using q-Vandermonde's formula, show that

$$\sum_{k=0}^{n}(-1)^k q^{\binom{k+1}{2}+k(y-n)}\begin{bmatrix} n \\ k \end{bmatrix}_q \frac{[x]_{k,q}}{[x+y]_{k,q}} = \frac{[y]_{n,q}}{[x+y]_{n,q}},$$

for $x+y \neq 0, 1, \ldots, n-1$, and conclude that

$$\sum_{k=0}^{n}(-1)^k q^{\binom{k+1}{2}}\begin{bmatrix} n \\ k \end{bmatrix}_q \frac{[x]_{k,q}}{[x+n]_{k,q}} = \sum_{k=0}^{n}(-1)^k q^{\binom{k+1}{2}}\begin{bmatrix} n \\ k \end{bmatrix}_q \begin{bmatrix} x \\ k \end{bmatrix}_q \Big/ \begin{bmatrix} x+n \\ k \end{bmatrix}_q$$

$$= 1\Big/\begin{bmatrix} x+n \\ n \end{bmatrix}_q.$$

1.6 Show that the sequence of sums

$$S_n = \sum_{k=1}^{n}(-1)^{k-1}q^{\binom{k+1}{2}}\frac{1}{[k]_q}\begin{bmatrix} n \\ k \end{bmatrix}_q, \quad n=1,2,\ldots,$$

satisfies the recurrence relation

$$S_n = S_{n-1} + \frac{q^n}{[n]_q}, \quad n=2,3,\ldots,$$

with $S_1 = q$, and conclude that

$$S_n = \sum_{k=1}^{n}\frac{q^k}{[k]_q}, \quad n=2,3,\ldots .$$

1.7 Show that the sequence of sums

$$S_n(x) = \sum_{k=0}^{n}(-1)^k q^{\binom{k+1}{2}-n(x+k)}\begin{bmatrix} n \\ k \end{bmatrix}_q \frac{[x]_q}{[x+k]_q}, \quad n=0,1,\ldots,$$

for $x \neq -1, -2, \ldots, -n$, satisfies the recurrence relation

$$S_n(x) = \frac{[n]_q}{[x+n]_q}S_{n-1}(x), \quad n=1,2,\ldots,$$

with $S_0(x) = 1$, and conclude that

$$S_n(x) = 1/\begin{bmatrix} x+n \\ n \end{bmatrix}_q.$$

1.8 Let n be a positive integer and let t, u, w, and q be real numbers, with $q \neq 1$. Show that

$$\sum_{k=0}^{n} \begin{bmatrix} n \\ k \end{bmatrix}_q w^k \prod_{i=1}^{n-k} (u + tq^{i-1}) = \sum_{r=0}^{n} \begin{bmatrix} n \\ r \end{bmatrix}_q u^r \prod_{i=1}^{n-r} (w + tq^{i-1})$$

and conclude that

$$\sum_{k=0}^{n} \begin{bmatrix} n \\ k \end{bmatrix}_q w^k \prod_{i=1}^{n-k} (1 - tq^{i-1}) = \sum_{r=0}^{n} \begin{bmatrix} n \\ r \end{bmatrix}_q \prod_{i=1}^{n-r} (w - tq^{i-1})$$

and

$$\sum_{k=0}^{n} \begin{bmatrix} n \\ k \end{bmatrix}_q t^k \prod_{i=1}^{n-k} (1 - tq^{i-1}) = 1.$$

1.9 *Additional negative q-binomial formulae.* For n a positive integer and t and q real numbers, with $0 < t < \infty$ and $0 < q < 1$ or $1 < q < \infty$, show that

$$\sum_{k=0}^{\infty} \begin{bmatrix} n+k-1 \\ k \end{bmatrix}_q \frac{t^k q^{\binom{k}{2}}}{\prod_{i=1}^{k}(1 + tq^{n+i-1})} = \prod_{i=1}^{n} (1 + tq^{i-1}),$$

or, equivalently, that

$$\sum_{k=0}^{\infty} \begin{bmatrix} n+k-1 \\ k \end{bmatrix}_q \frac{q^k}{\prod_{i=1}^{k}(1 + tq^{n+i-1})} = \frac{\prod_{i=1}^{n}(1 + tq^{i-1})}{t^n q^{\binom{n}{2}}}.$$

1.10 *A q-geometric series.* Consider the q-geometric progression

$$g_k = \prod_{j=1}^{k} (1 - tq^{j-1}) q^k, \quad k = 0, 1, \ldots,$$

for $0 < t < \infty$ and $0 < q < 1$ or $1 < q < \infty$, with $g_0 = 1$.
(a) Show that the sum of its first n terms is given by

$$\sum_{k=0}^{n-1} \prod_{j=1}^{k} (1 - tq^{j-1}) q^k = \frac{1 - \prod_{j=1}^{n}(1 - tq^{j-1})}{t}$$

and (b) deduce the limit of the q-geometric series as

$$\sum_{k=0}^{\infty} \prod_{j=1}^{k} (1 - tq^{j-1}) q^k = \frac{1 - E_q(-t/(1-q))}{t},$$

for $0 < t < \infty$ and $0 < q < 1$, where $E_q(t) = \prod_{i=1}^{\infty}(1 + t(1-q)q^{i-1})$ is a q-exponential function.

1.11 *Limit formulas for the q-exponential functions.* Show that the q-exponential functions $E_q(t) = \prod_{i=1}^{\infty}(1 + t(1-q)q^{i-1})$ and $e_q(t) = \prod_{i=1}^{\infty}(1 - t(1-q)q^{i-1})^{-1}$, for $|q| < 1$ and $|t| < 1/(1-q)$, may be obtained as

$$E_q(t) = \lim_{n\to\infty} \prod_{i=1}^{n}(1 + tq^{i-1}/[n]_q) \quad \text{and} \quad e_q(t) = \lim_{n\to\infty} \prod_{i=1}^{n}(1 - tq^{i-1}/[n]_q)^{-1},$$

respectively.

1.12 *Rogers–Szegö polynomial.* The polynomial

$$H_n(t;q) = \sum_{k=0}^{n} \begin{bmatrix} n \\ k \end{bmatrix}_q t^k, \quad -\infty < t < \infty, \quad 0 < q < 1, \quad n = 0, 1, \ldots,$$

is called Rogers-Szegö polynomial.

(a) Derive its q-exponential generating function as

$$\sum_{n=0}^{\infty} H_n(t;q)\frac{u^n}{[n]_q!} = e_q(u)e_q(ut),$$

where $e_q(u) = \prod_{i=1}^{\infty}(1 - u(1-q)q^{i-1})^{-1}$ is a q-exponential function.

(b) Show that

$$H_{n+1}(t;q) = (1+t)H_n(t;q) - t(1-q)[n]_q H_{n-1}(t;q), \quad n = 1, 2, \ldots,$$

with $H_0(t;q) = 1$ and $H_1(t;q) = t$.

1.13 *q-Multinomial coefficients.* The q-number

$$\begin{bmatrix} n \\ k_1, k_2, \ldots, k_{r-1} \end{bmatrix}_q = \frac{[n]_q!}{[k_1]_q![k_2]_q!\cdots[k_{r-1}]_q![k_r]_q!},$$

where $k_r = n - k_1 - k_2 - \cdots - k_{r-1}$, for $k_i = 0, 1, \ldots, n$, $i = 1, 2, \ldots, r$ and $n = 0, 1, \ldots,$ is called q-multinomial coefficient. The multivariate analogue of the Rogers-Szegö polynomial may be defined as

$$H_n(t_1, t_2, \ldots, t_{r-1}; q) = \sum \begin{bmatrix} n \\ k_1, k_2, \ldots, k_{r-1} \end{bmatrix}_q t_1^{k_1} t_2^{k_2} \cdots t_{r-1}^{k_{r-1}},$$

where the summation is extended over all $k_i = 0, 1, \ldots, n$, $i = 1, 2, \ldots, r$, such that $k_1 + k_2 + \cdots + k_{r-1} + k_r = n$. Derive its q-exponential generating function as

$$\sum_{n=0}^{\infty} H_n(t_1, t_2, \ldots, t_{r-1}; q)\frac{u^n}{[n]_q!} = e_q(u)e_q(ut_1)e_q(ut_2)\cdots e_q(ut_{r-1}),$$

where $e_q(u) = \prod_{i=1}^{\infty}(1 - u(1-q)q^{i-1})^{-1}$ is a q-exponential function.

1.14 *Noncentral q-Stirling numbers of the first kind.* Show that the noncentral q-Stirling numbers of the first kind are connected with the usual q-Stirling numbers of the first kind by

$$s_q(n,k;r) = \sum_{j=k}^{n} (-1)^{j-k} q^{r(n-j)} \binom{j}{k} [r]_q^{j-k} s_q(n,j)$$

and

$$s_q(n,k;r) = \sum_{j=k}^{n} q^{\binom{n-j}{2}+r(n-j)} \begin{bmatrix} n \\ j \end{bmatrix}_q [-r]_{n-j,q} s_q(j,k).$$

1.15 (*Continuation*). Show that

$$|s_q(n,j;r+\theta)| = q^{n-j} \sum_{k=j}^{n} \binom{k}{j} |s_q(n,k;r)| q^{(n-k)(\theta-1)} [\theta]_q^{k-j},$$

where $|s_q(n,k;r)|$ is the noncentral signless q-Stirling number of the first kind.

1.16 (*Continuation*). Show that

$$s_q(n,1) = (-1)^{n-1}[n-1]_q! \quad \text{and} \quad s_q(n,2) = (-1)^{n-2}[n-1]_q!\zeta_{n-1,q},$$

where $\zeta_{n,q} = \sum_{j=1}^{n} 1/[j]_q$.

1.17 *Noncentral q-Stirling numbers of the second kind.* Show that the noncentral q-Stirling numbers of the second kind are connected with the usual q-Stirling numbers of the second kind by

$$S_q(n,k;r) = \sum_{j=k}^{n} q^{\binom{j-k}{2}} \begin{bmatrix} j \\ k \end{bmatrix}_q [r]_{j-k,q} S_q(n,j)$$

and

$$S_q(n,k;r) = \sum_{j=k}^{n} q^{r(j-k)} \binom{n}{j} [r]_q^{n-j} S_q(j,k).$$

1.18 *Bivariate generating functions of the noncentral q-Stirling numbers.* Show that

$$\sum_{n=0}^{\infty} \sum_{k=0}^{n} s_q(n,k;r) t^k \frac{u^n}{[n]_q!} = \prod_{i=1}^{\infty} \frac{1+uq^{r+i-1}}{1+(1-(1-q)t)uq^{i-1}}$$

and

$$\sum_{n=0}^{\infty} \sum_{k=0}^{n} q^{\binom{k}{2}+rk} S_q(n,k;r) t^k \frac{u^n}{n!} = E_q(-t) \sum_{j=0}^{\infty} e^{[j+r]_q u} \frac{t^j}{[j]_q!},$$

where $E_q(t) = \prod_{i=1}^{\infty}(1+t(1-q)q^{i-1})$ is a q-exponential function.

1.19 (*Continuation*). Show that

$$\sum_{k=j}^{n}\binom{k}{j}(1-q)^{n-k}s_q(n,k;r)=q^{\binom{n-j}{2}+r(n-j)}\begin{bmatrix}n\\j\end{bmatrix}_q$$

and

$$\sum_{k=j}^{n}\begin{bmatrix}k\\j\end{bmatrix}_q q^{\binom{k-j}{2}+r(n-j)}(1-q)^{n-k}S_q(n,k;r)=\binom{n}{j}.$$

1.20 (*Continuation*). Show that

$$\sum_{n=k}^{\infty}\binom{n+j}{j}(1-q)^{n-k}S_q(n,k;r)=q^{-\binom{k+1}{2}-r(k+1)-j(r+k)}\begin{bmatrix}k+j\\j\end{bmatrix}_q.$$

1.21 *Noncentral generalized q-factorial coefficients*. Show that the noncentral generalized q-factorial coefficients are connected with the usual generalized q-factorial coefficients by

$$C_q(n,k;s,r)=q^{-r(n-k)}\sum_{j=k}^{n}q^{s\binom{j-k}{2}}\begin{bmatrix}j\\k\end{bmatrix}_{q^s}[r/s]_{j-k,q^s}C_q(n,j;s)$$

and

$$C_q(n,k;s,r)=\sum_{j=k}^{n}q^{\binom{n-j}{2}-r(n-j)}\begin{bmatrix}n\\j\end{bmatrix}_q[r]_{n-j,q}C_q(j,k;s).$$

1.22 *Noncentral q-Lah numbers*. Consider the expansion

$$[-(t-r)]_{n,q^{-1}}=q^{\binom{n}{2}-rn}\sum_{k=0}^{n}q^{\binom{k}{2}}L_q(n,k;r)[t]_{k,q}.$$

Since $[-(t-r)]_{n,q^{-1}}=[t-r+n-1]_{n,q}/[-1]_q^n$ and setting

$$|L_q(n,k;r)|=[-1]_q^nL_q(n,k;r),$$

it can be written as

$$[t-r+n-1]_{n,q}=q^{\binom{n}{2}-rn}\sum_{k=0}^{n}q^{\binom{k}{2}}|L_q(n,k;r)|[t]_{k,q}.$$

The coefficients $L_q(n,k;r)$ and $|L_q(n,k;r)|$ are called *noncentral q-Lah number* and *noncentral signless q-Lah number*, respectively. Note that for $r=0$ the noncentral q-Lah number and the noncentral signless q-Lah number reduce to $L_q(n,k;0)=L_q(n,k)$ and $|L_q(n,k;0)|=|L_q(n,k)|$, the usual (central) q-Lah number and the signless q-Lah number, respectively. Show that

$$|L_q(n,k;r)|=q^{-\binom{n}{2}+\binom{k}{2}+r(n-k)}\frac{[n]_q!}{[k]_q!}\begin{bmatrix}n-r-1\\k-r-1\end{bmatrix}_q.$$

1.23 *q-Eulerian numbers.* Consider the expansion of the nth power of a q-number into q-binomial coefficients of order n

$$[t]_q^n = \sum_{k=0}^{n} q^{\binom{k}{2}} A_q(n,k) \begin{bmatrix} t+n-k \\ n \end{bmatrix}_q, \quad n = 0, 1, \ldots .$$

The coefficient $A_q(n,k)$ is called *q-Eulerian number.* (a) Show that

$$A_q(n,k) = A_q(n, n-k+1), \quad k = 0, 1, \ldots, n, \quad n = 0, 1, \ldots,$$

and (b) derive the explicit expression

$$A_q(n,k) = q^{-\binom{k}{2}} \sum_{r=0}^{n} (-1)^r q^{\binom{r}{2}} \begin{bmatrix} n+1 \\ r \end{bmatrix}_q [k-r]_q^n,$$

for $k = 0, 1, \ldots, n$ and $n = 0, 1, \ldots$

1.24 (*Continuation*). Show that the q-Eulerian numbers $A_q(n,k)$, $k = 0, 1, \ldots, n$, $n = 0, 1, \ldots$, satisfy the triangular recurrence relation

$$A_q(n+1,k) = [k]_q A_q(n,k) + [n-k+2]_q A_q(n,k-1),$$

for $k = 1, 2, \ldots, n+1$, $n = 0, 1, \ldots$, with initial conditions

$$A_q(0,0) = 1, \quad A_q(n,0) = 0, n > 0, \quad A_q(n,k) = 0, k > n.$$

1.25 *Generalized Stirling numbers of the first kind.* Consider the expansion

$$\prod_{i=1}^{n} (t - a_i) = \sum_{k=0}^{n} s(n,k;\boldsymbol{a}) t^k,$$

or equivalently, the expansion

$$\prod_{i=1}^{n} (t + a_i) = \sum_{k=0}^{n} |s(n,k;\boldsymbol{a})| t^k,$$

where $|s(n,k;\boldsymbol{a})| = (-1)^{n-k} s(n,k;\boldsymbol{a})$, with $\boldsymbol{a} = (a_1, a_2, \ldots, a_n)$. The coefficient $s(n,k;\boldsymbol{a})$ is called *generalized Stirling number of the first kind* and the coefficient $|s(n,k;\boldsymbol{a})|$, which for $a_i \geq 0$, $i = 1, 2, \ldots, n$, is nonnegative, is called *generalized signless Stirling number of the first kind.*

(a) Show that

$$|s(n,k;\boldsymbol{a})| = \sum a_{i_1} a_{i_2} \cdots a_{i_{n-k}},$$

where the summation is extended over all $(n-k)$-combinations, $\{i_1, i_2, \ldots, i_{n-k}\}$, of the n indices $\{1, 2, \ldots, n\}$. Note that $|s(n, n-k; \boldsymbol{a})|$

is the *elementary symmetric function* (with respect to the n variables $a_1, a_2, \ldots, a_n$). Alternatively,

$$|s(n,k;\boldsymbol{a})| = \left(\prod_{i=1}^{n} a_i\right) \sum_{i=1}^{n} \frac{1}{a_{j_1} a_{j_2} \cdots a_{j_k}},$$

where the summation is extended over all k-combinations, $\{j_1, j_2, \ldots, j_k\}$, of the n indices $\{1, 2, \ldots, n\}$.

(b) Derive the triangular recurrence relation

$$|s(n,k;\boldsymbol{a})| = |s(n-1,k-1;\boldsymbol{a})| + a_n |s(n-1,k;\boldsymbol{a})|,$$

for $k = 1, 2, \ldots, n$, $n = 1, 2, \ldots$, with initial conditions

$$|s(0,0;\boldsymbol{a})| = 1, \quad |s(n,0;\boldsymbol{a})| = a_n a_{n-1} \cdots a_1, n > 0, \quad |s(n,k;\boldsymbol{a})| = 0, k > n.$$

(c) Show that

$$|s(n,k;\boldsymbol{a})| = \theta^{-(n-k)} q^{-\binom{n}{2}+\binom{k}{2}} \begin{bmatrix} n \\ k \end{bmatrix}_q, \quad \text{for } a_i = 1/(\theta q^{i-1}), \quad i = 1, 2, \ldots$$

and

$$|s(n,k;\boldsymbol{a})| = (\theta/q)^{n-k} |s_q(n,k;r)|, \quad \text{for } a_i = \theta[r+i-1]_q, \quad i = 1, 2, \ldots$$

1.26 *Generalized Stirling numbers of the second kind.* Consider the expansion

$$t^n = \sum_{k=0}^{n} S(n,k;\boldsymbol{a}) \prod_{i=1}^{k} (t - a_i).$$

The coefficient $S(n,k;\boldsymbol{a})$ is called *generalized Stirling number of the second kind.*

(a) Derive the triangular recurrence relation

$$S(n,k;\boldsymbol{a}) = S(n-1,k-1;\boldsymbol{a}) + a_{k+1} S(n-1,k;\boldsymbol{a}),$$

for $k = 1, 2, \ldots, n$, $n = 1, 2, \ldots$, with initial conditions

$$S(0,0;\boldsymbol{a}) = 1, \quad S(n,0;\boldsymbol{a}) = a_1^n, n > 0, \quad S(n,k;\boldsymbol{a}) = 0, k > n.$$

(b) Show that

$$\prod_{i=1}^{k} (1 - a_i u)^{-1} = \sum_{n=k}^{\infty} S(n-1,k-1;\boldsymbol{a}) u^{n-k}$$

and conclude that

$$S(n+k-1,n-1;\boldsymbol{a}) = \sum a_1^{r_1} a_2^{r_2} \cdots a_n^{r_n},$$

where the summation is extended over all $r_i = 0, 1, \ldots, k$, $i = 1, 2, \ldots, n$, such that $r_1 + r_2 + \cdots + r_n = k$. Note that $S(n+k-1,n-1;\boldsymbol{a})$ is the *homogeneous product sum symmetric function.*

(c) Also, show that

$$S(n,k;\boldsymbol{a}) = \theta^{n-k}\begin{bmatrix} n \\ k \end{bmatrix}_q, \quad \text{for } a_i = \theta q^{i-1}, \quad i = 1,2,\ldots$$

and

$$S(n,k;\boldsymbol{a}) = \theta^{n-k}S_q(n,k;r), \quad \text{for } a_i = \theta[r+i-1]_q, \quad i = 1,2,\ldots$$

1.27 (*Continuation*).

(a) Show that the generalized Stirling numbers of the first and second kind satisfy the following orthogonality relations

$$\sum_{j=k}^{n} s(n,j;\boldsymbol{a})S(j,k;\boldsymbol{a}) = \delta_{n,k}, \quad \sum_{j=k}^{n} S(n,j;\boldsymbol{a})s(j,k;\boldsymbol{a}) = \delta_{n,k},$$

where $\delta_{n,k} = 1$, if $k = n$ and $\delta_{n,k} = 0$, if $k \neq n$, is the Kronecker delta.

(b) Also, show that

$$\frac{1}{\prod_{i=1}^{k}(t-a_i)} = \sum_{n=k}^{\infty} S(n-1,k-1;\boldsymbol{a})\frac{1}{t^n}$$

and

$$\frac{1}{t^k} = \sum_{n=k}^{\infty} |s(n-1,k-1;\boldsymbol{a})| \frac{1}{\prod_{i=1}^{n}(t+a_i)}.$$

1.28 *Generalized Lah numbers*. Consider the expansion

$$\prod_{i=1}^{n}(t-a_i) = \sum_{k=0}^{n} C(n,k;\boldsymbol{a},\boldsymbol{b}) \prod_{j=1}^{k}(t-b_j),$$

with $\boldsymbol{a} = (a_1, a_2, \ldots, a_n)$ and $\boldsymbol{b} = (b_1, b_2, \ldots, b_k)$. The coefficient $C(n,k;\boldsymbol{a},\boldsymbol{b})$ is called *generalized Lah number.*

(a) Show that

$$C(n,k;\boldsymbol{a},\boldsymbol{b}) = \sum_{m=k}^{n} s(n,m;\boldsymbol{a})S(m,k;\boldsymbol{b})$$

and

$$C(n,k;\boldsymbol{a},\boldsymbol{b}) = [a]_q^n[b]_q^{-k}C_{q^a}(n,k;s,r), \quad s = b/a, \quad r = c/a,$$

for $a_i = [a(i-1)-c]_q$, $b_i = [b(i-1)]_q$, $i = 1,2,\ldots$, where $C_q(n,k;s,r)$ is the noncentral generalized q-factorial coefficient.

(b) Derive the triangular recurrence relation

$$C(n,k;\boldsymbol{a},\boldsymbol{b}) = C(n-1,k-1;\boldsymbol{a},\boldsymbol{b}) + (b_{k+1}-a_n)C(n-1,k;\boldsymbol{a},\boldsymbol{b}),$$

for $k = 1,2,\ldots,n$, $n = 1,2,\ldots$, with initial conditions

$$C(0,0;\boldsymbol{a},\boldsymbol{b})=1, C(n,0;\boldsymbol{a},\boldsymbol{b})=\prod_{i=1}^{n}(b_1-a_i), n>0, C(n,k;\boldsymbol{a},\boldsymbol{b})=0, k>n.$$

(c) Also, show that

$$\frac{1}{\prod_{j=1}^{k+1}(t-b_j)}=\sum_{n=k}^{\infty}C(n,k;\boldsymbol{a},\boldsymbol{b})\frac{1}{\prod_{i=1}^{n+1}(t-a_i)}.$$

1.29 *q-Derivative operator and q-exponential functions.* The q-derivative operator, denoted by $\mathcal{D}_q=d_q/d_qt$, is defined by

$$\mathcal{D}_qf(t)=\frac{d_qf(t)}{d_qt}=\frac{f(t)-f(qt)}{(1-q)t},$$

so that $\mathcal{D}_q1=0$. The higher-order q-derivatives are defined recursively by

$$\mathcal{D}_q^kf(t)=\mathcal{D}_q(\mathcal{D}_q^{k-1}f(t)),\quad k=2,3,\ldots .$$

(a) Show that

$$\mathcal{D}_qt^m=[m]_qt^{m-1},\mathcal{D}_{q^{-1}}t^m=[m]_{q^{-1}}t^{m-1}=q^{-(m-1)}[m]_qt^{m-1},\quad m\neq 0$$

and

$$\mathcal{D}_q^kt^m=[m]_{k,q}t^{m-k},\mathcal{D}_{q^{-1}}^kt^m=q^{-mk+\binom{k+1}{2}}[m]_{k,q}t^{m-k},\quad m\neq 0.$$

(b) Also, show that the q-exponential functions

$$e_q(t)=\prod_{i=1}^{\infty}(1-t(1-q)q^{i-1})^{-1}=\sum_{k=0}^{\infty}\frac{t^k}{[k]_q!},\quad |t|<1/(1-q)$$

and

$$E_q(t)=\prod_{i=1}^{\infty}(1+t(1-q)q^{i-1})=\sum_{k=0}^{\infty}q^{\binom{k}{2}}\frac{t^k}{[k]_q!},\quad -\infty<t<\infty.$$

satisfy the q-differential equations

$$\mathcal{D}_qe_q(t)=e_q(t),\quad \mathcal{D}_{q^{-1}}E_q(t)=E_q(t),\quad \mathcal{D}_qE_q(t)=E_q(qt).$$

1.30 (*Continuation*).

(a) Show that the noncentral q-Stirling numbers of the first kind may be written as

$$s_q(n,k;r)=q^{\binom{n}{2}+rn}\left[\frac{1}{[k]_q!}\mathcal{D}_q^k[t-r]_{n,q}\right]_{t=0},\quad k=0,1,\ldots,n,\ \ n=0,1,\ldots$$

(b) Also, derive the following *q-Leibnitz formula*:

$$\mathcal{D}_q^n(f(t)g(t)) = \sum_{k=0}^{n} \begin{bmatrix} n \\ k \end{bmatrix}_q \mathcal{D}_q^{n-k} f(q^k t)\mathcal{D}_q^k g(t).$$

1.31 *q-Integral and q-logarithmic function*. The q-integral is defined by

$$\int_0^x f(t)d_q t = x(1-q)\sum_{k=0}^{\infty} f(xq^k)q^k,$$

provided that the series converges, and

$$\int_a^b f(t)d_q t = \int_0^b f(t)d_q t - \int_0^a f(t)d_q t.$$

Note that, for a function $f(t)$ that is continuous on $[a, b]$, it holds

$$\int_a^b \mathcal{D}_q f(t)d_q t = f(b) - f(a).$$

Show that

$$\int_0^x t^n d_q t = \frac{x^{n+1}}{[n+1]_q}, \quad n \neq -1$$

and

$$\int_1^x \frac{d_q t}{t} = l_q(x),$$

with $l_q(x)$ the q-logarithmic function, for which

$$-l_q(1-x) = \sum_{k=1}^{\infty} \frac{x^k}{[k]_q}, \quad |x| < 1.$$

1.32 *A q-gamma function*. Consider the q-integral

$$I_{n,q} = \int_0^{\infty} t^n E_q(-qt)d_q t, \quad n = 0, 1, \ldots, \quad 0 < q < 1,$$

where $E_q(t) = \prod_{i=1}^{\infty}(1 + t(1-q)q^{i-1})$, $-\infty < t < \infty$ and $|q| < 1$, is a q-exponential function.

(a) Applying a q-integration by parts,

$$\int_a^b g(t)d_q f(t) = [f(t)g(t)]_a^b - \int_a^b f(qt)d_q g(t),$$

derive the first-order recurrence relation

$$I_{n,q} = [n]_q I_{n-1,q}, \quad n = 1, 2, \ldots, \quad 0 < q < 1,$$

with initial condition $I_{0,q} = 1$, and conclude that

$$I_{n,q} = \int_0^{\infty} t^n E_q(-qt)d_q t = [n]_q!, \qquad n = 0, 1, \ldots, \quad 0 < q < 1.$$

(b) The expression of q-factorial of n, for n a positive integer and $0 < q < 1$,

$$[n]_q! = \frac{\prod_{i=1}^{n}(1-q^i)\prod_{i=n+1}^{\infty}(1-q^i)}{(1-q)^n\prod_{i=n+1}^{\infty}(1-q^i)} = \frac{\prod_{i=1}^{\infty}(1-q^i)}{(1-q)^n\prod_{i=1}^{\infty}(1-q^{n+i})},$$

may be extended to a real number x as

$$\Gamma_q(x) = \frac{\prod_{i=1}^{\infty}(1-q^i)}{(1-q)^{x-1}\prod_{i=1}^{\infty}(1-q^{x+i-1})}, \quad |q| < 1.$$

The q-function $\Gamma_q(x)$ is called *q-gamma function*. Show that

$$\Gamma_q(x+1) = [x]_q\Gamma_q(x), \quad \Gamma_q(1) = 1$$

and $\lim_{q\to 1^-} \Gamma_q(x) = \Gamma(x)$, where $\Gamma(x)$ denotes the usual gamma function.

1.33 *Another q-gamma function.* Consider the q-integral

$$J_{n,q} = \int_0^{\infty} t^n e_q(-t)d_qt, \quad n = 0, 1, \ldots, \quad 0 < q < 1,$$

where $e_q(t) = \prod_{i=1}^{\infty}(1 - t(1-q)q^{i-1})^{-1}$, $|t| < 1/(1-q)$ and $|q| < 1$, is a q-exponential function.

(a) Applying a q-integration by parts, derive the first-order recurrence relation

$$J_{n,q} = q^{-n}[n]_qJ_{n-1,q}, \quad n = 1, 2, \ldots, \quad 0 < q < 1,$$

with initial condition $J_{0,q} = 1$, and conclude that

$$J_{n,q} = \int_0^{\infty} t^n e_q(-t)d_qt = q^{-\binom{n+1}{2}}[n]_q!$$

(b) The relation connecting the q-factorials of n, with bases (parameters) inverse to each other, $[n]_{q^{-1}}! = q^{-\binom{n}{2}}[n]_q!$, suggests the definition of a second *q-gamma function* as

$$\gamma_q(x) = \frac{q^{-\binom{x-1}{2}}\prod_{i=1}^{\infty}(1-q^i)}{(1-q)^{x-1}\prod_{i=1}^{\infty}(1-q^{x+i-1})}, \quad |q| < 1.$$

It should be noted that the definition of the first q-gamma function $\Gamma_q(x)$, which was given in Exercise 1.32 for $|q| < 1$, is extended for $|q| > 1$ through the second q-gamma function $\gamma_q(x)$, by the relation

$$\Gamma_q(x) \equiv \gamma_{q^{-1}}(x) = \frac{q^{\binom{x}{2}}\prod_{i=1}^{\infty}(1-q^{-i})}{(q-1)^{x-1}\prod_{i=1}^{\infty}(1-q^{-(x+i-1)})}, \quad |q| > 1.$$

Show that $\lim_{q\to 1^+} \Gamma_q(x) = \Gamma(x)$, where $\Gamma(x)$ denotes the gamma function.

1.34 A *q-Beta function.* Consider the q-integral

$$B_q(x,y) = \int_0^1 \prod_{i=1}^{\infty} \frac{1-tq^i}{1-tq^{x+i-1}} t^{y-1} d_q t = \int_0^1 t^{x-1} \prod_{i=1}^{\infty} \frac{1-tq^i}{1-tq^{y+i-1}} d_q t,$$

for $x > 0$, $y > 0$ and $0 < q < 1$.

(a) Show that

$$B_q(x,y) = \frac{\Gamma_q(x)\Gamma_q(y)}{\Gamma_q(x+y)},$$

where

$$\Gamma_q(x) = \frac{\prod_{i=1}^{\infty}(1-q^i)}{(1-q)^{x-1}\prod_{i=1}^{\infty}(1-q^{x+i-1})}, \quad |q| < 1.$$

is a q-gamma function, and conclude that

$$\lim_{q \to 1^-} B_q(x,y) = B(x,y),$$

where $B(x,y)$ is the beta function. The q-function $B_q(x,y)$ is called *q-beta function.*

(b) In particular, for $x = r+1$ and $y = n-r$ positive integers, deduce that

$$B_q(r+1, n-r) = \int_0^1 \prod_{i=1}^{n-r-1} (1-tq^{i-1}) t^r d_q t = \frac{[r]_q![n-r-1]_q!}{[n]_q!}.$$

1.35 *The q-operator $\Theta_q = tD_q$.* The operator $\Theta_q = tD_q$ is the q-analogue of the well-known operator $\Theta = tD$, to which it reduces for $q = 1$.

(a) Show that

$$\Theta_q = [\Theta]_q = \frac{1-q^{\Theta}}{1-q}.$$

(b) Express the operator Θ_q in terms of the operator D_q as

$$\Theta_q^n = \sum_{k=0}^{n} q^{\binom{k}{2}} S_q(n,k) t^k D_q^k,$$

and, inversely, express the operator D_q in terms of the operator Θ_q as

$$D_q^n = q^{-\binom{n}{2}} t^{-n} \sum_{k=0}^{n} s_q(n,k) \Theta_q^k,$$

where $s_q(n,k)$ and $S_q(n,k)$ are the q-Stirling numbers of the first and second kind, respectively.

1.36 *q-Difference operator.* The q-difference operator, denoted by Δ_q, is defined by

$$\Delta_q f(t) = f(t+1) - f(t).$$

The higher-order q-differences are defined recursively by

$$\Delta_q^k f(t) = \Delta_q^{k-1} f(t+1) - q^{k-1}\Delta_q^{k-1} f(t), \quad k = 2, 3, \ldots$$

Clearly, the kth-order q-difference operator is expressed in terms of the usual shift operator E by

$$\Delta_q^k = \prod_{i=1}^{k} (E - q^{i-1}), \quad k = 1, 2, \ldots .$$

(a) Show that

$$\Delta_q [t]_{m,q} = [m]_q [t]_{m-1,q} q^{t-m+1}, \quad m \neq 0$$

and

$$\Delta_q^k [t]_{m,q} = [m]_{k,q} [t]_{m-k,q} q^{k(t-m+k)}, \quad m \neq 0.$$

(b) Also, show that the noncentral q-Stirling numbers of the second kind and the noncentral generalized q-factorial coefficients may be written as

$$S_q(n,k;r) = q^{-\binom{k}{2}-rn} \left[\frac{1}{[k]_q!} \Delta_q^k [t+r]_q^n \right]_{t=0},$$

for $k = 0, 1, \ldots, n$, $n = 0, 1, \ldots$, and

$$C_q(n,k;s,r) = q^{\binom{n}{2}-s\binom{k}{2}-rn} \left[\frac{1}{[k]_{q^s}!} \Delta_{q^s}^k [st+r]_{n,q} \right]_{t=0},$$

for $k = 0, 1, \ldots, n$, $n = 0, 1, \ldots$, respectively.

1.37 *q-Factorial moments as q-derivatives of probability generating functions.* Let X be a nonnegative integer-valued random variable with probability generating function

$$P_X(t) = \sum_{x=0}^{\infty} f(x) t^x, \quad |t| \leq 1,$$

where $f(x) = P(X = x)$, $x = 0, 1, \ldots$, is the probability function. Show that (a)

$$\left[\frac{d^m P_X(t)}{dt^m} \right]_{t=1} = E[(X)_m], \quad m = 1, 2, \ldots,$$

where $E[(X)_m] = \sum_{x=m}^{\infty} (x)_m f(x)$ is the mth factorial moment and (b)

$$\left[\frac{d_q^m P_X(t)}{d_q t^m} \right]_{t=1} = E([X]_{m,q}), \quad m = 1, 2, \ldots,$$

where $E([X]_{m,q}) = \sum_{x=m}^{\infty} [x]_{m,q} f(x)$ is the mth q-factorial moment.

1.38 *q-Factorial moments of a discrete q-uniform distribution.* Let X_n be a discrete q-uniform random variable, with probability function

$$f_{X_n}(x) = P(X_n = x) = \frac{q^x}{[n]_q}, \quad x = 0, 1, \ldots, n-1.$$

(a) Find the mth q-factorial moment $E([X_n]_{m,q})$, $m = 1, 2, \ldots$, and deduce the q-expected value $E([X_n]_q)$ and the q-variance $V([X_n]_q)$.

(b) Derive the probability generating function $P_{X_n}(t) = \sum_{x=0}^{n-1} f_{X_n}(x)t^x$.

2

SUCCESS PROBABILITY VARYING WITH THE NUMBER OF TRIALS

2.1 q-BINOMIAL DISTRIBUTION OF THE FIRST KIND

Consider a random experiment with sample space Ω and an event $A \subseteq \Omega$. If $A' \subseteq \Omega$ is the complementary event of A, then the pair of events $\{A, A'\}$, since $A \cup A' = \Omega$ and $A \cap A' = \emptyset$, constitutes a *partition* of Ω. Event A is usually characterized as *success* and event A' as *failure*. Representing the success by s and the failure by f, the sample space can be expressed as $\Omega = \{f, s\}$. An experiment with such a sample space is called *Bernoulli trial*.

Poisson (1837) considered a sequence of independent Bernoulli trials, with the probability of success at the ith trial varying with the number of trials,

$$P_i(\{s\}) = p_i, \quad 0 < p_i < 1, \quad i = 1, 2, \ldots .$$

Clearly, the probability of failure at the ith trial is $P_i(\{f\}) = 1 - p_i \equiv q_i$, for $i = 1, 2, \ldots$. Note that the number X_n of successes in a sequence of n independent Bernoulli trials may be expressed as a sum of n independent zero-one Bernoulli random variables. Specifically, let Z_i be the number of successes at the ith trial, $i = 1, 2, \ldots, n$. Then, $X_n = \sum_{i=1}^{n} Z_i$, with

$$P(Z_i = 0) = 1 - p_i \equiv q_i, \quad P(Z_i = 1) = p_i, \quad i = 1, 2, \ldots, n.$$

Discrete q-Distributions, First Edition. Charalambos A. Charalambides.

Note also that, in the case the probability p_i, for $i = 1, 2, \ldots$, is of a general functional form very little can be inferred from it concerning the distributions of X_n and other random variables that may be defined in this model (see Exercise 2.1).

Furthermore, assuming that the probability of success p_i or, equivalently, the odds of success $\theta_i = p_i/(1 - p_i)$ at the ith trial varies geometrically, with rate (proportion) q, two interesting and useful families of q-binomial and negative q-binomial distributions are introduced. More precisely, consider a sequence of independent Bernoulli trials and assume that the odds of success at the ith trial is given by

$$\theta_i = \theta q^{i-1}, \quad i = 1, 2, \ldots, \quad 0 < \theta < \infty, \quad 0 < q < 1 \quad \text{or} \quad 1 < q < \infty, \tag{2.1}$$

which is a geometrically varying sequence, with rate q. Note that the case $q = 1$ corresponds to the classical case of constant odds (or constant probability) of success. Also, for $0 < q < 1$, the sequence (2.1) is geometrically decreasing, while for $1 < q < \infty$, it is geometrically increasing. Since $p_i = \theta_i/(1 + \theta_i)$, it follows that the probability of success at the ith trial is given by

$$p_i = \frac{\theta q^{i-1}}{1 + \theta q^{i-1}}, \quad i = 1, 2, \ldots, \quad 0 < \theta < \infty, \quad 0 < q < 1 \quad \text{or} \quad 1 < q < \infty. \tag{2.2}$$

The study of the distribution of the number of successes in a given number of trials is of theoretical and practical interest. In this respect, the following definition is introduced.

Definition 2.1. *Let X_n be the number of successes in a sequence of n independent Bernoulli trials, with probability of success at the ith trial given by (2.2). The distribution of the random variable X_n is called q-binomial distribution of the first kind, with parameters n, θ, and q.*

The probability function, the q-factorial moments, and the usual factorial moments of the q-binomial distribution of the first kind are obtained in the following theorem.

Theorem 2.1. *The probability function of the q-binomial distribution of the first kind, with parameters n, θ, and q, is given by*

$$P(X_n = x) = \begin{bmatrix} n \\ x \end{bmatrix}_q \frac{\theta^x q^{\binom{x}{2}}}{\prod_{i=1}^{n}(1 + \theta q^{i-1})}, \quad x = 0, 1, \ldots, n, \tag{2.3}$$

for $0 < \theta < \infty$ and $0 < q < 1$ or $1 < q < \infty$. Its q-factorial moments are given by

$$E([X_n]_{m,q}) = \frac{[n]_{m,q}\, \theta^m q^{\binom{m}{2}}}{\prod_{i=1}^{m}(1 + \theta q^{i-1})}, \quad m = 1, 2, \ldots, n, \tag{2.4}$$

and $E([X_n]_{m,q}) = 0$, for $m = n + 1, n + 2, \ldots$. Moreover, its (usual) factorial moments are given by

$$E[(X_n)_j] = j! \sum_{m=j}^{n} (-1)^{m-j} \begin{bmatrix} n \\ m \end{bmatrix}_q \frac{\theta^m q^{\binom{m}{2}} (1 - q)^{m-j} s_q(m, j)}{\prod_{i=1}^{m}(1 + \theta q^{i-1})}, \tag{2.5}$$

for $j = 1, 2, \ldots, n$, and $E[(X_n)_j] = 0$, for $j = n+1, n+2, \ldots$, where $s_q(m,j)$ is the q-Stirling number of the first kind. In particular, its mean and variance are given by

$$E(X_n) = \sum_{i=1}^{n} \frac{\theta q^{i-1}}{1+\theta q^{i-1}}, \quad V(X_n) = \sum_{i=1}^{n} \frac{\theta q^{i-1}}{(1+\theta q^{i-1})^2}. \tag{2.6}$$

Proof. Let A_i be the event of success at the ith trial, $i = 1, 2, \ldots, n$, and consider a permutation $(i_1, i_2, \ldots, i_x, i_{x+1}, \ldots, i_n)$ of the n positive integers $\{1, 2, \ldots, n\}$. Then, using the independence of the Bernoulli trials and the probabilities (2.2) together with the identity $\prod_{k=1}^{n}(1+\theta q^{i_k-1}) = \prod_{i=1}^{n}(1+\theta q^{i-1})$, we get

$$\begin{aligned} P(A_{i_1}A_{i_2}\cdots A_{i_x}A'_{i_{x+1}}\cdots A'_{i_n}) &= P(A_{i_1})\cdots P(A_{i_x})P(A'_{i_{x+1}})\cdots P(A'_{i_n}) \\ &= \frac{1}{\prod_{i=1}^{n}(1+\theta q^{i-1})}\theta^x q^{i_1+i_2+\cdots+i_x-x}. \end{aligned}$$

Summing these probabilities over all x-combinations $\{i_1, i_2, \ldots, i_x\}$ of the set $\{1, 2, \ldots, n\}$ and using (1.3),

$$\sum_{1\le i_1<i_2<\cdots<i_x\le n} q^{i_1+i_2+\cdots+i_x-x} = q^{\binom{x}{2}} \begin{bmatrix} n \\ x \end{bmatrix}_q,$$

we deduce the probability of x successes in n trials in the form (2.3).

Note that the probabilities (2.3), according the q-binomial formula (1.14), sum to unity,

$$\sum_{x=0}^{n} P(X_n = x) = \frac{1}{\prod_{i=1}^{n}(1+\theta q^{i-1})} \sum_{x=0}^{n} \begin{bmatrix} n \\ x \end{bmatrix}_q \theta^x q^{\binom{x}{2}} = 1,$$

which conforms with the definition of a probability function.

The mth q-factorial moment of X_n,

$$E([X_n]_{m,q}) = \sum_{x=m}^{n} [x]_{m,q} \begin{bmatrix} n \\ x \end{bmatrix}_q \frac{\theta^x q^{\binom{x}{2}}}{\prod_{i=1}^{n}(1+\theta q^{i-1})},$$

on using the relations

$$\begin{aligned} [x]_{m,q}\begin{bmatrix} n \\ x \end{bmatrix}_q &= \frac{[x]_q!}{[x-m]_q!} \cdot \frac{[n]_q!}{[x]_q![n-x]_q!} \\ &= \frac{[n]_q!}{[n-m]_q!} \cdot \frac{[n-m]_q!}{[x-m]_q![n-x]_q!} = [n]_{m,q}\begin{bmatrix} n-m \\ x-m \end{bmatrix}_q, \end{aligned}$$

and

$$\binom{x}{2} = \binom{m}{2} + \binom{x-m}{2} + m(x-m),$$

is expressed as

$$E([X_n]_{m,q}) = \frac{[n]_{m,q}\theta^m q^{\binom{m}{2}}}{\prod_{i=1}^{n}(1+\theta q^{i-1})} \sum_{x=m}^{n} \begin{bmatrix} n-m \\ x-m \end{bmatrix}_q q^{\binom{x-m}{2}} (\theta q^m)^{x-m},$$

which, by the q-binomial formula (1.14), yields the required expression (2.4),

$$E([X_n]_{m,q}) = \frac{[n]_{m,q}\theta^m q^{\binom{m}{2}} \prod_{i=1}^{n-m}(1+\theta q^{m+i-1})}{\prod_{i=1}^{m}(1+\theta q^{i-1}) \prod_{i=m+1}^{n}(1+\theta q^{i-1})} = \frac{[n]_{m,q}\theta^m q^{\binom{m}{2}}}{\prod_{i=1}^{m}(1+\theta q^{i-1})}.$$

Introducing the last expression into (1.61), the required formula (2.5) is deduced. In particular, from (2.5) with $j = 1$ and $j = 2$, the mean and variance of X_n may be obtained (see Exercises 2.3 and 2.4). Alternatively, using the expression $X_n = \sum_{i=1}^{n} Z_i$, where $Z_1, Z_2, \ldots, Z_n$ are independent zero–one Bernoulli random variables, with

$$E(Z_i) = \frac{\theta q^{i-1}}{1+\theta q^{i-1}}, \quad V(Z_i) = \frac{\theta q^{i-1}}{(1+\theta q^{i-1})^2}, \quad i = 1, 2, \ldots, n,$$

and since $E(X_n) = \sum_{i=1}^{n} E(Z_i)$ and $V(X_n) = \sum_{i=1}^{n} V(Z_i)$, the mean and variance (2.6) are readily deduced. □

Example 2.1. *Weldon's classical dice data.* Walter F. R. Weldon obtained the data from $m = 26,306$ throws of $n = 12$ dice. Among the $mn = 315,672$ recorded numbers from the set of the six faces of a die, $\{1, 2, 3, 4, 5, 6\}$, the number of dice showing face 5 or face 6 was $s = 106,602$.

A discrete probability distribution that fits to these data may be defined on a sequence of independent Bernoulli trials. Specifically, a throw of a die is considered as a Bernoulli trial, with success the event of showing face 5 or face 6. Then, each throw of the 12 dice constitutes a sequence of $n = 12$ independent Bernoulli trials.

Kemp and Kemp (1991) examined first the fair dice assumption, which leads to the usual binomial distribution, with $n = 12$ and $p = 1/3$. It was found out that this distribution does not fit to these data. After this conclusion, they replaced the success probability $p = 1/3$ by its moment estimate

$$\hat{p} = \frac{s}{mn} = \frac{106,602}{315,672} = 0.3377.$$

Although this equally unbalanced dice hypothesis may give a satisfactory fit to these data, the assumption that all $n = 12$ dice are identically unbalanced seems inherently implausible.

More realistic hypotheses for unfair dice, p_i, $i = 1, 2, \ldots, n$, were examined by Kemp and Kemp (1991). They assumed that there is a spectrum of unfairness among the dice. A log-linear odds assumption,

$$\log \theta_i = \log \theta + (i-1) \log q, \quad i = 1, 2, \ldots, n,$$

implies (2.1). Then, the number X_n of successes in n trials obeys the q-binomial distribution of the first kind with probability function (2.3).

Remark 2.1. *Stationary distribution in a birth and death process.* Consider a homogeneous birth and death process X_t, $t \geq 0$, with birth and death rates $\lambda_j > 0$, for $j = 0, 1, \ldots$, and $\mu_j > 0$, for $j = 1, 2, \ldots$, respectively. The probability function of the stationary distribution,

$$\lim_{t\to\infty} P(X_t = x) = P(X = x), \quad x = 0, 1, \ldots,$$

satisfies the recurrence relation

$$P(X = x) = \frac{\lambda_{x-1}}{\mu_x} P(X = x - 1), \quad x = 1, 2, \ldots,$$

and so

$$P(X = x) = P(X = 0) \prod_{j=1}^{x} \frac{\lambda_{j-1}}{\mu_j}, \quad x = 1, 2, \ldots,$$

with

$$P(X = 0) = \left(1 + \sum_{x=1}^{\infty} \prod_{j=1}^{x} \frac{\lambda_{j-1}}{\mu_j}\right)^{-1},$$

provided $\sum_{x=1}^{\infty} \prod_{j=1}^{x} (\lambda_{j-1}/\mu_j) < \infty$. Clearly, any distribution with probability function satisfying a recurrence relation of this form may be interpreted as a stationary distribution of a birth and death process. Thus, the q-binomial distribution of the first kind may be considered as the stationary distribution of a birth and death process with birth and death rates

$$\lambda_j = \theta q^j [n - j]_q = \theta([n]_q - [j]_q), \ j = 0, 1, \ldots, n, \quad \mu_j = [j]_q, \ j = 1, 2, \ldots, n.$$

Indeed,

$$\prod_{j=1}^{x} \frac{\lambda_{j-1}}{\mu_j} = \prod_{j=1}^{x} \frac{\theta q^{j-1}[n - j + 1]_q}{[j]_q} = \theta^x q^{\binom{x}{2}} \begin{bmatrix} n \\ x \end{bmatrix}_q$$

and by the q-binomial formula (1.14) $P(X = 0) = 1/\prod_{i=1}^{n}(1 + \theta q^{i-1})$, whence the probability function (2.3) is readily deduced.

An application of the q-binomial distribution of the first kind as the stationary distribution of a birth and death process is presented in the following example.

Example 2.2. *Stationary distribution for a dichotomized parasite population.* Let us consider a population of parasites of size n that is dichotomized to the subpopulations of type A parasites, which are on hosts without open wounds, and of type B parasites, which are on hosts with open wounds. The parasites may be either active or passive. A type A parasite, which slits an opening in the skin of the host, consumes an amount of blood and then remains passive for a while before actively seeking a new site, either on the same host or on another host never previously parasitized. A host very rarely has more than one parasite attached to it; to simplify the model, assume that no host has more than one parasite.

Let $X_{t,n}$ be the number of parasites on hosts without open wounds (type A parasites) at time $t \geq 0$. Assume that at time t, j of the n parasites are on hosts without open wounds and the other $n - j$ parasites are on hosts with open wounds (type B parasites). If a host, without open wounds (which has never previously been parasitized), is available, then one of the active parasites may transfer to it instead of relocating on its existing host; active type A parasites are assumed to take priority over active type B parasites.

Let p be the probability that a parasite is active, whence $q = 1 - p$ is the probability that it is passive. Also, let $\theta/(1 + \theta)$ be the probability for an active parasite to move to a host without open wounds.

The conditional probability that the number $X_{t,n}$ of parasites on hosts without open wounds, in a small time interval, is increased by one, given that $X_{t,n} = j$, equals the product of the following three probabilities. The probability q^j, that all the j type A parasites are passive, the probability $1 - q^{n-j}$ that at least one of the $n - j$ type B parasites is active and the probability $\theta/(1 + \theta)$ for an active parasite to move to a host without open wounds.

Also, the conditional probability that the number $X_{t,n}$ of parasites on hosts without open wounds, in a small time interval, is decreased by one, given that $X_{t,n} = j$, equals the product of the probability $1 - q^j$ that at least one of the j type A parasites is active and the probability $1/(1 + \theta)$ for an active parasite not to move to a host without open wounds.

Then, $X_{t,n}$ is a birth and death process with birth and death rates proportional to

$$b_j = \frac{\theta q^j(1 - q^{n-j})}{1 + \theta}, \quad j = 0, 1, \ldots, n, \quad d_j = \frac{1 - q^j}{1 + \theta}, \quad j = 1, 2, \ldots, n.$$

Therefore, according to Remark 2.1, the probability function of the stationary distribution $P(X_n = x)$, $x = 0, 1, \ldots, n$, is given by (2.3).

The number $Y_{t,n}$ of parasites on hosts with open wounds (type B parasites) at time $t \geq 0$ is also a birth and death process and its stationary distribution, $P(Y_n = x) = P(X_n = n - x)$, $x = 0, 1, \ldots, n$, is also a q-binomial distribution of the first kind; its probability function is of the form (2.3), with the parameters θ and q replaced by θ^{-1} and q^{-1}, respectively.

2.2 NEGATIVE q-BINOMIAL DISTRIBUTION OF THE FIRST KIND

Consider now a sequence of independent Bernoulli trials, which is terminated with the occurrence of the nth failure (or success). Also, assume that the odds of success at the ith trial is given by (2.1), or, equivalently, that the probability of success at the ith trial is given by (2.2). In this stochastic model, the interest is focused on the study of the number of successes (or failures) until the occurrence of the nth failure (or success). For this reason, the following definition is introduced.

Definition 2.2. *Let W_n be the number of failures until the occurrence of the nth success in a sequence of independent Bernoulli trials, with probability of success* *at the ith trial given by (2.2). The distribution of the random variable W_n is called negative q-binomial distribution of the first kind, with parameters n, θ, and q.*

The probability function, the q-factorial moments, and the usual factorial moments of the negative q-binomial distribution of the first kind are derived in the following theorem.

Theorem 2.2. *The probability function of the negative q-binomial distribution of the first kind, with parameters n, θ, and q, is given by*

$$P(W_n = w) = \begin{bmatrix} n+w-1 \\ w \end{bmatrix}_q \frac{\theta^n q^{\binom{n}{2}+w}}{\prod_{i=1}^{n+w}(1+\theta q^{i-1})}, \quad w = 0, 1, \ldots, \tag{2.7}$$

for $0 < \theta < \infty$ and $0 < q < 1$ or $1 < q < \infty$. Its q-factorial moments are given by

$$E([W_n]_{m,q}) = \frac{[n+m-1]_{m,q}}{\theta^m q^{\binom{m}{2}+(n-1)m}}, \quad m = 1, 2, \ldots. \tag{2.8}$$

Furthermore, its (usual) factorial moments are given by

$$E[(W_n)_j] = j! \sum_{m=j}^{\infty} (-1)^{m-j} \begin{bmatrix} n+m-1 \\ m \end{bmatrix}_q \frac{(1-q)^{m-j} s_q(m,j)}{\theta^m q^{\binom{m}{2}+(n-1)m}}, \tag{2.9}$$

for $j = 1, 2, \ldots$, where $s_q(m,j)$ is the q-Stirling number of the first kind.

Proof. The probability function of the negative q-binomial distribution of the first kind is closely connected to the probability function of the q-binomial distribution of the first kind. Precisely,

$$P(W_n = w) = P(X_{n+w-1} = n-1)p_{n+w},$$

where $P(X_{n+w-1} = n-1)$ is the probability of the occurrence of $n-1$ successes in $n+w-1$ trials and $p_{n+w} = \theta q^{n+w-1}/(1+\theta q^{n+w-1})$ is the probability of success at the $(n+w)$th trial. Thus, using (2.3), expression (2.7) is deduced.

Note that the probabilities (2.7), according to the negative q-binomial formula, which is given in Exercise 1.9, sum to unity

$$\sum_{w=0}^{\infty} P(W_n = w) = \frac{\theta^n q^{\binom{n}{2}}}{\prod_{i=1}^{n}(1+\theta q^{i-1})} \sum_{w=0}^{\infty} \begin{bmatrix} n+w-1 \\ w \end{bmatrix}_q \frac{q^w}{\prod_{i=1}^{w}(1+\theta q^{n+i-1})} = 1,$$

which conforms with the definition of a probability function.

The mth q-factorial moment of W_n,

$$E([W_n]_{m,q}) = \sum_{w=m}^{\infty} [w]_{m,q} \begin{bmatrix} n+w-1 \\ w \end{bmatrix}_q \frac{\theta^n q^{\binom{n}{2}+w}}{\prod_{i=1}^{n+w}(1+\theta q^{i-1})},$$

using the relation

$$[w]_{m,q}\begin{bmatrix} n+w-1 \\ w \end{bmatrix}_q = \frac{[w]_q!}{[w-m]_q!}\cdot\frac{[n+w-1]_q!}{[w]_q![n-1]_q!}$$
$$= \frac{[n+m-1]_q!}{[n-1]_q!}\cdot\frac{[n+w-1]_q!}{[w-m]_q![n+m-1]_q!}$$
$$= [n+m-1]_{m,q}\begin{bmatrix} n+w-1 \\ w-m \end{bmatrix}_q,$$

is expressed as

$$E([W_n]_{m,q}) = \frac{[n+m-1]_{m,q}\theta^n q^{\binom{n}{2}+m}}{\prod_{i=1}^{n+m}(1+\theta q^{i-1})}\sum_{w=m}^{\infty}\begin{bmatrix} n+w-1 \\ w-m \end{bmatrix}_q \frac{q^{w-m}}{\prod_{i=1}^{w-m}(1+\theta q^{n+m+i-1})}.$$

Therefore, setting $k = w - m$ and using the negative q-binomial formula given in Exercise 1.9, with $n + m$ instead of n, the mth q-factorial moment of W_n is obtained in the form

$$E([W_n]_{m,q}) = \frac{[n+m-1]_{m,q}\theta^n q^{\binom{n}{2}+m}}{\theta^{n+m}q^{\binom{n+m}{2}}}, \quad m = 1, 2, \ldots,$$

which, by the relation $\binom{n+m}{2} = \binom{n}{2} + \binom{m}{2} + nm$, implies (2.8). Introducing the last expression into (1.61), the required formula (2.9) is deduced. □

Remark 2.2. *Another negative q-binomial distribution of the first kind.* The probability function of the number U_n of successes until the occurrence of the nth failure is closely connected to (2.7). Precisely, from

$$P(U_n = u) = P(X_{n+u-1} = u)(1 - p_{n+u})$$

and

$$P(W_n = w) = P(X_{n+w-1} = n-1)p_{n+w}.$$

it follows that

$$P(U_n = u) = P(X_{n+u-1} = u)(1 - p_{n+u}) = P(W_{u+1} = n-1)\frac{1-p_{n+u}}{p_{n+u}}$$

and so

$$P(U_n = u) = \begin{bmatrix} n+u-1 \\ u \end{bmatrix}_q \frac{\theta^u q^{\binom{u}{2}}}{\prod_{i=1}^{n+u}(1+\theta q^{i-1})}, \quad u = 0, 1, \ldots, \qquad (2.10)$$

for $0 < \theta < \infty$ and $0 < q < 1$ or $1 < q < \infty$. Its q-factorial and usual factorial moments are given in Exercise 2.11.

Note that the probabilities (2.10), according to the negative q-binomial formula given in Exercise 1.9, sum to unity,

$$\sum_{u=0}^{\infty} P(U_n = u) = \frac{1}{\prod_{i=1}^{n}(1+\theta q^{i-1})} \sum_{u=0}^{\infty} \begin{bmatrix} n+u-1 \\ u \end{bmatrix}_q \frac{\theta^u q^{\binom{u}{2}}}{\prod_{i=1}^{u}(1+\theta q^{n+i-1})} = 1,$$

which conforms with the definition of a probability function. Also, notice that the probability function (2.10), on introducing the parameters $\lambda = \theta^{-1}$ and $p = q^{-1}$, with the same parametric space, reduces to probability function (2.7).

2.3 HEINE DISTRIBUTION

Definition 2.3. *Let X be a discrete random variable with probability function*

$$f(x) = P(X = x) = e_q(-\lambda) \frac{q^{\binom{x}{2}} \lambda^x}{[x]_q!}, \quad x = 0, 1, \ldots, \tag{2.11}$$

where $0 < \lambda < \infty$, $0 < q < 1$, *and* $e_q(t) = \prod_{i=1}^{\infty} (1 - t(1-q)q^{i-1})^{-1}$ *is the q-exponential function (1.24). The distribution of the random variable X is called Heine distribution, with parameters λ and q.*

The Heine distribution is a q-Poisson distribution since the probability function (2.11), for $q \to 1$, converges to the probability function of the Poisson distribution. Note that the function (2.11) is nonnegative,

$$f(x) > 0, \ x = 0, 1, \ldots, \quad f(x) = 0, \ x \neq 0, 1, \ldots,$$

and using the expansion of the q-exponential function $E_q(t)$ into a power series,

$$E_q(t) = \prod_{i=1}^{\infty} (1 + t(1-q)q^{i-1}) = \sum_{x=0}^{\infty} q^{\binom{x}{2}} \frac{t^x}{[x]_q!}, \quad -\infty < t < \infty,$$

together with the relation $e_q(-t)E_q(t) = 1$, it follows that it sums to unity,

$$\sum_{x=0}^{\infty} f(x) = e_q(-\lambda) \sum_{x=0}^{\infty} q^{\binom{x}{2}} \frac{\lambda^x}{[x]_q!} = e_q(-\lambda)E_q(\lambda) = 1,$$

as is required by the definition of a probability function.

The q-factorial moments and the usual factorial moments of the Heine distribution are derived in the following theorem.

Theorem 2.3. *The q-factorial moments of the Heine distribution are given by*

$$E([X]_{m,q}) = \frac{q^{\binom{m}{2}}\lambda^m}{\prod_{i=1}^{m}(1+\lambda(1-q)q^{i-1})}, \quad m = 1, 2, \ldots . \tag{2.12}$$

Moreover, its factorial moments are given by

$$E[(X)_j] = j! \sum_{m=j}^{\infty} (-1)^{m-j} \frac{q^{\binom{m}{2}}\lambda^m}{[m]_q!} \cdot \frac{(1-q)^{m-j} s_q(m,j)}{\prod_{i=1}^{m}(1+\lambda(1-q)q^{i-1})}, \tag{2.13}$$

for $j = 1, 2, \ldots$, where $s_q(m,j)$ is the q-Stirling number of the first kind. In particular, its mean and variance are given by

$$E(X) = \sum_{m=1}^{\infty} \frac{q^{\binom{m}{2}}\lambda^m(1-q)^{m-1}}{[m]_q \prod_{i=1}^{m}(1+\lambda(1-q)q^{i-1})} \tag{2.14}$$

and

$$V(X) = 2\sum_{m=2}^{\infty} \frac{q^{\binom{m}{2}}\lambda^m(1-q)^{m-2}\zeta_{m-1,q}}{[m]_q \prod_{i=1}^{m}(1+\lambda(1-q)q^{i-1})} + E(X) - [E(X)]^2, \tag{2.15}$$

where $\zeta_{m,q} = \sum_{j=1}^{m} 1/[j]_q$.

Proof. The mth q-factorial moment of X,

$$E([X]_{m,q}) = e_q(-\lambda) \sum_{x=m}^{\infty} [x]_{m,q} \frac{q^{\binom{x}{2}}\lambda^x}{[x]_q!} = e_q(-\lambda) \sum_{x=m}^{\infty} \frac{q^{\binom{x}{2}}\lambda^x}{[x-m]_q!},$$

on using the relation

$$\binom{x}{2} = \binom{x-m}{2} + \binom{m}{2} + m(x-m),$$

is expressed as

$$E([X]_{m,q}) = q^{\binom{m}{2}}\lambda^m e_q(-\lambda) \sum_{x=m}^{\infty} \frac{q^{\binom{x-m}{2}}(\lambda q^m)^{x-m}}{[x-m]_q!}.$$

Then, using the expansion of the q-exponential function $E_q(t)$ into a power series, we get

$$E([X]_{m,q}) = q^{\binom{m}{2}}\lambda^m e_q(-\lambda)E_q(\lambda q^m), \quad m = 1, 2, \ldots ,$$

and since

$$e_q(-\lambda)E_q(\lambda q^m) = \frac{\prod_{i=1}^{\infty}(1+\lambda(1-q)q^{m+i-1})}{\prod_{i=1}^{\infty}(1+\lambda(1-q)q^{i-1})} = \frac{1}{\prod_{i=1}^{m}(1+\lambda(1-q)q^{i-1})},$$

we conclude the required expression (2.12). Introducing the q-factorial moments (2.12) into (1.61), the required formula (2.13) is deduced. In particular, from (2.13) with $j = 1$ and $j = 2$, the mean (2.14) and variance and (2.15) are deduced, by using the expressions

$$s_q(m,1) = (-1)^{m-1}[m-1]_q!, \qquad s_q(m,2) = (-1)^{m-2}[m-1]_q!\zeta_{m-1,q},$$

with $\zeta_{m,q} = \sum_{j=1}^{m} 1/[j]_q$. □

The probability function (2.3) of the q-binomial distribution of the first kind, as the number of trials tends to infinity, and the probability function (2.10) of the negative q-binomial distribution of the first kind, as the number of failures tends to infinity, can be approximated by the probability function of Heine distribution, according to the following theorem.

Theorem 2.4. *The limit of the probability function (2.3) of the q-binomial distribution of the first kind, as $n \to \infty$, is the probability function of the Heine distribution,*

$$\lim_{n\to\infty} \begin{bmatrix} n \\ x \end{bmatrix}_q \frac{q^{\binom{x}{2}}\theta^x}{\prod_{i=1}^{n}(1+\theta q^{i-1})} = e_q(-\lambda)\frac{q^{\binom{x}{2}}\lambda^x}{[x]_q!}, \quad x = 0, 1, \ldots, \tag{2.16}$$

for $0 < \lambda < \infty$ and $0 < q < 1$, with $\lambda = \theta/(1-q)$.

Also, the limit of the probability function (2.10) of the negative q-binomial distribution of the first kind, as $n \to \infty$, is the probability function of the Heine distribution,

$$\lim_{n\to\infty} \begin{bmatrix} n+u-1 \\ u \end{bmatrix}_q \frac{q^{\binom{u}{2}}\theta^u}{\prod_{i=1}^{n+u}(1+\theta q^{i-1})} = e_q(-\lambda)\frac{q^{\binom{u}{2}}\lambda^u}{[u]_q!}, \quad u = 0, 1, \ldots, \tag{2.17}$$

for $0 < \lambda < \infty$ and $0 < q < 1$, with $\lambda = \theta/(1-q)$.

Proof. Since, for $0 < q < 1$,

$$\lim_{n\to\infty} \begin{bmatrix} n \\ x \end{bmatrix}_q = \frac{1}{(1-q)^x[x]_q!}\lim_{n\to\infty}\prod_{i=1}^{x}(1-q^{n-i+1}) = \frac{1}{(1-q)^x[x]_q!}$$

and

$$\lim_{n\to\infty}\prod_{i=1}^{n}(1+\lambda(1-q)q^{i-1}) = \prod_{i=1}^{\infty}(1+\lambda(1-q)q^{i-1}) = E_q(\lambda) = \frac{1}{e_q(-\lambda)},$$

the limiting expression (2.16) is readily deduced. Similarly, the limit (2.17) is obtained. □

An interesting application of the Heine distribution as feasible prior in a simple Bayesian model for oil exploration is presented in the following example.

Example 2.3. *Number of undiscovered oilfields in oil exploration I.* Assume that an oil company has an area in which to drill and the area contains an unknown number of oilfields. The probability function $f(x) = P(X = x)$, $x = 0, 1, \dots$, of the number X of undiscovered oilfields is required for finding optimal strategies for drilling.

Furthermore, assume that a single well can reach at most one of the undiscovered oilfields, so that it can be considered as a Bernoulli trial resulting in a success (S) or a failure (F). Let $P(S|X = j) = p_j$ and $P(F|X = j) = q_j = 1 - p_j$, for $j = 0, 1, \dots$, and assume that $q_0 = 1 > q_1 > q_2 > \cdots$. Then, the posterior probability functions $P(X = j + 1|S)$ and $P(X = j|F)$, given a success and a failure, respectively, by Bayes' theorem, are given by

$$P(X = j + 1|S) = \frac{P(X = j + 1)p_{j+1}}{\sum_{i=0}^{\infty} P(X = i)p_i}, \quad j = 0, 1, \dots, \tag{2.18}$$

and

$$P(X = j|F) = \frac{P(X = j)q_j}{\sum_{i=0}^{\infty} P(X = i)q_i}, \quad j = 0, 1, \dots, \tag{2.19}$$

respectively. Moreover, the posterior probability functions $P[X = j + 1|(S, F)]$ and $P[X = j + 1|(F, S)]$, given a success and a failure in either order, are deduced as

$$P[X = j + 1|(S, F)] = \frac{P(X = j + 1)p_{j+1}q_j}{\sum_{i=0}^{\infty} P(X = i + 1)p_{i+1}q_i}, \quad j = 0, 1, \dots,$$

and

$$P[X = j + 1|(F, S)] = \frac{P(X = j + 1)q_{j+1}p_{j+1}}{\sum_{i=0}^{\infty} P(X = i)q_i p_i}, \quad j = 0, 1, \dots,$$

respectively. Thus, the posterior probability function of the number X of undiscovered oilfields after a number of wells have been drilled depends on the order in which successes and failures occur. Let us impose the condition

$$P[X = j + 1|(S, F)] = P[X = j + 1|(F, S)], \quad j = 0, 1, \dots,$$

under which the number of successes (or failures) is a sufficient statistic. Then, since $p_0 = 1 - q_0 = 0$, it follows that

$$q_j \sum_{i=1}^{\infty} P(X = i)q_i p_i = q_{j+1} \sum_{i=0}^{\infty} P(X = i + 1)p_{i+1}q_i, \quad j = 0, 1, \dots,$$

and

$$\sum_{i=1}^{\infty} P(X = i)p_i(q_j q_i - q_{j+1}q_{i-1}) = 0, \quad j = 0, 1, \dots.$$

Therefore,

$$\frac{q_{j+1}}{q_j} = \frac{q_i}{q_{i-1}} < 1, \quad i = 1, 2, \dots, \; j = 0, 1, \dots,$$

and

$$q_j = q^j, \quad j = 0, 1, \dots, \quad 0 < q < 1.$$

The derivation of a specific distribution for the number X of undiscovered oilfields, requires an additional assumption. Assume that

$$P(X = j + 1|S) = P(X = j|F), \quad j = 0, 1, \dots .$$

Then, from (2.18) and (2.19) it follows that

$$P(X = j + 1)(1 - q^{j+1}) = \theta P(X = j)q^j, \quad j = 0, 1, \dots,$$

with

$$\theta = \frac{\sum_{i=0}^{\infty} P(X = i)(1 - q^i)}{\sum_{i=0}^{\infty} P(X = i)q^i} = \frac{1 - \sum_{i=0}^{\infty} P(X = i)q^i}{\sum_{i=0}^{\infty} P(X = i)q^i}.$$

Consequently,

$$P(X = x) = P(X = 0)\frac{q^{\binom{x}{2}}\lambda^x}{[x]_q!}, \quad x = 1, 2, \dots,$$

where $\lambda = \theta/(1 - q)$, with $0 < \theta < \infty$ and $0 < q < 1$. Since

$$E_q(t) = \prod_{i=1}^{\infty}(1 + t(1 - q)q^{i-1}) = \sum_{x=0}^{\infty} q^{\binom{x}{2}} \frac{t^x}{[x]_q!},$$

it follows that $P(X = 0) = 1/E_q(\lambda) = e_q(-\lambda)$ and so

$$P(X = x) = e_q(-\lambda)\frac{q^{\binom{x}{2}}\lambda^x}{[x]_q!}, \quad x = 0, 1, \dots,$$

which is the probability function of the Heine distribution.

2.4 HEINE STOCHASTIC PROCESS

In the stochastic model of a sequence of independent Bernoulli trials, the event of success, $A = \{s\}$, may occur at discrete points (trials) of its development. The possibility of an event to occur at continuous (time or space) points of the development of a stochastic model is of great theoretical and practical interest. In this respect, let us consider a stochastic model that is developing in time or space, in which successes or failures may occur at continuous points. Furthermore, let X_t be the number of successes (occurrences of event A) in the interval $(0, t]$. The family of random variables X_t, $t \geq 0$, is called *stochastic process*. A nonnegative integer-valued stochastic process X_t, $t \geq 0$, with independent and homogeneous increments is called Poisson process, if in a small time interval either a success, $A = \{s\}$, occurs, with probability analogous to the length of the interval, or a failure, $A' = \{f\}$. A Heine process, which constitutes a q-analogue of a Poisson process, is introduced in the following definition.

Definition 2.4. *Consider a stochastic model that is developing in time or space, and let X_t, $t \geq 0$, be the number of successes (occurrences of event A) in the interval $(0, t]$. Assume that X_t, $t \geq 0$, is a stochastic process, with independent and homogeneous increments, which starts at time $t = 0$ from state 0, $P(X_0 = 0) = 1$, and, in the small time interval $(qt, t]$, of length $\delta t = (1 - q)t$, satisfies the condition*

$$p_j(\delta t) = P(X_t - X_{qt} = j) = \begin{cases} \dfrac{1}{1 + \lambda(1 - q)t}, & j = 0, \\ \dfrac{\lambda(1 - q)t}{1 + \lambda(1 - q)t}, & j = 1, \\ 0, & j > 1, \end{cases} \tag{2.20}$$

with $0 < \lambda < \infty$ and $0 < q < 1$. Then, X_t, $t \geq 0$, is called Heine process, with parameters λ and q.

Theorem 2.5. *The probability function of the Heine process X_t, $t \geq 0$, with parameters λ and q, is given by*

$$p_x(t) = P(X_t = x) = e_q(-\lambda t) \frac{q^{\binom{x}{2}} (\lambda t)^x}{[x]_q!}, \quad x = 0, 1, \ldots, \tag{2.21}$$

where $0 < \lambda < \infty$, $0 < q < 1$, and $e_q(u) = \prod_{i=1}^{\infty} (1 - u(1 - q)q^{i-1})^{-1}$ is the q-exponential function (1.24).

Proof. The probability function $p_x(t)$, by the total probability theorem,

$$p_x(t) = p_x(qt + \delta t) = \sum_{k=0}^{x} p_{x-k}(qt) p_k(\delta t), \quad x = 0, 1, \ldots,$$

and condition (2.20), satisfies the system of equations

$$p_0(t) = \frac{1}{1 + \lambda(1 - q)t} p_0(qt),$$

$$p_x(t) = \frac{1}{1 + \lambda(1 - q)t} p_x(qt) + \frac{\lambda(1 - q)t}{1 + \lambda(1 - q)t} p_{x-1}(qt), \quad x = 0, 1, \ldots.$$

This system may be rewritten as

$$\frac{p_0(t) - p_0(qt)}{(1 - q)t} = -\lambda p_0(t),$$

$$\frac{p_x(t) - p_x(qt)}{(1 - q)t} = -\lambda p_x(t) + \lambda p_{x-1}(qt), \quad x = 0, 1, \ldots,$$

or equivalently, by introducing the q-derivative operator $\mathcal{D}_q$, as

$$\mathcal{D}_q p_0(t) = -\lambda p_0(t), \quad \mathcal{D}_q p_x(t) = -\lambda p_x(t) + \lambda p_{x-1}(qt),$$

for $x = 0, 1, \ldots$. Introducing the function $g(t)$ by

$$p_x(t) = g(t)\frac{q^{\binom{x}{2}}(\lambda t)^x}{[x]_q!}, \quad x = 0, 1, \ldots, \tag{2.22}$$

and since

$$\mathcal{D}_q p_x(t) = \frac{q^{\binom{x}{2}}(\lambda t)^x}{[x]_q!}\mathcal{D}_q g(t) + \lambda\frac{q^{\binom{x}{2}}(\lambda t)^{x-1}}{[x-1]_q!}g(qt),$$

the system of q-differential equations reduces to the q-differential equation

$$\mathcal{D}_q g(t) = -\lambda g(t),$$

with initial condition $g(0) = p_0(0) = 1$. Its solution is readily obtained as $g(t) = e_q(-\lambda t)$, and so, by (2.22), expression (2.21) is established. □

In a Heine process, the distribution of the waiting time until the occurrence of a fixed number of successes is connected to the distribution of the number of successes in a fixed time interval. In this respect, the following definition is introduced.

Definition 2.5. *Consider a stochastic model that is developing in time and successes occur according to a Heine process. Let W_n be the waiting time until the occurrence of the nth success. The distribution of W_n is called q-Erlang distribution of the first kind, with parameters n, λ, and q. In particular, the distribution of the waiting time until the occurrence of the first success, $W \equiv W_1$, is called q-exponential distribution of the first kind, with parameters λ and q.*

The distribution function, together with the q-density function and q-moments of the q-Erlang distribution of the first kind are derived in the following theorem.

Theorem 2.6. *The distribution function $F_n(w) = P(W_n \le w)$, $-\infty < w < \infty$, of the q-Erlang distribution of the first kind, with parameters n, λ, and q, is given by*

$$F_n(w) = 1 - \sum_{x=0}^{n-1} e_q(-\lambda w)\frac{q^{\binom{x}{2}}(\lambda w)^x}{[x]_q!}, \quad 0 < w < \infty, \tag{2.23}$$

and $F_n(w) = 0$, $-\infty < w < 0$, where n is a positive integer, $0 < \lambda < \infty$, and $0 < q < 1$. Its q-density function $f_n(w) = d_q F_n(w)/d_q w$ is given by

$$f_n(w) = \frac{q^{\binom{n}{2}}\lambda^n}{[n-1]_q!}w^{n-1}e_q(-\lambda w), \quad 0 < w < \infty. \tag{2.24}$$

Also, its jth q-moment is given by

$$\mu'_{j,q} = E(W_n^j) = \frac{[n+j-1]_{j,q}}{\lambda^j q^{\binom{j}{2}+nj}}, \quad j = 1, 2, \ldots. \tag{2.25}$$

Proof. The event $\{W_n > w\}$ that the nth success occurs after time w is equivalent to the event $\{X_w < n\}$ that the number of successes up to time w is less than n and so

$$P(W_n > w) = P(X_w < n) = \sum_{x=0}^{n-1} P(X_w = x).$$

Thus, the distribution function of the random variable W_n, on using the relation $F_n(w) = P(W_n \leq w) = 1 - P(W_n > w)$ and expression (2.21), is deduced as (2.23).

The q-density function of W_n, on taking the q-derivative of (2.23), by using the q-Leibnitz formula, is obtained in the form

$$\begin{aligned} f_n(w) =& \lambda e_q(-\lambda w) \sum_{x=0}^{n-1} \frac{q^{\binom{x}{2}}(\lambda q w)^x}{[x]_q!} - \lambda e_q(-\lambda w) \sum_{x=1}^{n-1} \frac{q^{\binom{x}{2}}(\lambda w)^{x-1}}{[x-1]_q!} \\ =& \lambda e_q(-\lambda w) \sum_{x=0}^{n-1} \frac{q^{\binom{x+1}{2}}(\lambda w)^x}{[x]_q!} - \lambda e_q(-\lambda w) \sum_{x=1}^{n-1} \frac{q^{\binom{x}{2}}(\lambda w)^{x-1}}{[x-1]_q!}, \end{aligned}$$

which reduces to (2.24). Note that, using the relation (see Exercise 1.33)

$$\int_0^\infty u^{n-1} e_q(-u) d_q u = q^{-\binom{n}{2}} [n-1]_q!, \tag{2.26}$$

it follows that

$$\int_0^\infty f_n(w) d_q w = 1,$$

which conforms with the definition of a q-density function.

The jth q-moment of W_n,

$$\mu'_{j,q} = E(W_n^j) = \frac{q^{\binom{n}{2}} \lambda^n}{[n-1]_q!} \int_0^\infty w^{n+j-1} e_q(-\lambda w) d_q w,$$

using the transformation $u = \lambda w$ and expression (2.26), is obtained as

$$\mu'_{j,q} = \frac{q^{\binom{n}{2}} \lambda^n}{[n-1]_q! \lambda^{n+j}} \int_0^\infty u^{n+j-1} e_q(-u) d_q u = \frac{[n+j-1]_q! q^{\binom{n}{2}}}{[n-1]_q! q^{\binom{n+j}{2}} \lambda^j}.$$

Since $[n+j-1]_q! = [n+j-1]_{j,q}[n-1]_q!$ and $\binom{n+j}{2} = \binom{n}{2} + \binom{j}{2} + nj$, the last relation implies the required expression (2.25). □

The q-density function and q-moments of the q-exponential distribution of the first kind are deduced in the following corollary of Theorem 2.6.

Corollary 2.1. *The q-density function of the q-exponential distribution of the first kind, with parameters λ and q, is given by*

$$f(w) = \lambda e_q(-\lambda w), \quad 0 < w < \infty,$$

where $0 < \lambda < \infty$ and $0 < q < 1$. Also, its jth q-moment is given by

$$\mu'_{j,q} = E(W_q^j) = \frac{[j]_q!}{\lambda^j q^{\binom{j+1}{2}}}, \quad j = 1, 2, \ldots .$$

Remark 2.3. The distribution function of the q-Erlang distribution of the first kind, in addition to expression (2.23), may be written as a q-integral of its q-density function as

$$F_n(w) = \int_0^w \frac{q^{\binom{n}{2}} \lambda^n}{[n-1]_q!} u^{n-1} e_q(-\lambda u) d_q u.$$

These two expressions of $F_n(w)$ imply the relation

$$\int_0^w \frac{q^{\binom{n}{2}} \lambda^n}{[n-1]_q!} u^{n-1} e_q(-\lambda u) d_q u = 1 - \sum_{x=0}^{n-1} e_q(-\lambda w) \frac{q^{\binom{x}{2}} (\lambda w)^x}{[x]_q!}.$$

2.5 q-STIRLING DISTRIBUTIONS OF THE FIRST KIND

Let us now consider a sequence of independent Bernoulli trials and assume that the probability of success at the ith trial is given by

$$p_i = \theta q^{i-1}, \quad i = 1, 2, \ldots, \quad 0 < \theta \leq 1, \quad 0 < q < 1 \text{ or } 1 < q < \infty, \tag{2.27}$$

where for $0 < \theta \leq 1$ and $1 < q < \infty$, the number i of trials is restricted by $\theta q^{i-1} < 1$, ensuring that $0 < p_i < 1$. This restriction imposes on i an upper bound, $i = 1, 2, \ldots, [-r]$, with $[-r]$ denoting the integral part of $-r$, where $r = \log\ \theta / \log\ q < 0$. The probabilities involved in this model are more conveniently expressed in terms of a new parameter r that replaces the parameter θ by $\theta = q^r$. Then, (2.27) is written as

$$p_i = q^{r+i-1}, \quad i = 1, 2, \ldots, \tag{2.28}$$

for $0 \leq r < \infty$ and $0 < q < 1$ or $-\infty < r < 0$ and $1 < q < \infty$, with $i \leq [-r]$.

The probability function and factorial moments of the number of successes in a specific number of trials are derived in the following theorem.

Theorem 2.7. *Let X_n be the number of successes in a sequence of n independent Bernoulli trials, with probability of success at the ith trial given by (2.28). The probability function of X_n is given by*

$$P(X_n = x) = q^{\binom{n}{2}+rn}(1-q)^{n-x}|s_{q^{-1}}(n,x;r)|, \quad x = 0,1,\ldots,n, \tag{2.29}$$

for $0 \leq r < \infty$ and $0 < q < 1$ and by

$$P(X_n = x) = q^{\binom{n}{2}+rn}(1-q^{-1})^{n-x}s_{q^{-1}}(n,x;r), \quad x = 0,1,\ldots,n, \tag{2.30}$$

for $-\infty < r < 0$ and $1 < q < \infty$, with $n \leq [-r]$, where $|s_q(n,x;r)|$ and $s_q(n,x;r)$ are the noncentral signless and the noncentral q-Stirling numbers of the first kind. Its factorial moments are given by

$$E[(X_n)_i] = i!q^{\binom{i}{2}+ri}\begin{bmatrix} n \\ i \end{bmatrix}_q, \quad i = 1,2,\ldots,n, \tag{2.31}$$

and $E[(X_n)_i] = 0$, for $i = n+1, n+2, \ldots$. In particular, its mean and variance are given by

$$E(X_n) = q^r[n]_q, \quad V(X_n) = q^r[n]_q - q^{2r}[n]_{q^2}. \tag{2.32}$$

Proof. Let A_i be the event of success at the ith trial, for $i = 1,2,\ldots$, and consider a permutation $(i_1, i_2, \ldots, i_x, i_{x+1}, \ldots, i_n)$ of $\{1,2,\ldots,n\}$. Then, using the independence of the Bernoulli trials and the probabilities (2.28), we get

$$\begin{aligned} P(A_{i_1}A_{i_2}\cdots A_{i_x}A'_{i_{x+1}}\cdots A'_{i_n}) &= P(A_{i_1})\cdots P(A_{i_x})P(A'_{i_{x+1}})\cdots P(A'_{i_n}) \\ &= \prod_{k=1}^{x} q^{r+i_k-1}\prod_{k=x+1}^{n}(1-q^{r+i_k-1}) \\ &= \prod_{k=1}^{n} q^{r+i_k-1}\prod_{k=x+1}^{n}(q^{-(r+i_k-1)}-1). \end{aligned}$$

Moreover since

$$\prod_{k=1}^{n} q^{r+i_k-1} = \prod_{i=1}^{n} q^{r+i-1} = q^{\binom{n}{2}+rn}$$

and

$$\prod_{k=x+1}^{n}(q^{-(r+i_k-1)}-1) = (q^{-1}-1)^{n-x}\prod_{k=x+1}^{n}[r+i_k-1]_{q^{-1}},$$

it reduces to

$$P(A_{i_1}A_{i_2}\cdots A_{i_x}A'_{i_{x+1}}\cdots A'_{i_n}) = q^{\binom{n}{2}+rn}(q^{-1}-1)^{n-x}\prod_{k=x+1}^{n}[r+i_k-1]_{q^{-1}}.$$

Summing these probabilities over all $(n-x)$-combinations $\{i_{x+1}, i_{x+2}, \ldots, i_n\}$ of the set $\{1, 2, \ldots, n\}$ and using (1.37), we deduce the probability of x successes in n trials in the form

$$P(X_n = x) = q^{\binom{n}{2}+rn}(1-q^{-1})^{n-x}s_{q^{-1}}(n, x; r), \quad x = 0, 1, \ldots, n.$$

Note that the factor $(1-q^{-1})^{n-x}$, for $0 < q < 1$, and the factor $s_{q^{-1}}(n, x; r)$, for $0 \leq r < \infty$, according to Remark 1.4, have the sign of $(-1)^{n-x}$. Introducing the noncentral signless q-Stirling numbers of the first kind by

$$|s_{q^{-1}}(n, x; r)| = (-1)^{n-x}q^{-(n-x)}s_{q^{-1}}(n, x; r),$$

expression (2.31) is deduced.

An alternative and useful expression of the probability function of X_n, on using (1.39) and the relations

$$\binom{n-j}{2} = \binom{n}{2} - \binom{j}{2} - j(n-j), \quad \begin{bmatrix} n \\ j \end{bmatrix}_{q^{-1}} = q^{-j(n-j)} \begin{bmatrix} n \\ j \end{bmatrix}_q,$$

may be obtained as

$$P(X_n = x) = \sum_{j=x}^{n} (-1)^{j-x} q^{\binom{j}{2}+rj} \begin{bmatrix} n \\ j \end{bmatrix}_q \binom{j}{x}, \quad x = 0, 1, \ldots, n,$$

for $0 \leq r < \infty$ and $0 < q < 1$ or $-\infty < r < 0$ and $1 < q < \infty$, with $n \leq [-r]$.

Multiplying the last expression of the probability function of X_n by $(x)_i = i!\binom{x}{i}$ and summing it for $x = i, i+1, \ldots, n$, we get for the ith factorial moment of X_n the expression

$$\begin{aligned} E[(X_n)_i] &= i! \sum_{x=i}^{n} \binom{x}{i} \sum_{j=x}^{n} (-1)^{j-x} q^{\binom{j}{2}+rj} \begin{bmatrix} n \\ j \end{bmatrix}_q \binom{j}{x} \\ &= i! \sum_{j=i}^{n} \left\{ \sum_{x=i}^{j} (-1)^{j-x} \binom{x}{i} \binom{j}{x} \right\} q^{\binom{j}{2}+rj} \begin{bmatrix} n \\ j \end{bmatrix}_q, \end{aligned}$$

which, on using the orthogonality relation

$$\sum_{x=i}^{j} (-1)^{j-x} \binom{x}{i} \binom{j}{x} = \delta_{i,j},$$

implies the required expression (2.31). The mean (2.32) is readily obtained from (2.31) by setting $i = 1$. Also, setting $i = 2$, we get

$$E[(X_n)_2] = 2q^{2r+1} \frac{[n]_q [n-1]_q}{[2]_q} = 2q^{2r} \frac{[n]_q([n]_q - 1)}{[2]_q}$$

and using the relation $V(X_n) = E[(X_n)_2] + E(X_n) - [E(X_n)]^2$, the variance is obtained as

$$V(X_n) = q^r[n]_q - q^{2r}\left([n]_q^2 - \frac{2[n]_q([n]_q - 1)}{[2]_q}\right).$$

Furthermore, since

$$\begin{aligned} [n]_q^2 - \frac{2[n]_q([n]_q - 1)}{[2]_q} &= \frac{[n]_q}{[2]_q}\left((1+q)[n]_q - 2[n]_q + 2\right) \\ &= \frac{(1-q^n)}{1-q^2}\cdot(1+q^n) = \frac{(1-q^{2n})}{1-q^2} = [n]_{q^2}, \end{aligned}$$

it reduces to the second of the expressions (2.32). □

The probability function of the number of trials until the occurrence of a given number of successes is obtained in the following corollary of Theorem 2.7.

Corollary 2.2. *Consider a sequence of independent Bernoulli trials, with probability of success at the ith trial given by (2.28), and let T_k be the number of trials until the occurrence of the kth success. The probability function of T_k is given by*

$$P(T_k = n) = q^{\binom{n}{2}+rn}(1-q)^{n-k}|s_{q^{-1}}(n-1, k-1; r)|, \tag{2.33}$$

for $n = k, k+1, \ldots$, with $0 \le r < \infty$ and $0 < q < 1$, where $|s_q(n-1, k-1; r)|$ is the noncentral signless q-Stirling number of the first kind.

Proof. The probability function of the number T_k of trials until the occurrence of the kth success is expressed in terms of the probability function of the number X_n of success in n trials by

$$P(T_k = n) = P(X_{n-1} = k-1)p_n,$$

where p_n is the probability of success at the nth trial. Therefore, using (2.28) and (2.29), formula (2.33) is readily deduced. □

Two interesting applications of the distribution (2.29) in a random graph model and in a defense model against an approaching attacker (missile) are presented in the following examples.

Example 2.4. *The number of sources (or sinks) in a random acyclic digraph.* A graph G is a pair (V, E), with $V = \{v_1, v_2, \ldots, v_n\}$ the set of vertices (nodes) and $E \subseteq V^2 = \{(v_k, v_r) : v_k \in V, v_r \in V\}$ the set of edges; a graph with $E = \{(v_k, v_r) : v_k < v_r, v_k \in V, v_r \in V\}$ or $E = \{(v_k, v_r) : v_k > v_r, v_k, v_r \in V\}$ is called digraph (or directed graph). A path in a graph $G = (V, E)$ is a subset $U = \{u_1, u_2, \ldots, u_r\}$ of the set of vertices V, such that $(u_k, u_{k+1}) \in E$ is an edge of the graph, for $k = 1, 2, \ldots, r-1$; a path U is a cycle if $u_r = u_1$. A graph without cycles is called acyclic.

A vertex (node) in an acyclic digraph is called source (or sink) if it does not have any predecessor (or successor).

Let us denote by $G_{n,q}$ a random acyclic digraph of n vertices, in which any edge occurs independently with probability $p = 1 - q$. Consider the sequential construction of $G_{i,q}$ from $G_{i-1,q}$ through the addition of the vertex i. This addition will create a new source if no vertex of $G_{i-1,q}$ is connected to i. Thus, the probability p_i that the addition of vertex i to of $G_{i-1,q}$ creates a new source is given by

$$p_i = q^{i-1}, \quad i = 1, 2, \ldots, \quad 0 < q < 1.$$

Since the sequential additions of nodes constitute a sequence of independent Bernoulli trials with success the creation of a new source, it follows from Theorem 2.7, with $r = 0$, that the probability function of the number X_n of sources in a random acyclic digraph $G_{n,q}$ is given by

$$P(X_n = x) = q^{\binom{n}{2}}(1 - q)^{n-x}|s_{q^{-1}}(n, x)|, \quad x = 0, 1, \ldots, n,$$

with $0 < q < 1$, where $|s_q(n, x)| = |s_q(n, x; 0)|$ is the signless q-Stirling number of the first kind.

Example 2.5. *Successful shots at an approaching attacker*. Consider a defense situation in which the motion of an attacker (missile) relative to the defender is assumed to be toward the defender with relative speed v. The attacker will annihilate the defender if it reaches him before he has disabled it.

The defender fires a sequence of shots at the attacker when the distance between them is $D_i = d - (i - 1)uv$, $i = 1, 2, \ldots, m$, where d is the distance when the first shot is fired, u is the time interval between shots and m is a fixed number such that $d - (m - 1)uv > 0$.

Assume that the distance D_i a shot would travel is an exponential random variable, with mean $E(D_i) = 1/\lambda$, that the shot is correctly aimed, and that it is successful if it reaches the attacker. Then, the probability that the ith shot is successful is

$$p_i = \exp\ \{\lambda(i - 1)uv - \lambda d\}, \quad i = 1, 2, \ldots, m.$$

Considering the sequence of shots as a sequence of independent Bernoulli trials, and setting $\theta = \exp\{-\lambda d\}$ and $q = \exp\{\lambda uv\}$, it follows that the probability of success (successful shot) at the ith trial is given by

$$p_i = \theta q^{i-1}, \quad i = 1, 2, \ldots, [-r], \ 0 < \theta < 1, \ 1 < q < \infty,$$

or equivalently by

$$p_i = q^{r+i-1}, \quad i = 1, 2, \ldots, [-r], \ -\infty < r < 0, \ 1 < q < \infty,$$

with $[-r]$ the integral part of $-r$ and $r = \log\ \theta/\log\ q < 0$. Therefore, according to Theorem 2.7, the probability function and factorial moments of the number X_n of successful shots, in a sequence of $n \leq [-r]$ shots by the defender, are given by (2.30) and (2.31), respectively.

Finally, consider a sequence of independent Bernoulli trials and assume that the odds of failure at the ith trial is given by

$$\lambda_i = q[r+i-1]_q = \frac{q}{1-q}(1-q^{r+i-1}), \quad i = 1, 2, \ldots, \tag{2.34}$$

with $0 \leq r < \infty$ and $0 < q < 1$ or $1 < q < \infty$, which is an increasing sequence. Note that the odds λ_i of failure at the ith trial is expressed in terms of the probability p_i of success at the ith trial by $\lambda_i = (1-p_i)/p_i$, whence $p_i = 1/(\lambda_i + 1)$. Consequently, assumption (2.34) is expressed in terms of the probability of success at the ith trial by

$$p_i = \frac{1}{[r+i]_q}, \quad i = 1, 2, \ldots, \tag{2.35}$$

with $0 \leq r < \infty$ and $0 < q < 1$ or $1 < q < \infty$, which is a decreasing sequence. The probability function and factorial moments of the number of successes in a specific number of trials are derived in the following theorem.

Theorem 2.8. *Let X_n be the number of successes in a sequence of n independent Bernoulli trials, with probability of success at the ith trial given by (2.35). The probability function of X_n is given by*

$$P(X_n = x) = \frac{|s_q(n, x; r)|}{[r+n]_{n,q}}, \quad x = 0, 1, \ldots, n, \tag{2.36}$$

with $0 \leq r < \infty$ and $0 < q < 1$ or $1 < q < \infty$, where $|s_q(n, x; r)|$ is the noncentral signless q-Stirling number of the first kind. Its factorial moments are given by

$$E[(X_n)_j] = \frac{j!q^{-(n-j)}|s_q(n, j; r+1)|}{[r+n]_{n,q}}, \quad j = 1, 2, \ldots, n, \tag{2.37}$$

and $E[(X_n)_j] = 0$, for $j = n+1, n+2, \ldots$. In particular, its mean and variance are given by

$$E(X_n) = \sum_{i=1}^{n} \frac{1}{[r+i]_q}, \quad V(X_n) = \sum_{i=1}^{n} \frac{q[r+i-1]_q}{[r+i]_q^2}. \tag{2.38}$$

Proof. Let A_i be the event of success at the ith trial, for $i = 1, 2, \ldots$, and consider a permutation $(i_1, i_2, \ldots, i_x, i_{x+1}, \ldots, i_n)$ of $\{1, 2, \ldots, n\}$. Then, using the independence of the trials and the probabilities (2.35), we get

$$\begin{aligned} P(A_{i_1}A_{i_2}\cdots A_{i_x}A'_{i_{x+1}}\cdots A'_{i_n}) &= P(A_{i_1})\cdots P(A_{i_x})P(A'_{i_{x+1}})\cdots P(A'_{i_n}) \\ &= \frac{q^{n-x}}{\prod_{k=1}^{n}[r+i_k]_q}\prod_{k=x+1}^{n}[r+i_k-1]_q. \end{aligned}$$

Furthermore, since

$$\prod_{k=1}^{n}[r+i_k]_q=\prod_{i=1}^{n}[r+i]_q=[r+n]_{n,q},$$

it is simplified to

$$P(A_{i_1}A_{i_2}\cdots A_{i_x}A'_{i_{x+1}}\cdots A'_{i_n})=\frac{q^{n-x}}{[r+n]_{n,q}}\prod_{k=x+1}^{n}[r+i_k-1]_q.$$

Summing these probabilities over all $(n-x)$-combinations $\{i_{x+1},i_{x+2},\dots,i_n\}$ of the n positive integers $\{1,2,\dots,n\}$ and using (1.37) together with (1.30), we deduce the probability of x successes in n trials as (2.36).

The jth factorial moment $E[(X_n)_j]$ is readily obtained as (2.37), by using the relation

$$|s_q(n,j;r+1)|=q^{n-j}\sum_{k=j}^{n}\binom{k}{j}|s_q(n,k;r)|,$$

which follows from the definition of the signless noncentral q-Stirling numbers of the first kind (see Exercise 1.15). In particular, from (2.37) with $j=1$ and $j=2$ and using by (1.37) with $k=1$ and $k=2$, the mean and variance of X_n may be deduced. Alternatively, using the expression $X_n=\sum_{i=1}^{n}Z_i$, where $Z_1,Z_2,\dots,Z_n$ are independent zero–one Bernoulli random variables, with

$$E(Z_i)=\frac{1}{[r+i]_q},\qquad V(Z_i)=\frac{q[r+i-1]_q}{[r+i]_q^2},\qquad i=1,2,\dots,n,$$

and since $E(X_n)=\sum_{i=1}^{n}E(Z_i)$ and $V(X_n)=\sum_{i=1}^{n}V(Z_i)$, the mean and variance (2.38) are readily deduced. □

The probability function of the number of trials until the occurrence of a given number of successes is obtained in the following corollary of Theorem 2.8.

Corollary 2.3. *Consider a sequence of independent Bernoulli trials, with probability of success at the ith trial given by (2.35), and let T_k be the number of trials until the occurrence of the kth success. The probability function of T_k is given by*

$$P(T_k=n)=\frac{|s_q(n-1,k-1;r)|}{[r+n]_{n,q}},\qquad n=k,k+1,\dots,\tag{2.39}$$

with $0\le r<\infty$ and $0<q<1$ or $1<q<\infty$, where $|s_q(n-1,k-1;r)|$ is the noncentral signless q-Stirling number of the first kind.

Proof. The probability function of the number T_k of trials until the occurrence of the kth success is expressed in terms of the probability function of the number X_n of successes in n trials by

$$P(T_k=n)=P(X_{n-1}=k-1)p_n,$$

where p_n is the probability of success at the nth trial. Therefore, using (2.35) and (2.36), formula (2.39) is readily deduced. □

An interesting application of these distributions in the theory of records is presented in the following example.

Example 2.6. *Records in a geometrically increasing population.* Consider a sequence of random variables Y_j, $j = 1, 2, \ldots$. The random variable Y_i, $i \geq 2$, is called a *record* if $Y_i > Y_j$ for all $j = 1, 2, \ldots, i-1$; by convention Y_1 is a record. Let X_n be the number of records up to time (index) n and T_k the time of the kth record.

Motivated by the increasing frequency of record breaking in the Olympic games, a model in which the breakings are attributed to the increase in the population size was proposed. Specifically, it was assumed that the random variable Y_i, $i = 1, 2, \ldots$, is the maximum of a number α_i of independent and identically distributed random variables; that is

$$Y_i = \max\{Y_{i,1}, Y_{i,2}, \ldots, Y_{i,\alpha_i}\}, \quad i = 1, 2, \ldots,$$

where $Y_{i,j}, j = 1, 2, \ldots, \alpha_i, i = 1, 2, \ldots$, is a double sequence of independent and identically distributed random variables, with an absolutely continuous distribution function $F(y)$, and α_i is the population size of the athletes of the world at the ith Olympic game, $i = 1, 2, \ldots$. Then, Y_i, $i = 1, 2, \ldots$, is a sequence of independent random variables with

$$F_{Y_i}(y) = P(Y_i \leq y) = [F(y)]^{\alpha_i}, \quad i = 1, 2, \ldots.$$

Yang (1975) examined the particular case of a geometrically increasing population, with size

$$\alpha_i = \theta q^{-(i-1)}, \quad i = 1, 2, \ldots, \quad 0 < \theta < \infty, \quad 0 < q < 1.$$

The case $1 < q < \infty$, of a geometrically decreasing population size, may be simultaneously treated without any additional problem. In order to find the distributions of the number X_n of records up to time (index) n and the time T_k of the kth record, consider the record indicator random variables Z_i, $i = 1, 2, \ldots$, defined by $Z_i = 1$, if Y_i is a record and $Z_i = 0$, if Y_i is not a record. Nevzorov (1984) proved that the record indicator random variables Z_i, $i = 1, 2, \ldots$, are independent and

$$p_i = P(Z_i = 1) = \frac{\alpha_i}{\alpha_1 + \alpha_2 + \cdots + \alpha_i}, \quad i = 1, 2, \ldots.$$

Thus, for $\alpha_i = \theta q^{-(i-1)}$, $i = 1, 2, \ldots$, and since

$$\sum_{j=1}^{i} \alpha_j = \sum_{j=1}^{i} \theta q^{-(j-1)} = \theta \frac{q^{-i} - 1}{q^{-1} - 1} = \theta q^{-(i-1)}[i]_q, \quad i = 1, 2, \ldots,$$

with $0 < q < 1$ or $1 < q < \infty$, it follows that

$$p_i = \frac{1}{[i]_q}, \quad i = 1, 2, \ldots, \quad 0 < q < 1 \text{ or } 1 < q < \infty.$$

Then, the probability function of the numbers X_n and T_k are deduced from (2.36) and (2.39), by setting $r = 0$, as

$$P(X_n = x) = \frac{|s_q(n, x)|}{[n]_q!}, \quad x = 0, 1, \ldots, n,$$

and

$$P(T_k = n) = \frac{|s_q(n-1, k-1)|}{[n]_q!}, \quad n = k, k+1, \ldots,$$

with $0 < q < 1$ or $1 < q < \infty$, respectively.

2.6 REFERENCE NOTES

Platonov (1976) considered the stochastic model of independent but not identically distributed Bernoulli trials, which was first introduced by Poisson (1837), and derived the probability function of the number of successes in a given number of trials. Balakrishnan and Nevzorov (1997) obtained this distribution as the distribution of the number of records up to a given time (index) in a general record model. These results are given in Exercises 2.1 and 2.2.

The q-binomial distribution of the first kind was introduced by Kemp and Kemp (1991) in their study of the Weldon's classical dice data, which was presented in Example 2.1. It was further examined by Kemp and Newton (1990), as a stationary distribution of a birth and death process; its application presented in Example 2.2 is taken from this paper.

The Heine distribution was derived by Benkherouf and Bather (1988) as feasible prior in a simple Bayesian model for oil exploration; this model is discussed in Example 2.3. In addition, its derivation as a limiting distribution of the q-binomial distribution of the first kind, presented in Theorem 2.4, was given by Kemp and Newton (1990). Also, the expression of the Heine distribution as an infinite convolution of zero-one Bernoulli distributions, examined in Exercise 2.13, was noticed by Benkherouf and Bather (1988). Finally, the derivation of the Heine distribution as a stationary distribution of a Markov chain, given in Exercise 2.16, is due to Kemp (1992b). Kyriakoussis and Vamvakari (2015) introduced the Heine process and derived its probability function. They also obtained the distribution and q-density functions together with the q-moments of the q-Erlang distribution of the first kind; these results are presented in Theorems 2.5 and 2.6, and Exercise 2.15. Ostrovska (2006) used the probability functions of the q-binomial distribution of the first kind and the Heine distribution in a probabilistic based study of the convergence of the Lupas q-analogue of the Bernstein operator.

The q-Stirling distribution of the first kind was obtained by Crippa et al. (1997) as the distribution of the number of sources (or sinks) in a random acyclic digraph, discussed in Example 2.4. A closely connected distribution was previously discussed by Kemp (1987) in connection with a defense model against an approaching attacker (missile). The same distribution, with different parameters, was derived by Rawlings (1998), under a stochastic model of Bernoulli trials with a discrete q-uniform success probability; Example 2.5 and Exercises 2.19 and 2.20 are based on these papers. The other q-Stirling distributions of the first kind (2.36) and (2.39), for $r = 0$, were obtained by Charalambides (2007) as the distributions of the number of records and the record times, respectively, in a geometrically increasing population model introduced by Yang (1975). Exercise 2.21 on the distribution of the inter-record times is taken from this paper. Analogous results were derived by Charalambides (2009) for a modified geometrically increasing population model together with a q-factorially increasing population model; Exercises 2.22–2.24 are based on this paper.

2.7 EXERCISES

2.1 Consider a sequence of independent Bernoulli trials, with the probability of success at the ith trial varying with the number of trials,

$$P_i(\{s\}) = p_i, \quad 0 < p_i < 1, \quad i = 1, 2, \ldots .$$

(a) Let X_n be the number of successes in n trials. Show that its probability function may be expressed as

$$P(X_n = x) = \frac{|s(n, x; \boldsymbol{a})|}{\prod_{i=1}^{n}(1 + a_i)}, \quad x = 0, 1, \ldots, n,$$

where $a_i = (1 - p_i)/p_i$, $i = 1, 2, \ldots, n$, and $|s(n, x; \boldsymbol{a})|$ is the generalized signless Stirling number of the first kind (see Exercise 1.25).

(b) Also, derive its jth factorial moment as

$$E[(X_n)_j] = \frac{j!|s(n, j; \boldsymbol{a} + 1)|}{\prod_{i=1}^{n}(1 + a_i)}, \quad j = 1, 2, \ldots, n,$$

and $E[(X_n)_j] = 0$, for $j = n + 1, n + 2, \ldots$. In particular, deduce its mean and variance as

$$E(X_n) = \sum_{i=1}^{n} \frac{1}{1 + a_i} = \sum_{i=1}^{n} p_i, \quad V(X_n) = \sum_{i=1}^{n} \frac{a_i}{(1 + a_i)^2} = \sum_{i=1}^{n} p_i(1 - p_i).$$

(c) Let T_k be the number of trials until the occurrence of the kth success. Show that its probability function may be expressed as

$$P(T_k = n) = \frac{|s(n - 1, k - 1; \boldsymbol{a})|}{\prod_{i=1}^{n}(1 + a_i)}, \quad n = k, k + 1, \ldots .$$

2.2 Consider the general record model discussed in Example 2.6, in which the record indicator random variables Z_i, $i = 1, 2, \ldots$, are independent with

$$p_i = P(Z_i = 1) = \frac{\alpha_i}{\alpha_1 + \alpha_2 + \cdots + \alpha_i}, \qquad i = 1, 2, \ldots .$$

Derive the probability function of the number X_n of records up to time (index) n and the probability function of the time T_k of the realization of the kth record.

2.3 The jth factorial moment of the q-binomial distribution of the first kind is given by

$$E[(X_n)_j] = j! \sum_{m=j}^{n} (-1)^{m-j} \begin{bmatrix} n \\ m \end{bmatrix}_q \frac{\theta^m q^{\binom{m}{2}} (1-q)^{m-j} s_q(m,j)}{\prod_{i=1}^{m}(1+\theta q^{i-1})},$$

for $j = 1, 2, \ldots, n$, and $E[(X_n)_j] = 0$, for $j = n+1, n+2, \ldots$, where $s_q(m,j)$ is the q-Stirling number of the first kind. Setting $j = 1$ and using the expression $s_q(m, 1) = (-1)^{m-1}[m-1]_q!$, deduce for the mean the expression

$$E(X_n) = \sum_{m=1}^{n} \begin{bmatrix} n \\ m \end{bmatrix}_q \frac{\theta^m q^{\binom{m}{2}} (1-q)^{m-1} [m-1]_q!}{\prod_{i=1}^{m}(1+\theta q^{i-1})}.$$

Using it, derive the recurrence relation

$$E(X_n) - E(X_{n-1}) = \frac{\theta q^{n-1}}{1+\theta q^{n-1}}, \quad n = 2, 3, \ldots, \qquad E(X_1) = \frac{\theta}{1+\theta},$$

and conclude that

$$E(X_n) = \sum_{i=1}^{n} \frac{\theta q^{i-1}}{1+\theta q^{i-1}}.$$

2.4 *(Continuation).* The second factorial moment, on setting $j = 2$ and using the expression $s_q(m, 2) = (-1)^{m-2}[m-1]_q!\zeta_{m-1,q}$, with $\zeta_{m,q} = \sum_{j=1}^{m} 1/[j]_q$, is deduced as

$$E[(X_n)_2] = 2 \sum_{m=2}^{n} \begin{bmatrix} n \\ m \end{bmatrix}_q \frac{\theta^m q^{\binom{m}{2}} (1-q)^{m-2} [m-1]_q! \zeta_{m-1,q}}{\prod_{i=1}^{m}(1+\theta q^{i-1})}.$$

Using it, derive the recurrence relation

$$E[(X_n)_2] - E[(X_{n-1})_2] = \frac{2\theta q^{n-1}}{1+\theta q^{n-1}} \cdot E(X_{n-1}), \quad n = 3, 4, \ldots,$$

with initial condition

$$E[(X_2)_2] = 2 \frac{\theta}{1+\theta} \cdot \frac{\theta q}{1+\theta q}.$$

Applying it repeatedly, derive the expression

$$E[(X_n)_2] = 2\sum_{j=2}^{n} \frac{\theta q^{j-1}}{1+\theta q^{j-1}} \sum_{i=1}^{j-1} \frac{\theta q^{i-1}}{1+\theta q^{i-1}}$$

and conclude that

$$V(X_n) = \sum_{i=1}^{n} \frac{\theta q^{i-1}}{(1+\theta q^{i-1})^2}.$$

2.5 Let X_n be a nonnegative integer-valued random variable with q-binomial moments

$$E\left(\begin{bmatrix} X_n \\ m \end{bmatrix}_q\right) = \begin{bmatrix} n \\ m \end{bmatrix}_q \frac{\theta^m q^{\binom{m}{2}}}{\prod_{i=1}^{m}(1+\theta q^{i-1})}, \qquad m = 0, 1, \ldots, n,$$

for $0 < \theta < 1$ and $0 < q < 1$ or $1 < q < \infty$. Find the probability function of X_n,

$$f_n(x) = P(X_n = x), \quad x = 0, 1, \ldots, n.$$

2.6 (a) Derive the probability generating function $P_{X_n}(t) = E(t^{X_n})$ of the q-binomial distribution of the first kind and

(b) using it obtain the mth q-factorial moment $E([X_n]_{m,q})$, $m = 1, 2, \ldots$.

2.7 *q-Variance of the q-binomial distribution of the first kind.*

(a) Show that the q-variance of the q-binomial distribution of the first kind is given by

$$V([X]_q) = \frac{[n]_q\theta}{(1+\theta)(1+\theta q)} + \frac{[n]_q^2\theta^2(q-1)}{(1+\theta)^2(1+\theta q)}.$$

(b) Derive the mth q^{-1}-factorial moment of the q-binomial distribution of the first kind as

$$E([X_n]_{m,q^{-1}}) = \frac{[n]_{m,q}\theta^m}{\prod_{i=1}^{m}(1+\theta q^{n-m+i-1})}, \qquad m = 1, 2, \ldots .$$

(c) Deduce its q^{-1}-variance as

$$V([X_n]_{q^{-1}}) = \frac{[n]_q\theta}{(1+\theta q^{n-1})(1+\theta q^{n-2})} + \frac{[n]_q^2\theta^2(1-q)}{q(1+\theta q^{n-1})^2(1+\theta q^{n-2})}.$$

2.8 Let X_n be the number of successes in n independent Bernoulli trials, with probability of success at the ith trial given by (2.2). Show that the probability of the occurrence of at most r successes,

$$P(X_n \le r) = \sum_{x=0}^{r} \begin{bmatrix} n \\ x \end{bmatrix}_q \frac{q^{\binom{x}{2}}\theta^x}{\prod_{i=1}^{n}(1+\theta q^{i-1})},$$

may be expressed by a q-integral as

$$P(X_n \le r) = 1 - \frac{[n]_q! q^{\binom{r+1}{2}}}{[r]_q![n-r-1]_q!} \int_0^\theta \frac{t^r d_q t}{\prod_{i=1}^{n+1}(1+tq^{i-1})}.$$

2.9 Let W_n be a nonnegative integer-valued random variable with q-binomial moments

$$E\left(\begin{bmatrix} W_n \\ m \end{bmatrix}_q\right) = \begin{bmatrix} n+m-1 \\ m \end{bmatrix}_q \theta^{-m} q^{-\binom{m}{2}-(n-1)m}, \quad m = 0, 1, \dots,$$

for $0 < \theta < \infty$ and $0 < q < 1$ or $1 < q < \infty$. Find the probability function of W_n,

$$f_n(w) = P(W_n = w), \quad w = 0, 1, \dots .$$

2.10 *Distribution of the number of inter-success failures.* Consider a sequence of independent Bernoulli trials and assume that the probability of success at the ith trial is given by (2.2). Let W_n be the number of failures until the occurrence of the nth success, and let $U_n = W_n - W_{n-1}$, for $n = 1, 2, \dots$, be the sequence of the numbers of failures between consecutive successes, with $P(W_0 = 0) = 1$.

(a) Determine the conditional probability function

$$P(U_{n+1} = u | W_n = w), \quad u = 0, 1, \dots ,$$

(b) compute the conditional q-factorial moments

$$E([U_{n+1}]_{m,q} | W_n = w), \quad m = 1, 2, \dots .$$

(c) and deduce the conditional factorial moments

$$E[(U_{n+1})_j | W_n = w], \quad j = 1, 2, \dots .$$

The conditional distribution of U_{n+1}, given that $W_n = w$, is a *q-geometric distribution of the first kind*, with parameters $\lambda = \theta q^{n+w}$ and q. Notice that the random variable W_n, which follows a negative q-binomial distribution of the first kind, may be written as a sum $W_n = \sum_{j=1}^n U_j$, of dependent random variables U_j, $j = 1, 2, \dots, n$, with conditional distribution, given W_{j-1}, a q-geometric distribution of the first kind.

2.11 Let U_n be the number of successes until the occurrence of the nth failure, in a sequence of independent Bernoulli trials, with probability of success at the ith trial given by (2.2). The distribution of U_n is a negative q-binomial distribution of the first kind with probability function (see Remark 2.2)

$$P(U_n = u) = \begin{bmatrix} n+u-1 \\ u \end{bmatrix}_q \frac{\theta^u q^{\binom{u}{2}}}{\prod_{i=1}^{n+u}(1+\theta q^{i-1})}, \quad u = 0, 1, \dots ,$$

for $0 < \theta < \infty$ and $0 < q < 1$ or $1 < q < \infty$.

(a) Show that the mth q^{-1}-factorial moment is given by

$$E([U_n]_{m,q^{-1}}) = [n+m-1]_{m,q}\theta^m, \quad m = 0, 1, \ldots,$$

and deduce the jth factorial moment as

$$E[(U_n)_j] = j! \sum_{m=j}^{\infty} (-1)^{m-j} \begin{bmatrix} n+m-1 \\ m \end{bmatrix}_q \theta^m q^{\binom{m}{2}} (1-q^{-1})^{m-j} s_{q^{-1}}(m,j),$$

for $j = 1, 2, \ldots$.

(b) Also, show that

$$V([U_n]_{q^{-1}}) = [n]_q \theta(1 + \theta q^{-1}).$$

2.12 *Distribution of the number of inter-failure successes.* Consider a sequence of independent Bernoulli trials and assume that the probability of success at the ith trial is given by (2.2). Let U_n be the number of successes until the occurrence of the nth failure, and let $S_n = U_n - U_{n-1}$, for $n = 1, 2, \ldots$, be the sequence of the numbers of successes between consecutive failures, with $P(U_0 = 0) = 1$.

(a) Determine the conditional probability function

$$P(S_{n+1} = s | U_n = u), \quad s = 0, 1, \ldots,$$

(b) compute the conditional q^{-1}-factorial moments

$$E([S_{n+1}]_{m,q^{-1}} | U_n = u), \quad m = 1, 2, \ldots,$$

(c) and deduce the conditional factorial moments

$$E[(S_{n+1})_j | U_n = u], \quad j = 1, 2, \ldots.$$

The conditional distribution of S_{n+1}, given that $U_n = u$, is a *q-geometric distribution of the first kind*, with parameters $\lambda = \theta q^{n+u}$ and q. Notice that the random variable U_n, which follows a negative q-binomial distribution of the first kind, may be written as a sum $U_n = \sum_{j=1}^{n} S_j$, of dependent random variables $S_j, j = 1, 2, \ldots, n$, with conditional distribution, given U_{j-1}, a q-geometric distribution of the first kind.

2.13 Let X be a nonnegative integer-valued random variable obeying a Heine distribution, with parameters λ and q.

(a) Derive the probability generating function $P_X(t) = E(t^X)$ and conclude that $X = \sum_{i=1}^{\infty} Z_i$, with Z_i, $i = 1, 2, \ldots$, independent zero-one Bernoulli random variables.

(b) Show that the mean and variance of X may be expressed as

$$E(X) = \sum_{i=1}^{\infty} \frac{\lambda(1-q)q^{i-1}}{1+\lambda(1-q)q^{i-1}}, \quad V(X) = \sum_{i=1}^{\infty} \frac{\lambda(1-q)q^{i-1}}{(1+\lambda(1-q)q^{i-1})^2}.$$

2.14 *q-Mean and q-variance of the Heine distribution*. Consider a random variable X that follows a Heine distribution with parameters λ and q.

(a) Show that

$$E([X]_q) = \frac{\lambda}{1+\lambda(1-q)}, \quad V([X]_q) = \frac{\lambda}{(1+\lambda(1-q))^2(1+\lambda(1-q)q)}.$$

(b) Also, derive the mth q^{-1}-factorial moment as

$$E([X]_{m,q^{-1}}) = \lambda^m, \quad m = 1, 2, \ldots,$$

(c) and deduce the q^{-1}-mean and q^{-1}-variance as

$$E([X]_{q^{-1}}) = \lambda, \quad V([X]_{q^{-1}}) = \lambda + \lambda^2(q^{-1} - 1).$$

2.15 *Elementary derivation of the probability function of the Heine process*. Consider a stochastic model in which successes or failures (events A or A') may occur at continuous time (or space) points. Furthermore, consider a time interval $(0, t]$ and its partition in n subintervals

$$\left(\frac{[i-1]_q t}{[n]_q}, \frac{[i]_q t}{[n]_q}\right], \quad i = 1, 2, \ldots, n, \quad 0 < q < 1,$$

with lengths $\delta_{n,i}(t) = tq^{i-1}/[n]_q$, $i = 1, 2, \ldots, n$. Note that $q^{i-1}/[n]_q$, for $i = 1, 2, \ldots, n$, is a discrete q-uniform distribution on the set $\{1, 2, \ldots, n\}$. Assume that in each subinterval either a success or a failure may occur. Also, assume that the odds of success is analogous to the length of the subinterval, $\theta_{n,i}(t) = \lambda\delta_{n,i}(t) = \lambda tq^{i-1}/[n]_q$, $i = 1, 2, \ldots, n$, with $0 < \lambda < \infty$. Let $X_{t,n}$ be the number of successes that occur in the n subintervals of $(0, t]$. Derive the probability function $P(X_{t,n} = x)$, $x = 0, 1, \ldots, n$, and show that

$$P(X_t = x) = \lim_{n\to\infty} P(X_{t,n} = x) = e_q(-\lambda t)\frac{q^{\binom{x}{2}}(\lambda t)^x}{[x]_q!}, \quad x = 0, 1, \ldots,$$

where $0 < \lambda < \infty$, $0 < t < \infty$, and $0 < q < 1$.

2.16 Let X_n be the accumulated number of money units saved at the nth unit time interval, $n = 1, 2, \ldots$, under the following savings scheme. A surplus money unit acquired each time interval. Furthermore, at any time interval and given that the current total savings is i, the conditional probability that the surplus money unit is saved is $p_{i,i+1} = \theta q^i$ and is spent is $p_{i,i} = q^i$, $i = 1, 2, \ldots$, and $p_{0,0} = 1 - \theta$. Also, the conditional probability that the total savings are spent is $p_{i,0} = 1 - (1+\theta)q^i$, $i = 1, 2, \ldots$. Show that the stationary distribution,

$$P(X = x) = \lim_{n\to\infty} P(X_n = x), \quad x = 0, 1, \ldots,$$

is the Heine distribution, with parameters $\lambda = \theta/(1-q)$ and q.

2.17 Let X_n be a nonnegative integer-valued random variable with binomial moments

$$E\left[\binom{X_n}{j}\right] = q^{\binom{j}{2}} \begin{bmatrix} n \\ j \end{bmatrix}_q, \quad j = 0, 1, \ldots, n, \quad 0 < q < 1 \text{ or } 1 < q < \infty,$$

and $E\left[\binom{X_n}{j}\right] = 0, j = n, n+1, \ldots$ Find the probability function of X_n,

$$f_n(x) = P(X_n = x), \quad x = 0, 1, \ldots, n.$$

2.18 Consider a sequence of independent Bernoulli trials and assume that the probability of success at the ith trial is given by

$$p_i = q^{r+i-1}, \quad i = 1, 2, \ldots,$$

for $0 \leq r < \infty$ and $0 < q < 1$ or $-\infty < r < 0$ and $1 < q < \infty$; in the second case the number of trials is bounded: $i \leq [-r]$. The probability function of the number X_n of successes in n trials was obtained in Theorem 2.7.

(a) Show that the mth q-factorial moment of X_n is given by

$$E([X_n]_{m,q}) = [m]_q! \sum_{j=m}^{n} (-1)^{j-m} S_q(j, m) q^{\binom{j}{2}+rj} \begin{bmatrix} n \\ j \end{bmatrix}_q, \quad m = 1, 2, \ldots,$$

where $S_q(j, m)$ is the q-Stirling number of the second kind.

(b) Applying expression (1.61) and using the orthogonality relation of the q-Stirling numbers, deduce the jth factorial moment of X_n as

$$E[(X_n)_j] = j! q^{\binom{j}{2}+rj} \begin{bmatrix} n \\ j \end{bmatrix}_q, \quad j = 1, 2, \ldots,$$

in agreement with the expression obtained directly in Theorem 2.7. Note that this is one of the rare cases in which the expression of the factorial moments is simpler than that of the q-factorial moments.

2.19 Consider a sequence of independent Bernoulli trials and assume that the probability of success at the ith trial is given by

$$p_i = \theta q^{i-1}, \quad i = 1, 2, \ldots, \quad 0 < \theta \leq 1, \quad 0 < q < 1 \quad \text{or} \quad 1 < q < \infty,$$

where for $0 < \theta \leq 1$ and $1 < q < \infty$, the number of trials is restricted by $i \leq [-r]$, the integral part of $-r$, with $r = \log \theta / \log q < 0$. Let X_n be the number of successes in n trials. Show that

$$E\left[\binom{X_n}{j}\right] = \theta^j q^{\binom{j}{2}} \begin{bmatrix} n \\ j \end{bmatrix}_q, \quad j = 0, 1, \ldots, n,$$

and $E\left[\binom{X_n}{j}\right] = 0$, for $j = n+1, n+2, \ldots$, and conclude that

$$P(X_n = x) = \sum_{j=x}^{n} (-1)^{j-x} \theta^j q^{\binom{j}{2}} \begin{bmatrix} n \\ j \end{bmatrix}_q \binom{j}{x}, \quad x = 0, 1, \ldots, n.$$

2.20 *Bernoulli trials with discrete q-uniform success probability.* Assume that a ball (white or black) is randomly selected, with probability θ for a white and $1-\theta$ for a black ball. The selected ball is placed in the first (bottom) cell of the first column of an $n \times n$ array of cells. Then, a coin, with probability q of tails, is sequentially tossed until heads occurs. Each time tails occurs, the ball moves up a cell, with one exception: If tails occurs when the ball is in the nth (top) cell, then the ball moves back to the first cell. When heads occurs, the ball comes to rest. This procedure is successively repeated for the second to the nth column. Let X_n be the number of white balls that come to rest on the diagonal running from the first cell of the first column to the nth cell of the nth column.

(a) Derive the probability function of X_n as

$$P(X_n = x) = \sum_{j=x}^{n} (-1)^{j-x} \binom{j}{x} \begin{bmatrix} n \\ j \end{bmatrix}_q \frac{\theta^j q^{\binom{j}{2}}}{[n]_q^j}, \quad x = 0, 1, \ldots, n.$$

(b) Show that the probability function of the limiting distribution, as $n \to \infty$, is given by

$$P(X = x) = \lim_{n\to\infty} P(X_n = x) = \sum_{j=x}^{\infty} (-1)^{j-x} \binom{j}{x} \frac{\theta^j q^{\binom{j}{2}}}{[j]_q!}, \quad x = 0, 1, \ldots .$$

(c) Moreover, show that

$$\lim_{q\to 1} P(X = x) = e^{-\theta} \frac{\theta^x}{x!}, \quad x = 0, 1, \ldots .$$

2.21 *Inter-record times in a geometrically increasing population.* Consider, again, the stochastic model of the geometrically increasing population discussed in Example 2.6, and let $W_1 = T_1$ and $W_k = T_k - T_{k-1}$, $k = 2, 3, \ldots$, be the inter-record times.

(a) Show that

$$P(W_k = n) = q^{n-1} \sum_{i=0}^{n} (-1)^i q^{\binom{i}{2}} \begin{bmatrix} n \\ i \end{bmatrix}_q \frac{q^{i(k-1)}}{[i+1]_q^{k-1}} \left(\frac{[n-i]_q}{[n]_q} - q \right),$$

for $n = 1, 2, \ldots$, and deduce that

$$\lim_{k\to\infty} P(W_k = n) = (1-q)q^{n-1}, \quad n = 1, 2, \ldots .$$

(b) Also, show that

$$E(W_k) = \sum_{i=0}^{\infty} (-1)^i q^{\binom{i}{2}} \frac{(1-q)^{-i-1}}{[i]_q!} \cdot \frac{q^{ik}}{[i+1]_q^k}.$$

2.22 *Records in a modified geometrically increasing population.* Consider again the general record model discussed in Example 2.6, in which the record indicator random variables Z_i, $i = 1, 2, \ldots$, are independent. Furthermore, assume that

$$p_i = P(Z_i = 1) = \frac{\alpha_i}{\alpha_0 + \alpha_1 + \cdots + \alpha_i}, \quad i = 1, 2, \ldots,$$

with the population size of the athletes given by

$$\alpha_i = \theta q^{-(i-1)}, \quad i = 1, 2, \ldots, \quad 0 < \theta < 1, \quad 0 < q < 1,$$

which is a geometrically increasing sequence. An initial term $\alpha_0 = \theta q[r]_q$, with $0 \leq r < \infty$, attached to the sequence α_i, $i = 1, 2, \ldots$, modifies Yang's model by allowing the probability $p_1 = P(Z_1 = 1)$ to be not necessarily one. Show that the probability function and factorial moments of the number X_n of records up to time (index) n are given by (2.36) and (2.37). Also, show that the probability function of the time T_k of the realization of the kth record is given by (2.39).

2.23 *(Continuation).* Let $W_1 = T_1$ and $W_k = T_k - T_{k-1}$, for $k = 2, 3, \ldots$, be the inter-record times.

(a) Show that

$$P(W_k = n) = q^{n-1} \sum_{i=0}^{n} (-1)^i q^{\binom{i}{2}+ri} \frac{[n]_{i,q} q^{i(k-1)}}{[r+i]_{i,q}[i+1]_q^{k-1}} \left(\frac{[n-i]_q}{[n]_q} - q \right),$$

for $n = 1, 2, \ldots$, and deduce that

$$\lim_{k \to \infty} P(W_k = n) = (1-q)q^{n-1}, \quad n = 1, 2, \ldots.$$

(b) Also, show that

$$E(W_k) = \sum_{i=0}^{\infty} (-1)^i q^{\binom{i}{2}+ri} \frac{(1-q)^{-i-1}}{[r+i]_{i,q}} \cdot \frac{q^{ik}}{[i+1]_q^k}.$$

2.24 *Records in a q-factorially increasing population.* Consider, once more, the general record model discussed in Example 2.6, in which the record indicator random variables Z_i, $i = 1, 2, \ldots$, are independent. Moreover, assume that

$$p_i = P(Z_i = 1) = \frac{\alpha_i}{\alpha_0 + \alpha_1 + \cdots + \alpha_i}, \quad i = 1, 2, \ldots,$$

with the population size of the athletes given by

$$\alpha_i = q^{-(i-1)\theta} \frac{[\theta + r + i - 1]_{i-1,q}}{[r + i - 1]_{i-1,q}}, \quad i = 1, 2, \ldots,$$

where $0 < \theta < \infty$, $0 \leq r < \infty$, $0 < q < 1$, and $\alpha_0 = q^\theta [r]_q / [\theta]_q$.

(a) Show that the probability function of the number X_n of records up to time (index) n is given by

$$P(X_n = x) = \frac{q^{(n-x)(\theta-1)}|s_q(n, x; r)|[\theta]_q^x}{[\theta + r + n - 1]_{n,q}}, \quad x = 0, 1, \dots, n,$$

where $|s_q(n, x; r)|$ is the noncentral signless q-Stirling number of the first kind.

(b) Derive the jth factorial moment of X_n as

$$E[(X_n)j] = \frac{j!q^{-(n-j)}|s_q(n, j; r + \theta)|[\theta]_q^j}{[\theta + r + n - 1]_{n,q}}, \quad j = 1, 2, \dots, n,$$

and $E[(X_n)_j] = 0$, for $j = n + 1, n + 2, \dots$, and deduce the mean and variance as

$$E(X_n) = \sum_{i=1}^{n} \frac{[\theta]_q}{[\theta + r + i - 1]_q}, \quad V(X_n) = \sum_{i=1}^{n} \frac{q^{\theta}[r + i - 1]_q}{[\theta + r + i - 1]_q^2}.$$

(c) Show that the probability function of the time T_k of the realization of the kth record is given by

$$P(T_k = n) = \frac{q^{(n-k)(\theta-1)}|s_q(n - 1, k - 1; r)|[\theta]_q^k}{[\theta + r + n - 1]_{n,q}}, \quad n = k, k + 1, \dots .$$

3

SUCCESS PROBABILITY VARYING WITH THE NUMBER OF SUCCESSES

3.1 NEGATIVE q-BINOMIAL DISTRIBUTION OF THE SECOND KIND

Consider a random experiment with sample space $\Omega = \{f, s\}$, where the sample points (events) f and s are characterized as *failure* and *success*, respectively. An experiment with such a sample space is called *Bernoulli trial*. Furthermore, a sequence of independent Bernoulli trials, with constant success probability, which is terminated with the occurrence of the first success, $g = (f, f, \ldots, f, s)$, is called *geometric sequence of trials*.

Consider a sequence of independent geometric sequences of trials with probability of success at the jth geometric sequence of trials,

$$Q_j(\{s\}) = p_j, \ \ 0 < p_j < 1, \ \ j = 1, 2, \ldots,$$

varying with the number of geometric sequences of trials, or, equivalently, with the number of successes in the sequence of Bernoulli trials. Clearly, the probability of failure at the jth geometric sequence of trials is $Q_j(\{f\}) = 1 - p_j \equiv q_j$, $j = 1, 2, \ldots$. Notice that the probability p_j is essentially the conditional probability of success at any Bernoulli trial, given that $j - 1$ successes occur in the previous trials. It is interesting and useful to note that the number W_n of failures until the occurrence of the nth success, in a sequence of independent geometric sequences of trials, may be expressed as a sum of n independent geometric random variables. Specifically,

Discrete q-Distributions, First Edition. Charalambos A. Charalambides.

let U_j be the number of failures at the jth geometric sequence of trials, $j = 1, 2, \ldots, n$. Then, $W_n = \sum_{j=1}^{n} U_j$, with

$$P(U_j = u) = p_j q_j^u, \quad u = 0, 1, \ldots, \quad j = 1, 2, \ldots, n.$$

Also note that, in the case of a general probability sequence $p_j, j = 1, 2, \ldots$, very little can be deduced from it about the distributions of W_n and other random variables that may be defined in this model (see Exercises 3.1 and 3.2). The restriction to particular cases, which allow a thorough study of probability distributions, is of interest.

Let us first consider the case with probability of success at the jth geometric sequence of trials given by

$$p_j = 1 - \theta q^{j-1}, \quad j = 1, 2, \ldots, \quad 0 < \theta < 1, \quad 0 < q < 1 \text{ or } 1 < q < \infty, \tag{3.1}$$

where, for $0 < \theta < 1$ and $1 < q < \infty$, the number j of geometric sequences of trials is restricted by $\theta q^{j-1} < 1$, ensuring that $0 < p_j < 1$. This restriction imposes on j an upper bound, $j = 1, 2, \ldots, [r]$, with $[r]$ denoting the integral part of $r = -\log\,\theta / \log\,q > 0$. Notice that the probability p_j is geometrically varying, with rate q.

The study of the distribution of the number of failures until the occurrence of a specific number of successes is of theoretical and practical interest. In this respect, the following definition is introduced.

Definition 3.1. *Let W_n be the number of failures until the occurrence of the nth success, in a sequence of independent geometric sequences of trials, with probability of success at the jth geometric sequence of trials given by (3.1). The distribution of the random variable W_n is called negative q-binomial distribution of the second kind, with parameters n, θ, and q.*

The probability function, the q-factorial moments, and the usual factorial moments of the negative q-binomial distribution of the second kind are given in the following theorem.

Theorem 3.1. *The probability function of the negative q-binomial distribution of the second kind, with parameters n, θ, and q, is given by*

$$P(W_n = w) = \begin{bmatrix} n+w-1 \\ w \end{bmatrix}_q \theta^w \prod_{j=1}^{n} (1 - \theta q^{j-1}), \quad w = 0, 1, \ldots, \tag{3.2}$$

for $0 < \theta < 1$ and $0 < q < 1$ or $1 < q < \infty$ with $\theta q^{n-1} < 1$. Its q-factorial moments are given by

$$E([W_n]_{m,q}) = \frac{[n+m-1]_{m,q}\theta^m}{\prod_{j=1}^{m}(1 - \theta q^{n+j-1})}, \quad m = 1, 2, \ldots . \tag{3.3}$$

Moreover, its factorial moments are given by

$$E[(W_n)_j] = j! \sum_{m=j}^{\infty} (-1)^{m-j} \begin{bmatrix} n+m-1 \\ m \end{bmatrix}_q \frac{\theta^m (1-q)^{m-j} s_q(m,j)}{\prod_{i=1}^{m} (1-\theta q^{n+i-1})}, \tag{3.4}$$

for $j = 1, 2, \ldots,$ *where* $s_q(m,j)$ *is the q-Stirling number of the first kind. In particular, its mean and variance are given by*

$$E(W_n) = \sum_{j=1}^{n} \frac{\theta q^{j-1}}{1-\theta q^{j-1}}, \quad V(W_n) = \sum_{j=1}^{n} \frac{\theta q^{j-1}}{(1-\theta q^{j-1})^2}. \tag{3.5}$$

Proof. Let A_{j,k_j} be the event that k_j failures precede the first success in the jth geometric sequence of trials, for $j = 1, 2, \ldots, n$. Then, using the independence of the events $A_{1,k_1}, A_{2,k_2}, \ldots, A_{n,k_n}$ and the (geometric) probabilities

$$P(A_{j,k_j}) = (\theta q^{j-1})^{k_j} (1-\theta q^{j-1}), \quad j = 1, 2, \ldots, n,$$

we get

$$\begin{aligned} P(A_{1,k_1} A_{2,k_2} \cdots A_{n,k_n}) =& P(A_{1,k_1}) P(A_{2,k_2}) \cdots P(A_{n,k_n}) \\ =& (\theta q^{-1})^{k_1+k_2+\cdots+k_n} q^{k_1+2k_2+\cdots+nk_n} \prod_{j=1}^{n} (1-\theta q^{j-1}), \end{aligned}$$

with $k_j \geq 0$, $j = 1, 2, \ldots, n$. Summing these probabilities over all $k_j \geq 0$, $j = 1, 2, \ldots, n$, with fixed $k_1 + k_2 + \cdots + k_n = w$, and using (1.5),

$$\sum_{\substack{k_j \geq 0, j=1,2,\ldots,n \\ k_1+k_2+\cdots+k_n = w}} q^{k_1+2k_2+\cdots+nk_n} = q^w \begin{bmatrix} n+w-1 \\ w \end{bmatrix}_q,$$

the probability that w failures precede the nth success is obtained as (3.2).

Notice that, according to the negative q-binomial formula (1.15), the probabilities (3.2) sum to unity,

$$\begin{aligned} \sum_{w=0}^{\infty} P(W_n = w) &= \prod_{j=1}^{n} (1-\theta q^{j-1}) \sum_{w=0}^{\infty} \begin{bmatrix} n+w-1 \\ w \end{bmatrix}_q \theta^w \\ &= \prod_{j=1}^{n} (1-\theta q^{j-1}) \prod_{j=1}^{n} (1-\theta q^{j-1})^{-1} = 1, \end{aligned}$$

which conforms to the definition of a probability function.

The mth q-factorial moment of W_n,

$$E([W_n]_{m,q}) = \sum_{w=m}^{\infty} [w]_{m,q} \begin{bmatrix} n+w-1 \\ w \end{bmatrix}_q \theta^w \prod_{j=1}^{n} (1-\theta q^{j-1}),$$

on using the relation

$$[w]_{m,q}\begin{bmatrix} n+w-1 \\ w \end{bmatrix}_q = \frac{[w]_q!}{[w-m]_q!}\cdot\frac{[n+w-1]_q!}{[w]_q![n-1]_q!}$$
$$= \frac{[n+m-1]_q!}{[n-1]_q!}\cdot\frac{[n+w-1]_q!}{[w-m]_q![n+m-1]_q!}$$
$$= [n+m-1]_{m,q}\begin{bmatrix} n+w-1 \\ w-m \end{bmatrix}_q,$$

is expressed as

$$E([W_n]_{m,q}) = [n+m-1]_{m,q}\theta^m\prod_{j=1}^{n}(1-\theta q^{j-1})\sum_{w=m}^{\infty}\begin{bmatrix} n+w-1 \\ w-m \end{bmatrix}_q \theta^{w-m}.$$

Therefore, using the negative q-binomial formula (1.15), it reduces to

$$E([W_n]_{m,q}) = [n+m-1]_{m,q}\theta^m\frac{\prod_{j=1}^{n}(1-\theta q^{j-1})}{\prod_{j=1}^{n+m}(1-\theta q^{j-1})}$$

and since

$$\prod_{j=1}^{n+m}(1-\theta q^{j-1}) = \prod_{j=1}^{n}(1-\theta q^{j-1})\prod_{j=1}^{m}(1-\theta q^{n+j-1}),$$

expression (3.3) is deduced.

Introducing the mth q-factorial moment (3.3) into (1.61), the required formula (3.4) is deduced. In particular, from (3.4) with $j = 1$ and $j = 2$, the mean and variance of W_n may be obtained (see Exercises 3.3 and 3.4). Alternatively, using the expression $W_n = \sum_{j=1}^{n} U_j$, where $U_1, U_2, \ldots, U_n$ are independent geometric random variables, with

$$E(U_j) = \frac{\theta q^{j-1}}{1-\theta q^{j-1}},\quad V(U_j) = \frac{\theta q^{j-1}}{(1-\theta q^{j-1})^2},\quad j = 1, 2, \ldots, n,$$

and $E(W_n) = \sum_{j=1}^{n} E(U_j)$ and $V(W_n) = \sum_{j=1}^{n} V(U_j)$, the mean and variance (3.5) are readily deduced. □

Remark 3.1. *Inverse absorption distribution*. The success probability (3.1), in the case $1 < q < \infty$, may be preferably expressed as follows. Replacing the parameter q by q^{-1}, with $0 < q < 1$, and then substituting $\theta = q^r$, we get

$$p_j = 1 - q^{r-j+1},\quad j = 1, 2, \ldots, [r],\quad 0 < r < \infty,\quad 0 < q < 1, \tag{3.6}$$

which is a geometrically decreasing sequence of a finite number of terms. In this case, the probability function (3.2) reduces to

$$P(W_n = w) = \begin{bmatrix} n+w-1 \\ w \end{bmatrix}_q q^{(r-n+1)w}(1-q)^n[r]_{n,q},\quad w = 0, 1, \ldots, \tag{3.7}$$

for $0 < r < \infty$, $0 < q < 1$ and $n \leq [r]$. This negative q-binomial distribution of the second kind is particularly known as *inverse absorption distribution*.

Example 3.1. *Proofreading a manuscript.* Assume that a proofreader reads a manuscript, which has a fixed number of errors, m, and when he/she finds an error, corrects it and starts reading the manuscript from the beginning. Also, the proofreader starts reading the manuscript from the beginning when he/she reaches its end. A scan (reading) of the manuscript is successful if the proofreader finds (and corrects) an error and is a failure otherwise. Thus, a scan of the manuscript constitutes a Bernoulli trial. Assume that the probability of finding any particular error is $p = 1 - q$. Then, the conditional probability that a scan (trial) is successful, given that $j - 1$ scans (trials) were successful in the previous scans, is

$$p_j = 1 - q^{m-j+1}, \quad j = 1, 2, \ldots, m, \quad 0 < q < 1.$$

Consequently, the distribution of the number W_n of unsuccessful scans until the occurrence of the nth successful scan, with $n \leq m$, is the inverse absorption distribution, with probability function (3.7), where $r = m$ is a positive integer.

Remark 3.2. *A q-geometric distribution of the second kind.* The probability function of the number Z_1 of successes until the occurrence of the first failure is of interest and may be obtained as follows. Considering the event A_j of success at the jth trial, for $j = 1, 2, \ldots$, the probability $P_z = P(A_1A_2 \cdots A_zA'_{z+1})$, that z successes precede the occurrence of the first failure, on using the multiplication formula, is deduced as

$$\begin{aligned} P_z &= P(A_1)P(A_2|A_1) \cdots P(A_z|A_1A_2 \cdots A_{z-1})P(A'_{z+1}|A_1A_2 \cdots A_z) \\ &= \prod_{j=1}^{z}(1 - \theta q^{j-1})\theta q^z, \end{aligned}$$

for $z = 0, 1, \ldots$. Also, the probability Q of the occurrence of at least one failure in an infinite number of Bernoulli trials is readily deduced as

$$\begin{aligned} Q &= \lim_{n\to\infty} P(A'_1 \cup A'_2 \cup \cdots \cup A'_n) = 1 - \lim_{n\to\infty} P(A_1A_2 \cdots A_n) \\ &= 1 - \lim_{n\to\infty} \prod_{j=1}^{n}(1 - \theta q^{j-1}) = 1 - E_q(-\theta/(1-q)), \end{aligned}$$

where $E_q(t) = \prod_{j=1}^{\infty}(1 + t(1-q)q^{j-1})$ is the q-exponential function (1.23). Clearly, the probability function of the random variable Z_1 is the conditional probability that z successes precede the occurrence of the first failure, given the occurrence of at least one failure in an infinite number of Bernoulli trials, $P(Z_1 = z) = P_z/Q$, and so

$$P(Z_1 = z) = (1 - E_q(-\theta/(1-q)))^{-1} \prod_{j=1}^{z}(1 - \theta q^{j-1})\theta q^z, \tag{3.8}$$

for $z = 0, 1, \ldots,$ where $0 < \theta < 1$, and $0 < q < 1$.

Note that the probabilities (3.8), according to a q-geometric series given in Exercise 1.10, sum to unity,

$$\sum_{z=0}^{\infty} P(Z_1 = z) = (1 - E_q(-\theta/(1-q)))^{-1} \sum_{z=0}^{\infty} \prod_{j=1}^{z} (1 - \theta q^{j-1}) \theta q^z = 1,$$

which conforms to the definition of a probability function.

3.2 q-BINOMIAL DISTRIBUTION OF THE SECOND KIND

Consider again a sequence of independent geometric sequences of trials with probability of success at the jth geometric sequence of trials given by (3.1). In this stochastic model, the interest now is focused on the study of the number of successes (or failures) in a given number of trials. For this reason, the following definition is introduced.

Definition 3.2. *Let X_n be the number of failures in a sequence of n independent Bernoulli trials, with probability of success at the jth geometric sequence of trials given by (3.1). The distribution of the random variable X_n is called q-binomial distribution of the second kind, with parameters n, θ, and q.*

Note that the probability function of the number $Y_n = n - X_n$ of successes in a sequence of n independent Bernoulli trials is closely connected to that of X_n, $P(Y_n = y) = P(X_n = n - y)$, $y = 0, 1, \ldots, n$. The reason for choosing the distribution of X_n as the q-binomial distribution of the second kind is that its moments are more easily derived.

The probability function, the q-factorial moments, and the usual factorial moments of the q-binomial distribution of the second kind, with parameters n, θ, and q, are obtained in the following theorem.

Theorem 3.2. *The probability function of the q-binomial distribution of the second kind, with parameters n, θ, and q, is given by*

$$P(X_n = x) = \begin{bmatrix} n \\ x \end{bmatrix}_q \theta^x \prod_{j=1}^{n-x} (1 - \theta q^{j-1}), \quad x = 0, 1, \ldots, n, \tag{3.9}$$

for $0 < \theta < 1$ and $0 < q < 1$ or $1 < q < \infty$, with $\theta q^{n-1} < 1$. Its q-factorial moments are given by

$$E([X_n]_{m,q}) = [n]_{m,q} \theta^m, \quad m = 1, 2, \ldots, n, \tag{3.10}$$

and $E([X_n]_{m,q}) = 0$, for $m = n+1, n+2, \ldots$. Moreover, its factorial moments are given by

$$E[(X_n)_j] = j! \sum_{m=j}^{n} (-1)^{m-j} \begin{bmatrix} n \\ m \end{bmatrix}_q \theta^m (1-q)^{m-j} s_q(m, j), \tag{3.11}$$

for $j = 1, 2, \ldots, n$, *and* $E[(X_n)_j] = 0$, *for* $j = n+1, n+2, \ldots$, *where* $s_q(m,j)$ *is the* q*-Stirling number of the first kind. In particular, its mean and variance are given by*

$$E(X_n) = \sum_{m=1}^{n} \frac{[n]_{m,q}(1-q)^{m-1}\theta^m}{[m]_q} \tag{3.12}$$

and

$$V(X_n) = 2\sum_{m=2}^{n} \frac{[n]_{m,q}(1-q)^{m-2}\theta^m \zeta_{m-1,q}}{[m]_q} + E(X_n) - [E(X_n)]^2, \tag{3.13}$$

where $\zeta_{m,q} = \sum_{j=1}^{m} 1/[j]_q$.

Proof. The probability function of the q-binomial distribution of the second kind is closely connected to the probability function of the negative q-binomial distribution of the second kind. Precisely,

$$P(W_{n-x+1} = x) = P(X_n = x)p_{n-x+1},$$

where $P(X_n = x)$ is the probability of the occurrence of x failures and $n - x$ successes in n trials, $p_{n-x+1} = 1 - \theta q^{n-x}$ is the conditional probability of success at a Bernoulli trial given that $n - x$ successes occur in the previous trials and $P(W_{n-x+1} = x)$ is the probability that x failures precede the occurrence of the $n - x + 1$st success. Thus, using (3.2), expression (3.9) is deduced.

Note that the probabilities (3.9), according to the q-binomial formula (1.19), sum to unity,

$$\sum_{x=0}^{n} P(X_n = x) = \sum_{x=0}^{n} \begin{bmatrix} n \\ x \end{bmatrix}_q \theta^x \prod_{j=1}^{n-x}(1-\theta q^{j-1}) = 1,$$

as is required by the definition of a probability function.

The mth q-factorial moment of X_n,

$$E([X_n]_{m,q}) = \sum_{x=m}^{n} [x]_{m,q} \begin{bmatrix} n \\ x \end{bmatrix}_q \theta^x \prod_{j=1}^{n-x}(1-\theta q^{j-1}),$$

on using the relation

$$\begin{aligned} [x]_{m,q}\begin{bmatrix} n \\ x \end{bmatrix}_q &= \frac{[x]_q!}{[x-m]_q!} \cdot \frac{[n]_q!}{[x]_q![n-x]_q!} \\ &= \frac{[n]_q!}{[n-m]_q!} \cdot \frac{[n-m]_q!}{[x-m]_q![n-x]_q!} = [n]_{m,q}\begin{bmatrix} n-m \\ x-m \end{bmatrix}_q, \end{aligned}$$

is expressed as

$$E([X_n]_{m,q}) = [n]_{m,q}\theta^m \sum_{x=m}^{n} \begin{bmatrix} n-m \\ x-m \end{bmatrix}_q \theta^{x-m} \prod_{j=1}^{n-x}(1-\theta q^{j-1}),$$

which, by (1.19), reduces to (3.10). Introducing it into (1.61), the required formula (3.11) is deduced. The mean and variance of X_n may be deduced from (3.11), using the expressions

$$s_q(m,1) = (-1)^{m-1}[m-1]_q!, \qquad s_q(m,2) = (-1)^{m-2}[m-1]_q!\zeta_{m-1,q},$$

with $\zeta_{m,q} = \sum_{j=1}^{m} 1/[j]_q$. □

Remark 3.3. *Absorption distribution.* The probability function of the number $Y_n = n - X_n$ of successes in n independent Bernoulli trials, with probability of success at the jth geometric sequence of trials given by (3.1), on using the relation $P(Y_n = y) = P(X_n = n-y)$, $y = 0, 1, \ldots, n$, and the expression (3.9), is deduced as

$$P(Y_n = y) = \begin{bmatrix} n \\ y \end{bmatrix}_q \theta^{n-y} \prod_{j=1}^{y} (1 - \theta q^{j-1}), \quad y = 0, 1, \ldots, n,$$

Also, the success probability (3.1), in the case $1 < q < \infty$, may be expressed as in (3.6). In this case, the probability function of Y_n reduces to

$$P(Y_n = y) = \begin{bmatrix} n \\ y \end{bmatrix}_q q^{(n-y)(r-y)}(1-q)^y [r]_{y,q}, \quad y = 0, 1, \ldots, n, \tag{3.14}$$

for $0 < r < \infty$, $0 < q < 1$, and $n \leq [r]$. This q-binomial distribution of the second kind is particularly known as *absorption distribution.*

Example 3.2. *Sequential capture of endangered animals.* Consider a closed population of m endangered animals and assume that each day a search is organized to find an animal. When an animal is found, it is transferred to captive breeding program and search is abandoned for that day. A search day in which an animal is captured is considered as successful. Assume that on any particular day a particular animal has constant probability q of evading capture. Then, the conditional probability that a search day is successful, given that $j-1$ search days were successful in the previous search days, is

$$p_j = 1 - q^{m-j+1}, \quad j = 1, 2, \ldots, m, \quad 0 < q < 1.$$

Therefore, the distribution of the number X_n of captures in n search days, with $n \leq m$, is the absorption distribution, with probability function (3.14), where $r = m$ is a positive integer. Also, the distribution of the number W_n of unsuccessful search days until the nth animal capture, with $n \leq m$, is the inverse absorption distribution, with probability function (3.7), where $r = m$ is a positive integer.

Example 3.3. *An absorption process.* Assume that batches of k particles are sequentially propelled into a chamber of l consecutive lined cells, with the capacity of each cell limited to one particle. Initially, a batch of k particles occupies the k leftmost cells. Then, a coin, with probability p of heads and $q = 1 - p$ of tails, is successively tossed. When tails occurs each of the k particles of the batch moves one cell to the right, while when heads occurs the batch of the k particles is absorbed and the cells with the absorbed particles are removed. A batch of k particles, which successfully reaches the k rightmost cells, is said to have escaped and its particles are removed from the chamber without removing these cells. Subsequent batches of r particles are propelled into the chamber of the remaining cells. Clearly, the conditional probability of an absorption of a batch of k particles, given that $j-1$ absorptions occur, is given by

$$p_j = 1 - q^{l-kj+1}, \quad j = 1, 2, \ldots .$$

Substituting $l = (r+1)k - 1$, with $r > 0$ not necessarily an integer, it follows that

$$p_j = 1 - q^{k(r-j+1)}, \quad j = 1, 2, \ldots, [r].$$

Therefore, the probability function of the number X_n of absorbed batches of k particles, when n batches are propelled into the chamber of l cells, is given (3.14), with q^k instead of q.

3.3 EULER DISTRIBUTION

Definition 3.3. *Let X be a discrete random variable with probability function*

$$f(x) = P(X = x) = E_q(-\lambda)\frac{\lambda^x}{[x]_q!}, \quad x = 0, 1, \ldots, \tag{3.15}$$

where $0 < \lambda < 1/(1-q)$, $0 < q < 1$, *and* $E_q(t) = \prod_{i=1}^{\infty}(1 + t(1-q)q^{i-1})$ *is the q-exponential function (1.23). The distribution of the random variable X is called Euler distribution, with parameters* λ *and q.*

The Euler distribution is a q-Poisson distribution since the probability function (3.15), for $q \to 1$, converges to the probability function of the Poisson distribution. Note that the function (3.15) is nonnegative,

$$f(x) > 0, \quad x = 0, 1, \ldots, \quad f(x) = 0, \quad x \neq 0, 1, \ldots,$$

and using the expansion of the q-exponential function $e_q(t)$ into a power series,

$$e_q(t) = \prod_{i=1}^{\infty}(1 - t(1-q)q^{i-1})^{-1} = \sum_{x=0}^{\infty}\frac{t^x}{[x]_q!}, \quad |t| < 1/(1-q), \quad 0 < q < 1,$$

together with the relation $E_q(-t)e_q(t) = 1$, it follows that it sums to unity,

$$\sum_{x=0}^{\infty} P(X = x) = E_q(-\lambda)\sum_{x=0}^{\infty}\frac{\lambda^x}{[x]_q!} = E_q(-\lambda)e_q(\lambda) = 1,$$

as is required by the definition of a probability function.

The q-factorial and the usual factorial moments of the Euler distribution are derived in the following theorem.

Theorem 3.3. *The q-factorial moments of the Euler distribution are given by*

$$E([X]_{m,q}) = \lambda^m, \quad m = 1, 2, \ldots \tag{3.16}$$

Moreover, its factorial moments are given by

$$E[(X)_j] = j! \sum_{m=j}^{\infty} (-1)^{m-j} \frac{\lambda^m}{[m]_q!} (1-q)^{m-j} s_q(m,j), \quad j = 1, 2, \ldots, \tag{3.17}$$

where $s_q(m,j)$ is the q-Stirling number of the first kind. In particular, its mean and variance are given by

$$E(X) = \frac{-l_q(1-\lambda(1-q))}{1-q}, \tag{3.18}$$

where $-l_q(1-t) = \sum_{m=1}^{\infty} t^m/[m]_q$ is a q-logarithmic function and

$$V(X) = 2\sum_{m=2}^{\infty} \frac{\lambda^m (1-q)^{m-2} \zeta_{m-1,q}}{[m]_q} + E(X) - [E(X)]^2, \tag{3.19}$$

where $\zeta_{m,q} = \sum_{j=1}^{m} 1/[j]_q$.

Proof. The mth q-factorial moment of X,

$$E([X]_{m,q}) = E_q(-\lambda) \sum_{x=m}^{\infty} [x]_{m,q} \frac{\lambda^x}{[x]_q!} = \lambda^m E_q(-\lambda) \sum_{x=m}^{\infty} \frac{\lambda^{x-m}}{[x-m]_q!},$$

on using the relations

$$\sum_{x=m}^{\infty} \frac{\lambda^{x-m}}{[x-m]_q!} = e_q(\lambda), \quad E_q(-\lambda) e_q(\lambda) = 1,$$

is deduced in the form (3.16). Introducing it into (1.61), the required formula (3.17) is deduced. The mean and variance of X_n may be deduced from (3.17), by using the expressions

$$s_q(m,1) = (-1)^{m-1}[m-1]_q!, \qquad s_q(m,2) = (-1)^{m-2}[m-1]_q! \zeta_{m-1,q},$$

with $\zeta_{m,q} = \sum_{j=1}^{m} 1/[j]_q$. □

The probability function (3.9) of the q-binomial distribution of the second kind, as the number of trials tends to infinity, and the probability function (3.2) of the negative q-binomial distribution of the second kind, as the number of successes tends to infinity, can be approximated by the probability function of Euler distribution, according to the following theorem.

Theorem 3.4. *The limit of the probability function (3.9) of the q-binomial distribution of the second kind, as $n \to \infty$, is the probability function of the Euler distribution,*

$$\lim_{n\to\infty} \begin{bmatrix} n \\ x \end{bmatrix}_q \theta^x \prod_{i=1}^{n-x} (1-\theta q^{i-1}) = E_q(-\lambda)\frac{\lambda^x}{[x]_q!}, \quad x = 0, 1, \ldots, \tag{3.20}$$

for $0 < \lambda < 1/(1-q)$ and $0 < q < 1$, with $\lambda = \theta/(1-q)$.

Also, the limit of the probability function (3.2) of the negative q-binomial distribution of the second kind, as $n \to \infty$, is the probability function of the Euler distribution,

$$\lim_{n\to\infty} \begin{bmatrix} n+w-1 \\ w \end{bmatrix}_q \theta^w \prod_{i=1}^{n} (1-\theta q^{i-1}) = E_q(-\lambda)\frac{\lambda^w}{[w]_q!}, \quad w = 0, 1, \ldots, \tag{3.21}$$

for $0 < \lambda < 1/(1-q)$ and $0 < q < 1$, with $\lambda = \theta/(1-q)$.

Proof. Since

$$\lim_{n\to\infty} \begin{bmatrix} n \\ x \end{bmatrix}_q = \frac{1}{(1-q)^x[x]_q!} \lim_{n\to\infty} \prod_{i=1}^{x} (1-q^{n-i+1}) = \frac{1}{(1-q)^x[x]_q!}$$

and

$$\lim_{n\to\infty} \prod_{i=1}^{n-x} (1-\lambda(1-q)q^{i-1}) = \prod_{i=1}^{\infty} (1-\lambda(1-q)q^{i-1}) = E_q(-\lambda),$$

for $0 < q < 1$, the limiting expression (3.20) is readily deduced. Also, the limiting expression (3.21) is similarly obtained. □

Remark 3.4. *q-Poisson distributions.* As has already been noticed, the Euler and Heine distributions are both q-Poisson distributions. Their probability mass functions may be expressed by the same functional formula, with different parametric spaces. Specifically, the probability function of the Heine distribution (2.11),

$$f(x) = e_q(-\lambda)\frac{q^{\binom{x}{2}}\lambda^x}{[x]_q!}, \quad x = 0, 1, \ldots,$$

with $0 < \lambda < \infty$ and $0 < q < 1$, on replacing q by the q^{-1}, with $1 < q < \infty$, and using the relations $[x]_{q^{-1}}! = q^{-\binom{x}{2}}[x]_q$ and $e_{q^{-1}}(t) = E_q(t)$, may be expressed as

$$f(x) = E_q(-\lambda)\frac{\lambda^x}{[x]_q!}, \quad x = 0, 1, \ldots,$$

with $0 < \lambda < \infty$ and $1 < q < \infty$. Note that this is the same expression as that of the Euler probability function, (3.15), with a different parametric space.

An interesting application of the Euler distribution as feasible prior in a simple Bayesian model for oil exploration is presented in the following example, which is a continuation of Example 2.3.

Example 3.4. *Number of undiscovered oilfields in oil exploration II.* Returning to Example 2.3, assume that

$$P(X = j) = P(X = j + 1|S), \ \ j = 0, 1, \dots .$$

Then, from (2.18) it follows that

$$P(X = j + 1)(1 - q^{j+1}) = \theta P(X = j), \ \ j = 0, 1, \dots ,$$

with

$$\theta = \sum_{i=0}^{\infty} P(X = i)(1 - q^i) = 1 - \sum_{i=0}^{\infty} P(X = i)q^i.$$

Consequently,

$$P(X = x) = P(X = 0)\frac{\lambda^x}{[x]_q!}, \ \ x = 1, 2, \dots ,$$

where $\lambda = \theta/(1 - q)$, with $0 < \theta < 1$ and $0 < q < 1$. Since

$$e_q(t) = \prod_{i=1}^{\infty} (1 - t(1 - q)q^{i-1})^{-1} = \sum_{x=0}^{\infty} \frac{t^x}{[x]_q!},$$

it follows that $P(X = 0) = 1/e_q(\lambda) = E_q(-\lambda)$ and so

$$P(X = x) = E_q(-\lambda)\frac{\lambda^x}{[x]_q!}, \ \ x = 0, 1, \dots ,$$

which is the probability function of the Euler distribution.

Example 3.5. *A general negative q-binomial distribution of the second kind.* In the Bayesian model for oil exploration, presented in Example 3.4, a general negative q-binomial distribution of the second kind is obtained, by replacing the assumption

$$P(X = j) = P(X = j + 1|S), \ \ j = 0, 1, \dots ,$$

with the more general assumption

$$P(X = j) = \rho P(X = j + 1|S) + (1 - \rho)P(X = j|F), \ \ j = 0, 1, \dots ,$$

where $0 < \rho \leq 1$. Then, using (2.18) and (2.19), it follows that

$$P(X = j) = P(X = j + 1)(1 - q^{j+1})\rho\theta^{-1} + P(X = j)q^j(1 - \rho)(1 - \theta)^{-1},$$

for $j = 0, 1, \ldots,$ where $\theta = \sum_{i=0}^{\infty} P(X = i)(1 - q^i)$, with $\theta < \rho \leq 1$ and $0 < q < 1$. Introducing the parameters,

$$\lambda = \frac{\theta}{\rho}, \quad r = \frac{\log\,(1-\rho) - \log\,(1-\theta)}{\log\,q}, \quad 0 < \theta < \rho \leq 1, \quad 0 < q < 1,$$

the first-order recurrence relation

$$P(X = j+1)(1 - q^{j+1}) = P(X = j)(1 - q^{r+j})\lambda, \quad j = 0, 1, \ldots,$$

where $0 < \lambda < 1, 0 < q < 1$, and $r > 0$, is deduced. Consequently,

$$\begin{aligned} P(X = x) &= P(X = 0)\frac{(1-q^r)(1-q^{r+1})\cdots(1-q^{r+x-1})}{(1-q)(1-q^2)\cdots(1-q^x)}\lambda^x \\ &= P(X = 0)\begin{bmatrix} r+x-1 \\ x \end{bmatrix}_q \lambda^x, \quad x = 0, 1, \ldots, \end{aligned}$$

where $0 < \lambda < 1$, $0 < q < 1$, and $r > 0$. Thus, using the general q-binomial formula (1.22),

$$\prod_{i=1}^{\infty} \frac{1 - tq^{r+i-1}}{1 - tq^{i-1}} = \sum_{k=0}^{\infty} \begin{bmatrix} r+x-1 \\ x \end{bmatrix}_q t^k, \quad |t| < 1, \quad 0 < q < 1, \quad r > 0,$$

we get

$$P(X = 0) = \frac{1}{\prod_{i=1}^{\infty}(1 - \lambda q^{r+i-1})/(1 - \lambda q^{i-1})}$$

and so

$$P(X = x) = \begin{bmatrix} r+x-1 \\ x \end{bmatrix}_q \frac{\lambda^x}{\prod_{i=1}^{\infty}(1 - \lambda q^{r+i-1})/(1 - \lambda q^{i-1})},$$

for $x = 0, 1, \ldots,$ where $0 < \lambda < 1, 0 < q < 1$, and $r > 0$.

Note that the probability function of the general negative q-binomial distribution of the second kind for $r = n$, a positive integer, reduces to the probability function (3.2) of the negative q-binomial distribution of the second kind.

3.4 EULER STOCHASTIC PROCESS

Consider a stochastic model that is developing in time and successes occur at continuous time or space points of an interval $(0, t]$. An Euler process, which constitutes a q-analogue of a Poisson process, is introduced in the following definition, by considering the geometrically decreasing sequence of time differences

$$\delta t_i = (1 - q)q^{i-1}t, \quad i = 1, 2, \ldots, \quad 0 < q < 1,$$

with $\sum_{i=1}^{\infty} \delta t_i = t$, to partition the time interval $(0, t]$.

Definition 3.4. *Consider a stochastic model that is developing in time or space and let X_t, $t \geq 0$, be the number of successes (occurrences of event A) in the interval $(0, t]$. Assume that X_t, $t \geq 0$, is a stochastic process, with dependent and homogeneous increments, which starts at $t = 0$ from state 0, $P(X_0 = 0) = 1$, and, in the small time interval $(q^i t, q^{i-1} t]$, of length $\delta t_i = (1 - q)q^{i-1}t$, for $i = 1, 2, \ldots,$ satisfies the condition*

$$p_{j,k}(\delta t_i) = P(X_{q^{i-1}t} = k | X_{q^i t} = j) = \begin{cases} 1 - \lambda(1-q)q^{i-j-1}t, & k = j, \\ \lambda(1-q)q^{i-j-1}t, & k = j+1, \\ 0, & k > j+1, \end{cases} \quad (3.22)$$

for $j = 0, 1, \ldots, i-1$ and $i = 1, 2, \ldots,$ with $0 < \lambda t < 1/(1-q)$ and $0 < q < 1$. Then, X_t, $t \geq 0$, is called Euler process, with parameters λ and q.

It is worth noticing that, in contrast to a Poisson process, an Euler process does not have independent increments. Also, the condition of the occurrence of at most one success in a small time interval is expressed in terms of a series of small time intervals of varying (q-decreasing) lengths.

Theorem 3.5. *The probability function of the Euler process X_t, $t \geq 0$, with parameters λ and q, is given by*

$$p_x(t) = P(X_t = x) = E_q(-\lambda t)\frac{(\lambda t)^x}{[x]_q!}, \quad x = 0, 1, \ldots, \quad (3.23)$$

where $0 < \lambda t < 1/(1-q)$, $0 < q < 1$, and $E_q(u) = \prod_{i=1}^{\infty}(1 + u(1-q)q^{i-1})$ is a q-exponential function.

Proof. The probability function $p_x(q^{i-1}t)$ of the Euler process, by the total probability theorem,

$$p_x(q^{i-1}t) = p_x(q^i t + \delta t_i) = \sum_{k=0}^{x} p_{x-k}(q^i t)p_{x-k,x}(\delta t_i),$$

for $x = 0, 1, \ldots, i-1$, and condition (3.22), satisfies the system of equations

$$p_0(q^{i-1}t) = (1 - \lambda(1-q)q^{i-1}t)p_0(q^i t),$$
$$p_x(q^{i-1}t) = (1 - \lambda(1-q)q^{i-x-1}t)p_x(q^i t) + \lambda(1-q)q^{i-x}tp_{x-1}(q^i t),$$

for $x = 1, 2, \ldots, i-1$. Substituting $u = q^{i-1}t$, this system of equations may be rewritten as

$$p_0(u) = (1 - \lambda(1-q)u)p_0(qu),$$
$$p_x(u) = (1 - \lambda(1-q)q^{-x}u)p_x(qu) + \lambda(1-q)q^{-(x-1)}up_{x-1}(u),$$

for $x = 1, 2, \ldots$, or as

$$\frac{p_0(u) - p_0(qu)}{(1-q)u} = -\lambda p_0(qu),$$
$$\frac{p_x(u) - p_x(qu)}{(1-q)u} = -\lambda q^{-x} p_x(qu) + \lambda q^{-(x-1)} p_{x-1}(qu),$$

for $x = 1, 2, \ldots$. Introducing the q-derivative operator $\mathcal{D}_q$, with respect to u, we deduce the system of q-differential equations

$$\mathcal{D}_q p_0(u) = -\lambda p_0(qu),$$
$$\mathcal{D}_q p_x(u) = -\lambda q^{-x} p_x(qu) + \lambda q^{-(x-1)} p_{x-1}(qu), \quad x = 1, 2, \ldots.$$

Introducing the function $g(u)$ by

$$p_x(u) = g(u)\frac{(\lambda u)^x}{[x]_q!}, \quad x = 0, 1, \ldots, \tag{3.24}$$

and since

$$\mathcal{D}_q p_x(u) = \frac{(\lambda u)^x}{[x]_q!}\mathcal{D}_q g(u) + \lambda \frac{(\lambda u)^{x-1}}{[x-1]_q!} g(qu),$$

the system of q-differential equations reduces to the q-differential equation

$$\mathcal{D}_q g(u) = -\lambda g(qu),$$

with initial condition $g(0) = p_0(0) = 1$. Its solution is readily obtained as $g(u) = E_q(-\lambda u)$, and so, by (3.24), expression (3.23) is established, with u instead of t. □

Remark 3.5. *q-Poisson stochastic processes.* As has already been noted, the Euler and Heine stochastic processes constitute q-analogues of the Poisson stochastic process. Their probability functions may be expressed by the same functional formula, with different parametric spaces. Specifically, the probability function (2.21), of the Heine stochastic process,

$$p_x(t) = e_q(-\lambda t)\frac{q^{\binom{x}{2}}(\lambda t)^x}{[x]_q!}, \quad x = 0, 1, \ldots,$$

with $0 < \lambda t < \infty$ and $0 < q < 1$, on replacing q by the q^{-1}, with $1 < q < \infty$, and using the relations $[x]_{q^{-1}}! = q^{-\binom{x}{2}}[x]_q$ and $e_{q^{-1}}(-\lambda t) = E_q(-\lambda t)$, may be expressed as

$$p_x(t) = E_q(-\lambda t)\frac{(\lambda t)^x}{[x]_q!}, \quad x = 0, 1, \ldots,$$

with $0 < \lambda t < \infty$ and $1 < q < \infty$. Note that this is the same expression as that of the probability function, (3.23), of the Euler stochastic process, with a different parametric space. It should also be remarked the significant difference in the definitions of the two q-Poisson stochastic processes; the increments of a Heine process are independent, while those of an Euler process are dependent.

In a time-dependent stochastic model, in which successes occur according to an Euler process, the distribution of the waiting time until the occurrence of a fixed number of successes is connected to the distribution of the number of successes in a fixed time interval. In this respect, the following definition is introduced.

Definition 3.5. *Consider a stochastic model that is developing in time and successes occur according to an Euler process. Let W_n be the waiting time until the occurrence of the nth success. The distribution of W_n is called q-Erlang distribution of the second kind, with parameters n, λ, and q. In particular, the distribution of the waiting time until the occurrence of the first success, $W \equiv W_1$, is called q-exponential distribution of the second kind, with parameters λ and q.*

The distribution function, together with the q-density function and q-moments of the q-Erlang distribution of the second kind are derived in the following theorem.

Theorem 3.6. *The distribution function $F_n(w) = P(W_n \leq w)$, $-\infty < w < \infty$, of the q-Erlang distribution of the second kind, with parameters n, λ, and q, is given by*

$$F_n(w) = 1 - \sum_{x=0}^{n-1} E_q(-\lambda w)\frac{(\lambda w)^x}{[x]_q!}, \quad 0 < w < \infty, \tag{3.25}$$

and $F_n(w) = 0$, $-\infty < w < 0$, where n is a positive integer, $0 < \lambda < \infty$ and $0 < q < 1$. Its q-density function $f_n(w) = d_qF_n(w)/d_qw$ is given by

$$f_n(w) = \frac{\lambda^n}{[n-1]_q!} w^{n-1} E_q(-\lambda q w), \quad 0 < w < \infty. \tag{3.26}$$

Also, its jth q-moment is given by

$$\mu'_{j,q} = E(W_n^j) = \frac{[n+j-1]_{j,q}}{\lambda^j}, \quad j = 1, 2, \ldots. \tag{3.27}$$

Proof. The event $\{W_n > w\}$ in which the nth success occurs after time w is equivalent to the event $\{X_w < n\}$ in which the number of successes up to time w is less than n and so

$$P(W_n > w) = P(X_w < n) = \sum_{x=0}^{n-1} P(X_w = x).$$

Thus, the distribution function of the random variable W_n, on using the relation $F_n(w) = P(W_n \leq w) = 1 - P(W_n > w)$ and expression (3.23), is deduced as (3.25).

The q-density function of W_n, on taking the q-derivative of (3.25), by using the q-Leibnitz formula, is obtained in the form

$$f_n(w) = \lambda E_q(-\lambda q w) \sum_{x=0}^{n-1} \frac{(\lambda w)^x}{[x]_q!} - \lambda E_q(-\lambda q w) \sum_{x=1}^{n-1} \frac{(\lambda w)^{x-1}}{[x-1]_q!},$$

which reduces to (3.26). Note that, using the relation (see Exercise 1.32)

$$\int_0^\infty u^{n-1} E_q(-qu) d_q u = [n-1]_q!, \tag{3.28}$$

it follows that

$$\int_0^\infty f_n(w) d_q w = 1,$$

which conforms to the definition of a q-density function.

The jth q-moment of W_n,

$$\mu'_{j,q} = E(W_n^j) = \frac{\lambda^n}{[n-1]_q!} \int_0^\infty w^{n+j-1} E_q(-\lambda q w) d_q w,$$

using the transformation $u = \lambda w$ and expression (3.28), is obtained as

$$\mu'_{j,q} = \frac{\lambda^n}{[n-1]_q! \lambda^{n+j}} \int_0^\infty u^{n+j-1} E_q(-qu) d_q u = \frac{[n+j-1]_q!}{[n-1]_q! \lambda^j}.$$

Since $[n+j-1]_q! = [n-1]_q! [n+j-1]_{j,q}$, the last relation implies the required expression (3.27). □

The q-density function and q-moments of the q-exponential distribution of the second kind are deduced in the following corollary of Theorem 3.6.

Corollary 3.1. *The q-density function of the q-exponential distribution of the second kind, with parameters λ and q, is given by*

$$f(w) = \lambda E_q(-\lambda q w), \quad 0 < w < \infty,$$

where $0 < \lambda < \infty$ and $0 < q < 1$. Also, its jth q-moment is given by

$$\mu'_{j,q} = E(W^j) = \frac{[j]_q!}{\lambda^j}, \quad j = 1, 2, \ldots .$$

Remark 3.6. The distribution function of the q-Erlang distribution of the second kind, in addition to expression (3.25), may be obtained as a q-integral of its q-density function as

$$F_n(w) = \int_0^w \frac{\lambda^n}{[n-1]_q!} u^{n-1} E_q(-\lambda q u) d_q u.$$

These two expressions of $F_n(w)$ imply the relation

$$\int_0^w \frac{\lambda^n}{[n-1]_q!} u^{n-1} E_q(-\lambda q u) d_q u = 1 - \sum_{x=0}^{n-1} E_q(-\lambda q w) \frac{(\lambda w)^x}{[x]_q!}.$$

3.5 q-LOGARITHMIC DISTRIBUTION

Definition 3.6. *Let X be a discrete random variable with probability function*

$$f(x) = P(X = x) = [-l_q(1-\theta)]^{-1} \frac{\theta^x}{[x]_q}, \quad x = 1, 2, \ldots, \tag{3.29}$$

where $0 < \theta < 1$, $0 < q < 1$, *and* $-l_q(1-\theta) = \sum_{j=1}^{\infty} \theta^j / [j]_q$ *is the q-logarithmic function (1.25). The distribution of the random variable X is called q-logarithmic distribution, with parameters* θ *and q.*

Note that the limit of this distribution, for $q \to 1$, is the logarithmic distribution. Also, note that the function (3.29) is nonnegative,

$$f(x) > 0, \quad x = 1, 2, \ldots, \quad f(x) = 0, \quad x \neq 1, 2, \ldots,$$

and since $-l_q(1-\theta) = \sum_{j=1}^{\infty} \theta^j / [j]_q$, it sums to unity,

$$\sum_{x=1}^{\infty} f(x) = [-l_q(1-\theta)]^{-1} \sum_{x=1}^{\infty} \frac{\theta^x}{[x]_q} = [-l_q(1-\theta)]^{-1}[-l_q(1-\theta)] = 1,$$

as is required by the definition of a probability function.

The q-factorial moments and the usual factorial moments of the q-logarithmic distribution are derived in the following theorem.

Theorem 3.7. *The q-factorial moments of the q-logarithmic distribution are given by*

$$E([X]_{m,q}) = \frac{[m-1]_q! \theta^m}{-l_q(1-\theta) \prod_{i=1}^{m} (1-\theta q^{i-1})}, \quad m = 1, 2, \ldots . \tag{3.30}$$

Moreover, its factorial moments are given by

$$E[(X)_j] = \frac{j!(1-q)^{-j}}{-l_q(1-\theta)} \sum_{m=j}^{\infty} (-1)^{m-j} \frac{[\theta(1-q)]^m s_q(m,j)}{[m]_q \prod_{i=1}^{m} (1-\theta q^{i-1})}, \tag{3.31}$$

for $j = 1, 2, \ldots$, *where* $s_q(m,j)$ *is the q-Stirling number of the first kind. In particular, its mean and variance are given by*

$$E(X) = \frac{(1-q)^{-1}}{-l_q(1-\theta)} \sum_{m=1}^{\infty} \frac{[\theta(1-q)]^m [m-1]_q!}{[m]_q \prod_{i=1}^{m} (1-\theta q^{i-1})} \tag{3.32}$$

and

$$V(X) = \frac{2(1-q)^{-2}}{-l_q(1-\theta)} \sum_{m=2}^{\infty} \frac{[\theta(1-q)]^m [m-1]_q! \zeta_{m-1,q}}{[m]_q \prod_{i=1}^{m}(1-\theta q^{i-1})} + E(X) - [E(X)]^2, \quad (3.33)$$

where $\zeta_{m,q} = \sum_{j=1}^{m} 1/[j]_q$.

Proof. The mth q-factorial moment of the q-logarithmic distribution,

$$E([X]_{m,q}) = \frac{1}{-l_q(1-\theta)} \sum_{x=m}^{\infty} [x]_{m,q} \frac{\theta^x}{[x]_q},$$

on using successively the relations $[x]_{m,q} = [x]_q [x-1]_{m-1,q}$ and

$$[x-1]_{m,q} = [m-1]_q! \frac{[x-1]_q!}{[x-m]_q![m-1]_q!} = [m-1]_q! \begin{bmatrix} x-1 \\ x-m \end{bmatrix}_q,$$

and then substituting $y = x - m$, is expressed as

$$E([X]_{m,q}) = \frac{[m-1]_q!}{-l_q(1-\theta)} \sum_{x=m}^{\infty} \begin{bmatrix} x-1 \\ x-m \end{bmatrix}_q \theta^x = \frac{[m-1]_q!\, \theta^m}{-l_q(1-\theta)} \sum_{y=0}^{\infty} \begin{bmatrix} m+y-1 \\ y \end{bmatrix}_q \theta^y.$$

Thus, using the negative q-binomial formula (1.15), it is obtained in the form (3.30). Introducing the mth q-factorial moment (3.30) into (1.61), the required formula (3.31) is deduced. The mean and variance of X_n may be deduced from (3.31), using the expressions

$$s_q(m,1) = (-1)^{m-1}[m-1]_q!, \qquad s_q(m,2) = (-1)^{m-2}[m-1]_q! \zeta_{m-1,q},$$

with $\zeta_{m,q}(1) = \sum_{j=1}^{m} 1/[j]_q$. □

Consider the zero-truncated random variable $Z_n = W_n | W_n > 0$, where W_n obeys the negative q-binomial distribution of the second kind, with probability function (3.2). The probability function of Z_n,

$$P(Z_n = z) = P(W_n = z | W_n > 0), \quad z = 1, 2, \ldots,$$

is readily obtained as

$$P(Z_n = z) = \begin{bmatrix} n+z-1 \\ z \end{bmatrix}_q \theta^z \left(\prod_{i=1}^{n} (1-\theta q^{i-1})^{-1} - 1 \right)^{-1}, \quad (3.34)$$

for $z = 1, 2, \ldots$, with $0 < \theta < 1$ and $0 < q < 1$. This distribution, for $n \to 0$, can be approximated by the q-logarithmic distribution, according to the following theorem.

Theorem 3.8. *The limit of the probability function (3.34) of the zero-truncated negative q-binomial distribution of the second kind, for $n \to 0$, is the q-logarithmic distribution,*

$$\lim_{n\to 0} \begin{bmatrix} n+z-1 \\ z \end{bmatrix}_q \theta^z \left(\prod_{i=1}^{n} (1-\theta q^{i-1})^{-1} - 1 \right)^{-1} = [-l_q(1-\theta)]^{-1} \frac{\theta^z}{[z]_q}, \qquad (3.35)$$

for $z = 1, 2, \ldots,$ with $0 < \theta < 1$ and $0 < q < 1$.

Proof. The limit of the probability function (3.34), since

$$\lim_{n\to 0} \frac{1}{[n]_q} \begin{bmatrix} n+z-1 \\ z \end{bmatrix}_q = \frac{1}{[z]_q} \cdot \frac{\lim_{n\to 0} [n+1]_q [n+2]_q \cdots [n+z-1]_q}{[z-1]_q!} = \frac{1}{[z]_q}$$

and

$$\lim_{n\to 0} \frac{1}{[n]_q} \left(\prod_{i=1}^{n} (1-\theta q^{i-1})^{-1} - 1 \right) = \lim_{n\to 0} \frac{1}{[n]_q} \sum_{j=1}^{\infty} \begin{bmatrix} n+j-1 \\ j \end{bmatrix}_q \theta^j$$

$$= \sum_{j=1}^{\infty} \frac{\theta^j}{[j]_q} = -l_q(1-\theta),$$

is readily obtained as (3.35). □

Example 3.6. *A group size distribution.* A typical statistical model considers the group size distribution as the equilibrium distribution arising from a birth and death process X_t, $t \geq 0$. Adopting this approach, let us assume that the minimum size of a group is unity, so that the process starts with a group of size $x \geq 1$, and that the birth and death rates are given by

$$\lambda_j = [j]_q \lambda, \ \ j = 1, 2, \ldots, \ \ \mu_j = [j]_q \mu, \ \ j = 2, 3, \ldots, \ \ \mu_1 = 0.$$

Then, according to Remark 2.1, the probability function of the equilibrium (stationary) distribution,

$$\lim_{t\to\infty} P[X_t = x] = P(X = x), \ \ x = 1, 2, \ldots,$$

is given by

$$P(X = x) = P(X = 1) \prod_{j=2}^{x} \frac{\lambda_{j-1}}{\mu_j} = P(X = 1) \frac{\theta^{x-1}}{[x]_q}, \ \ x = 2, 3, \ldots,$$

where $\theta = \lambda/\mu$, provided $0 < \theta < 1$ and $0 < q < 1$. Since $\sum_{x=1}^{\infty} P(X = x) = 1$ and

$$-l_q(1-\theta) = \sum_{x=1}^{\infty} \frac{\theta^x}{[x]_q}, \ \ 0 < \theta < 1, \ \ 0 < q < 1,$$

it follows that $P(X = 1) = \theta[-l_q(1-\theta)]^{-1}$ and so

$$P(X = x) = [-l_q(1-\theta)]^{-1}\frac{\theta^x}{[x]_q},\quad x = 1, 2, \ldots,$$

for $0 < \theta < 1$ and $0 < q < 1$, which is the probability mass function of a q-logarithmic distribution.

3.6 q-STIRLING DISTRIBUTIONS OF THE SECOND KIND

Consider now a sequence of independent geometric sequences of trials and assume that the probability of success at the jth geometric sequence of trials is given by

$$p_j = \theta q^{j-1},\quad j = 1, 2, \ldots,\quad 0 < \theta \leq 1,\quad 0 < q < 1 \text{ or } 1 < q < \infty, \tag{3.36}$$

where for $0 < \theta \leq 1$ and $1 < q < \infty$, the number j of geometric sequences of trials is restricted by $\theta q^{j-1} < 1$, ensuring that $0 < p_j < 1$. This restriction imposes on j an upper bound, $j = 1, 2, \ldots, [-r]$, with $[-r]$ denoting the integral part of $-r$, where $r = \log\ \theta/\log\ q < 0$. The probabilities involved in this model are more conveniently written in terms of a new parameter r that replaces the parameter θ by $\theta = q^r$. Then,

$$p_j = q^{r+j-1},\quad j = 1, 2, \ldots, \tag{3.37}$$

for $0 \leq r < \infty$ and $0 < q < 1$ or $-\infty < r < 0$ and $1 < q < \infty$, with $j \leq [-r]$.

The probability function and factorial moments of the number of trials until the occurrence of a given number of successes is obtained in the following theorem. In the case $-\infty < r < 0$ and $1 < q < \infty$, the given number of successes is bounded by $[-r]$.

Theorem 3.9. *Consider a sequence of independent geometric sequences of trials, with probability of success at the jth geometric sequence of trials given by (3.37), and let T_k be the number of trials until the occurrence of the kth success. The probability function of T_k is given by*

$$P(T_k = n) = q^{\binom{k}{2}+rk}(1-q)^{n-k}S_q(n-1, k-1; r), \tag{3.38}$$

for $n = k, k+1, \ldots,$ with $0 \leq r < \infty$ and $0 < q < 1$ or $-\infty < r < 0$ and $1 < q < \infty$, with $r + k < 0$, where $S_q(n, k; r)$ is the noncentral q-Stirling number of the second kind. Its ascending factorial moments are given by

$$E[(T_k + j - 1)_j] = j!q^{-j(r+k-1)}\begin{bmatrix} k+j-1 \\ j \end{bmatrix}_q,\quad j = 1, 2, \ldots . \tag{3.39}$$

In particular, its mean and variance are given by

$$E(T_k) = q^{-(r+k-1)}[k]_q,\quad V(T_k) = q^{-2(r+k-1)}[k]_{q^2} - q^{-(r+k-1)}[k]_q. \tag{3.40}$$

Proof. Let A_{j,m_j} be the event that m_j failures precede the first success in the jth geometric sequence of trials, for $j = 1, 2, \ldots, k$. Then, using the independence of the events $A_{1,m_1}, A_{2,m_2}, \ldots, A_{k,m_k}$ and the (geometric) probabilities

$$P(A_{j,m_j}) = (1 - q^{r+j-1})^{m_j} q^{r+j-1}, \; j = 1, 2, \ldots, k,$$

we get

$$\begin{aligned} P(A_{1,m_1} A_{2,m_2} \cdots A_{k,m_k}) &= P(A_{1,m_1}) P(A_{2,m_2}) \cdots P(A_{k,m_k}) \\ &= q^{\binom{k}{2}+rk} (1-q)^{m_1+m_2+\cdots+m_k} \prod_{j=1}^{k} [r+j-1]_q^{m_j}, \end{aligned}$$

with $m_j \geq 0, \; j = 1, 2, \ldots, k$. Summing these probabilities over all $m_j \geq 0$, $j = 1, 2, \ldots, k$, with fixed $m_1 + m_2 + \cdots + m_k = n - k$, and using the expression

$$S_q(n-1, k-1; r) = \sum_{\substack{m_j \geq 0, \, j=1,2,\ldots,k \\ m_1+m_2+\cdots+m_k=n-k}} [r]_q^{m_1} [r+1]_q^{m_2} \cdots [r+k]_q^{m_k},$$

which is equivalent to (1.38), the probability that $n - k$ failures precede the kth success is obtained as (3.38).

The jth ascending factorial moment, since, by Exercise 1.20,

$$\sum_{n=k}^{\infty} (n+j-1)_j (1-q)^{n-k} S_q(n-1, k-1; r) = j! q^{-\binom{k}{2}-rk-j(r+k-1)} \begin{bmatrix} k+j-1 \\ j \end{bmatrix}_q,$$

is readily deduced as (3.39). Furthermore, substituting in (3.39) $j = 1$ and $j = 2$, we get

$$E(T_k) = q^{-(r+k-1)} [k]_q, \;\; E[T_k(T_k+1)] = 2q^{2(r+k-1)} \frac{[k]_q [k+1]_q}{[2]_q}$$

and

$$\begin{aligned} V(T_k) &= E[T_k(T_k+1)] - E(T_k) - [E(T_k)]^2 \\ &= 2q^{2(r+k-1)} \frac{[k]_q [k+1]_q}{[2]_q} - q^{-(r+k-1)} [k]_q - q^{-2(r+k-1)} [k]_q^2 \\ &= q^{2(r+k-1)} \left(2 \frac{[k+1]_q}{[2]_q} - [k]_q \right) - q^{-(r+k-1)} [k]_q \\ &= q^{-2(r+k-1)} [k]_{q^2} - q^{-(r+k-1)} [k]_q, \end{aligned}$$

which completes the proof of the theorem. □

The probability function, the q-factorial moments, and the factorial moments of the number of successes in a given number of trials are derived in the following theorem.

Theorem 3.10. *Consider a sequence of independent geometric sequences of trials, with probability of success at the jth geometric sequence of trials given by (3.37), and let X_n be the number of successes in n trials. The probability function of X_n is given by*

$$P(X_n = x) = q^{\binom{x}{2}+rx}(1-q)^{n-x}S_q(n,x;r),\quad x=0,1,\ldots,n, \tag{3.41}$$

with $0 \le r < \infty$ and $0 < q < 1$ or $-\infty < r < 0$, $1 < q < \infty$ and $r+n<0$, where $S_q(n,x;r)$ is the noncentral q-Stirling number of the second kind. Its q-factorial moments are given by

$$E([X_n]_{m,q^{-1}}) = [m]_{q^{-1}}!q^{\binom{m}{2}+rm}\binom{n}{m},\quad m=1,2,\ldots,n, \tag{3.42}$$

and $E([X_n]_{m,q^{-1}})=0$, for $m=n+1,n+2,\ldots$. Also, its factorial moments are given by

$$E[(X_n)_j] = j!\sum_{m=j}^{n}(-1)^{m-j}\binom{n}{m}q^{\binom{m}{2}+rm}(1-q^{-1})^{m-j}s_{q^{-1}}(m,j), \tag{3.43}$$

for $j=1,2,\ldots,n$ and $E[(X_n)_j]=0$, for $j=n+1,n+2,\ldots$, where $s_q(m,j)$ is the q-Stirling number of the first kind.

Proof. The probability function of the number X_n of successes in n trials is closely connected to the probability function of the number T_k of trials until the occurrence of the kth success. Precisely, $P(T_{x+1}=n+1)=P(X_n=x)p_{x+1}$, where $P(X_n=x)$ is the probability of the occurrence of x successes in n trials and $p_{x+1}=q^{r+x}$ is the conditional probability of success at a Bernoulli trial given that x successes occur in the previous trials. Thus, using (3.38), expression (3.41) is obtained. Introducing expression (1.40) of the noncentral generalized q-Stirling numbers of the second kind into (3.41), we get

$$P(X_n=x) = q^{\binom{x}{2}}\sum_{j=x}^{n}(-1)^{j-x}q^{rj}\binom{n}{j}\begin{bmatrix} j \\ x \end{bmatrix}_q,\quad x=0,1,\ldots,n.$$

The mth q-factorial moments $E([X_n]_{m,q^{-1}})$, $m=1,2,\ldots$, upon using the last expression of the probability function together with the relations

$$\begin{bmatrix} x \\ m \end{bmatrix}_{q^{-1}} = q^{-m(x-m)}\begin{bmatrix} x \\ m \end{bmatrix}_q,\quad \binom{x}{2} = \binom{x-m}{2} + \binom{m}{2} + m(x-m)$$

and interchanging the order of summation, is obtained as

$$\begin{aligned} E([X_n]_{m,q^{-1}}) &= [m]_{q^{-1}}!\sum_{j=m}^{n}(-1)^{j-m}q^{rj}\binom{n}{j}\sum_{x=m}^{j}(-1)^{x-m}q^{\binom{x}{2}}\begin{bmatrix} x \\ m \end{bmatrix}_{q^{-1}}\begin{bmatrix} j \\ x \end{bmatrix}_q \\ &= [m]_{q^{-1}}!\sum_{j=m}^{n}(-1)^{j-m}q^{\binom{m}{2}+rj}\binom{n}{j}\sum_{x=m}^{j}(-1)^{x-m}q^{\binom{x-m}{2}}\begin{bmatrix} x \\ m \end{bmatrix}_q\begin{bmatrix} j \\ x \end{bmatrix}_q. \end{aligned}$$

Since, by (1.16),

$$\sum_{x=m}^{j}(-1)^{x-m}q^{\binom{x-m}{2}}\begin{bmatrix} x \\ m \end{bmatrix}_q \begin{bmatrix} j \\ x \end{bmatrix}_q = \delta_{j,m},$$

the mth q-factorial moment (3.42) is readily deduced. Thus, applying (1.61) with q^{-1} instead of q, expression (3.43) is obtained. □

The q-Stirling distribution of the second kind, with probability function (3.41), plays a central role in many algorithmic analyses. A probabilistic (approximate) counting algorithm is presented in the following example.

Example 3.7. *A probabilistic algorithm for counting events in a small counter.* An n-bit register can ordinarily be used to count up to $2^n - 1$ events. If the requirement of accuracy is dropped, the following probabilistic (approximate) counting algorithm was proposed. If C_n is the number of events counted after n trials (occurrences of events), the approximate counting starts with the initial value $C_0 = 1$. At each trial, the occurrence of an event is counted with probability

$$P(C_{i+1} = j+1 | C_i = j) = q^j, \ \ j = 1, 2, \ldots, i, \ \ i = 1, 2, \ldots, n-1,$$

where $q = 1/a$, with a the base in the increment procedure of the algorithm. Considering the counting of an event at any trial as success, $X_n = C_n - 1$ is the number of successes in n independent Bernoulli trials. Clearly, the conditional probability of success at the ith trial, given that $j-1$ successes occur at the $i-1$ previous trials, is given by

$$P(X_i = j | X_{i-1} = j-1) = q^j, \ \ j = 1, 2, \ldots, i, \ \ i = 1, 2, \ldots, n,$$

which is of the form (3.37), with $r = 1$ and $0 < q < 1$. Therefore, using (3.41), with $r = 1$ and $0 < q < 1$, since $\binom{x+1}{2} = \binom{x}{2} + x$ and, by (1.27),

$$S_q(n, x; 1) = S_q(n+1, x+1; 0) = S_q(n+1, x+1),$$

the probability function of X_n is obtained as

$$P(X_n = x) = q^{\binom{x+1}{2}}(1-q)^{n-x}S_q(n+1, x+1), \ \ x = 0, 1, \ldots, n,$$

with $0 < q < 1$. Clearly, the probability function $P(C_n = k) = P(X_n = k-1)$, of $C_n = X_n + 1$, is readily deduced as

$$P(C_n = k) = q^{\binom{k}{2}}(1-q)^{n-k+1}S_q(n+1, k), \ \ k = 1, 2, \ldots, n+1,$$

with $0 < q < 1$. Also, from (3.42) with $r = 1$ and $m = 1$, it follows that

$$E(q^{-1}[C_n - 1]_{q^{-1}}) = q^{-1}E([X_n]_{q^{-1}}) = n$$

and so

$$\hat{n} = q^{-1}[C_n - 1]_{q^{-1}} = a[C_n - 1]_a$$

is an unbiased estimator of n, $E(\hat{n}) = n$, which was the objective for choosing the probabilities of counting or not counting an event.

Example 3.8. *The width of a chain decomposition of a random acyclic digraph.* A graph G is a pair (V, E), with $V = \{v_1, v_2, \ldots, v_n\}$ the set of vertices (nodes) and $E \subseteq V^2 = \{(v_k, v_r) : v_k \in V, v_r \in V\}$ the set of edges; a graph with $E = \{(v_k, v_r) : v_k < v_r, v_k \in V, v_r \in V\}$ or $E = \{(v_k, v_r) : v_k > v_r, v_k, v_r \in V\}$ is called digraph (or directed graph). A path in a graph $G = (V, E)$ is a subset $U = \{u_1, u_2, \ldots, u_r\}$ of the set of vertices V, such that $(u_k, u_{k+1}) \in E$ is an edge of the graph, for $k = 1, 2, \ldots, r - 1$; a path U is a cycle if $u_r = u_1$. A graph without cycles is called acyclic. A partition $\{U_1, U_2, \ldots, U_j\}$ of the set of vertices V is called a *chain decomposition of width (size) j* of the graph $G = (V, E)$ if and only if $U_i = \{v_{1,i}, v_{2,i}, \ldots, v_{r_i,i}\}$ is a path in G, for $i = 1, 2, \ldots, j$.

Let us denote by $G_{n,q}$ a random acyclic digraph of n vertices, in which any edge occurs independently with probability $p = 1 - q$. Consider the sequential construction of $G_{i+1,q}$ from $G_{i,q}$ through the addition of the vertex v_{i+1} and assume that the width of a chain decomposition of the graph $G_{i,q}$ equals j. This addition will create a new chain (path) $U_{j+1} = \{v_{i+1}\}$ containing only the new vertex if no one of the j vertices $\{v_{r_1,1}, v_{r_2,2}, \ldots, v_{r_j,j}\}$ is connected to v_{i+1}. Thus, if S_n denotes the size (width) of the chain decomposition of $G_{n,q}$, then the conditional probability that the addition of vertex v_{i+1} creates a new chain (path) is given by

$$P(S_{i+1} = j + 1 | S_i = j) = q^j, \ \ j = 1, 2, \ldots, i, \ \ i = 1, 2, \ldots, 0 < q < 1.$$

Furthermore, the sequential additions of vertices may be considered as a sequence of independent Bernoulli trials with success the creation of a new chain. Then, $X_{n-1} = S_n - 1$ is the number of success in $n - 1$ independent Bernoulli trials with the conditional probability of a success at the ith trial given that $j - 1$ successes occur in the previous $i - 1$ trials is given by

$$P(X_i = j | X_{i-1} = j - 1) = q^j, \ \ j = 1, 2, \ldots, i, \ \ i = 1, 2, \ldots, 0 < q < 1,$$

which is of the form (3.37), with $r = 1$ and $0 < q < 1$. Therefore, applying Theorem 3.10, with $r = 1$ and $0 < q < 1$, since $\binom{k}{2} = \binom{k-1}{2} + (k - 1)$ and, by (1.27),

$$S_q(n - 1, k - 1; 1) = S_q(n, k; 0) = S_q(n, k),$$

the probability function of the width $S_n = X_{n-1} + 1$ of a chain decomposition of a random acyclic digraph $G_{n,q}$ is deduced as

$$P(S_n = k) = P(X_{n-1} = k - 1) = q^{\binom{k}{2}}(1 - q)^{n-k} S_q(n, k), \ \ k = 1, 2, \ldots, n,$$

with $0 < q < 1$, where $S_q(n, k)$ is the q-Stirling number of the second kind.

3.7 REFERENCE NOTES

The stochastic model of a sequence of independent Bernoulli trials with the probability of success at any trial varying with the number of previous successes was examined by Woodbury (1949). In this paper, the probability function of the number of successes in a given number of trials was obtained. Also the probability function of the number of trials until the occurrence of a given number of successes was derived by Sen and Balakrishnan (1999). Exercises 3.1 and 3.2 are based on these papers.

The stochastic model of a sequence of independent Bernoulli trials with the probability of success at any trial varying geometrically with the number of previous successes was studied, as a reliability growth model, by Dubman and Sherman (1969). The positive and negative q-binomial distributions of the second kind were defined in this model and some of their properties were examined in Charalambides (2010a). It should be noted that the probability function of the q-binomial distribution of the second kind was also obtained by Il'inskii (2004) using a different and more laborious method. Furthermore, Il'inskii and Ostrovska (2002) and Ostrovska (2003), in their probabilistic-based study of the convergence of the q-Bernstein polynomials, used the probability functions of the q-binomial distribution of the second kind, the absorption distribution and the Euler distribution. Goodman et al. (1999) merely derived the q-power moments of the q-binomial distribution of the second kind, using the expression of the q-Bernstein polynomial in terms of the q-difference operator. It is worth mentioning that Phillips (1997) introduced the q-Bernstein polynomials by using the probability function of the q-binomial distribution of the second kind, which forms a normalized totally positive basis, as q-Bernstein basis. This paper initiated an intensive research on these polynomials and their use in approximation theory. Jing and Fan (1993) and Jing (1994), in order to construct a q-binomial state for the q-boson, examined the q-binomial distribution of the second kind as a q-deformed binomial distribution. The absorption and the inverse absorption distributions were studied by Blomqvist (1952), Borenius (1953), Dunkl (1981) and Kemp (1998) under different but equivalent scenarios. Examples 3.1 and 3.2 and Exercise 3.12 on proofreading a manuscript, capture of endangered animals and crossing a minefield are extracted from these papers. Newby (1999) discussed a shift operator technique for developing recurrence relations for the factorial moments of these distributions. An extension of the absorption stochastic model, which is studied by Rawlings (1997), is presented in Example 3.3. Zacks and Goldfard (1966), Barakat (1985) and Charalambides (2012b) discussed another extension of the absorption model by considering a random number of absorption points. Exercises 3.13–3.15 are based on these papers. Kemp (2002b) developed further q-analogues of the binomial distribution.

The Euler distribution was derived (together with the Heine distribution) by Benkherouf and Bather (1988) as feasible prior in a simple Bayesian model for oil exploration; this model is discussed in Example 3.4. The expression of the Euler distribution as an infinite convolution of geometric distributions and the modality and the failure rate of the q-Poisson distribution (Heine and Euler distributions) presented in Exercises 3.15 and 3.17 were discussed by Kemp (1992a). Also, characterizations of various discrete q-distributions were discussed by Kemp (2001, 2003); Exercises

3.19 and 3.20 are based on this paper. An interesting study of the q-binomial and the negative q-binomial distributions of the second kind, together with the Euler distribution and other discrete q-distributions was provided by Kupershmidt (2000); Exercises 3.8 and 3.9 are extracted from this paper. The characterization of the Euler distributions presented in Exercise 3.21 was shown in Charalambides and Papadatos (2005). The negative q-binomial distribution of the second kind and its limit to the Euler distribution, which is given in Exercise 3.23, was derived by Rawlings (1998), under a stochastic model of geometric sequences of trials with discrete q-uniform success probability. A generalized Euler distribution, which is presented in Example 3.5 as a general negative q-binomial distribution of the second kind, was examined by Benkherouf and Alzaid (1993).

Kemp (1992b) obtained the Heine and Euler distributions as steady-state distributions of Markov chains; the q-Foster process, presented in Exercise 3.24, is discussed in this paper. Furthermore, Kemp (2002a) examined the existence conditions and properties for the generalized Euler family of distributions. It is worth noticing that this family, in addition to the generalized Euler distribution of Benkherouf and Alzaid (1993), includes a variety of q-distributions with probability functions of the same mathematical form but usually with different parameter constraints. A. Kemp (1997) expressed the probability functions of the differences of (i) two Heine random variables and (ii) two Euler random variables in terms of two modified q-Bessel functions.

The q-logarithmic distribution was introduced and studied by C.D. Kemp (1997) as the group size distribution discussed in Example 3.6. Its q-factorial and usual factorial moments together with its derivation as an approximation of a zero-truncated negative q-binomial distribution were derived in Charalambides (2010b). An interesting generalization of the q-logarithmic distribution, obtained as a cluster distribution, was discussed by Kemp and Kemp (2009).

The q-Stirling distribution of the second kind emerged as the distribution of the number of events counted in a small counter in accordance to a probabilistic (stochastic) algorithm. A detailed analysis of this approximate counting, briefly presented in Example 3.7, was given by Flajolet (1985). The distribution of the width of a chain decomposition of a random acyclic digraph, discussed in Example 3.8, was studied by Simon (1988) and Crippa et al. (1997). The distribution of the size of the transitive closure in a random acyclic digraph, given as Exercise 3.11, was examined by Simon et al. (1993).

3.8 EXERCISES

3.1 Consider a sequence of independent geometric sequences of trials, with probability of success at the jth geometric sequence of trials,

$$Q_j(\{s\}) = p_j, \quad j = 1, 2, \ldots,$$

varying with the number of geometric sequences of trials, or equivalently, varying with the number of successes in the sequence of Bernoulli trials.

(a) Let T_k be the number of trials until the occurrence of the kth success. Show that its probability function is given by

$$P(T_k = n) = \left(\prod_{j=1}^{k} p_j\right) S(n-1, k-1; \boldsymbol{q}), \quad n = k, k+1, \ldots,$$

where $q_j = 1 - p_j, j = 1, 2, \ldots$, and $S(n-1, k-1; \boldsymbol{q})$ is the generalized Stirling number of the second kind.

(b) Also, derive its ith ascending factorial moment as

$$E[(T_k + i - 1)_i] = i!S(k + i - 1, k - 1; \boldsymbol{a}), \quad i = 1, 2, \ldots,$$

where $a_j = 1/p_j, j = 1, 2, \ldots$. In particular, deduce its mean and variance as

$$E(T_k) = \sum_{j=1}^{k} a_j = \sum_{j=1}^{k} \frac{1}{p_j}, \quad V(T_k) = \sum_{j=1}^{k} a_j(a_j - 1) = \sum_{j=1}^{k} \frac{1 - p_j}{p_j^2}.$$

(c) Let X_n be the number of successes in n trials. Show that its probability function is given by

$$P(X_n = x) = \left(\prod_{j=1}^{x} p_j\right) S(n, x; \boldsymbol{q}), \quad x = 0, 1, \ldots, n,$$

where $q_j = 1 - p_j$, $j = 1, 2, \ldots, n$ and $S(n, x; \boldsymbol{q})$ is the generalized Stirling number of the second kind.

3.2 Consider a sequence of independent Bernoulli trials and assume that the conditional probability of success at a trial, given that $j - 1$ successes occur in the previous trials, is given by

$$Q_j(\{s\}) = p_j, \quad j = 1, 2, \ldots .$$

Let X_n be the number of successes in n trials. Show that the probability function $P(X_n = x)$, $x = 0, 1, \ldots, n$, satisfies the recurrence relation

$$P(X_n = x) = (1 - q_{x-1})P(X_{n-1} = x - 1) + q_x P(X_{n-1} = x),$$

for $x = 1, 2, \ldots, n$, where $q_j = 1 - p_j$, $j = 1, 2, \ldots$, with $P(X_0 = 0) = 1$ and $P(X_n = x) = 0$, for $x > n$. Solving it, derive the expression

$$P(X_n = x) = \left(\prod_{j=1}^{x} p_j\right) S(n, x; \boldsymbol{q}), \quad x = 0, 1, \ldots, n,$$

where

$$S(n, x; \boldsymbol{q}) = \sum_{j=0}^{x} \frac{q_{j+1}^n}{\prod_{i=1}^{j}(q_{j+1} - q_i) \prod_{i=1}^{x-j}(q_{j+1} - q_{j+i+1})},$$

is the generalized Stirling number of the second kind.

3.3 The jth factorial moment of the negative q-binomial distribution of the second kind is given by

$$E[(W_n)_j] = j! \sum_{m=j}^{\infty} (-1)^{m-j} \begin{bmatrix} n+m-1 \\ m \end{bmatrix}_q \frac{\theta^m (1-q)^{m-j} s_q(m,j)}{\prod_{i=1}^{m}(1-\theta q^{n+i-1})},$$

for $j = 1, 2, \ldots,$ where $s_q(m,j)$ is the q-Stirling number of the first kind. Setting $j = 1$ and using the expression $s_q(m, 1) = (-1)^{m-1}[m-1]_q!$, deduce for the mean the expression

$$E(W_n) = \sum_{m=1}^{\infty} \begin{bmatrix} n+m-1 \\ m \end{bmatrix}_q \frac{\theta^m (1-q)^{m-1} [m-1]_q!}{\prod_{i=1}^{m}(1-\theta q^{n+i-1})}.$$

Using it, derive the recurrence relation

$$E(W_n) - E(W_{n-1}) = \frac{\theta q^{n-1}}{1-\theta q^{n-1}}, \quad n = 2, 3, \ldots,$$

with initial condition

$$E(W_1) = \sum_{w=1}^{\infty} w\theta^w (1-\theta) = \frac{\theta}{1-\theta},$$

and conclude that

$$E(W_n) = \sum_{j=1}^{n} \frac{\theta q^{j-1}}{1-\theta q^{j-1}}.$$

3.4 (*Continuation*). The second factorial moment, on setting $j = 2$ and using the expression $s_q(m, 2) = (-1)^{m-2}[m-1]_q!\zeta_{m-1,q}$, with $\zeta_{m,q} = \sum_{j=1}^{m} 1/[j]_q$, is deduced as

$$E[(W_n)_2] = 2 \sum_{m=2}^{\infty} \begin{bmatrix} n+m-1 \\ m \end{bmatrix}_q \frac{\theta^m (1-q)^{m-2} [m-1]_q! \zeta_{m-1,q}}{\prod_{i=1}^{m}(1-\theta q^{n+i-1})}.$$

Using it, derive the recurrence relation

$$E[(W_n)_2] - E[(W_{n-1})_2] = \frac{2\theta q^{n-1}}{1-\theta q^{n-1}} \cdot E(W_n), \quad n = 2, 3, \ldots,$$

with initial condition

$$E[(W_1)_2] = \sum_{w=2}^{\infty} (w)_2 \theta^w (1-\theta) = \frac{2\theta}{(1+\theta)^2}.$$

Applying it repeatedly, derive the expression

$$E[(W_n)_2] = 2 \sum_{j=1}^{n} \frac{\theta q^{j-1}}{1-\theta q^{j-1}} \sum_{i=1}^{j} \frac{\theta q^{i-1}}{1-\theta q^{i-1}}$$

and conclude that

$$V(W_n) = \sum_{i=1}^{n} \frac{\theta q^{i-1}}{(1-\theta q^{i-1})^2}.$$

3.5 Let W_n be a nonnegative integer valued random variable with q-binomial moments

$$E\left(\begin{bmatrix} W_n \\ m \end{bmatrix}_q\right) = \begin{bmatrix} n+m-1 \\ m \end{bmatrix}_q \frac{\theta^m}{\prod_{j=1}^{m}(1-\theta q^{n+j-1})}, \quad m = 0, 1, \ldots,$$

for $0 < \theta < 1$ and $0 < q < 1$ or $1 < q < \infty$ with $\theta q^{n-1} < 1$. Find the probability function of W_n,

$$f_n(w) = P(W_n = w), \quad w = 0, 1, \ldots .$$

3.6 *q-Factorial moments via probability generating function.*

(a) Derive the probability generating function $P_{W_n}(t) = E(t^{W_n})$ of the negative q-binomial distribution of the second kind and

(b) using it obtain the mth q-factorial moment $E([W_n]_{m,q})$, $m = 1, 2, \ldots$.

3.7 *Moments of the inverse absorption distribution.*

(a) Derive the q-factorial moments of the inverse absorption distribution as

$$E([W_n]_{m,q}) = \frac{[n+m-1]_{m,q} q^{(r-n+1)m}}{(1-q)^m [r+m]_{m,q}}, \quad m = 1, 2, \ldots,$$

for $0 < r < \infty$ and $0 < q < 1$, with $n \leq [r]$.

(b) Deduce the usual factorial moments as

$$E[(W_n)_j] = \frac{j!}{(1-q)^j} \sum_{m=j}^{\infty} (-1)^{m-j} \begin{bmatrix} n+m-1 \\ m \end{bmatrix}_q \frac{q^{(r-n+1)m} s_q(m,j)}{[r+m]_{m,q}},$$

for $j = 1, 2, \ldots$, where $s_q(m,j)$ is the q-Stirling number of the first kind.

(c) Show, in particular, that

$$E(W_n) = \sum_{j=1}^{n} \frac{q^{r-j+1}}{1-q^{r-j+1}}, \quad V(W_n) = \sum_{j=1}^{n} \frac{q^{r-j+1}}{(1-q^{r-j+1})^2}.$$

3.8 *q-Binomial distribution of the second kind as sum of two-point distributions.* Consider a sequence of two-point (zero and nonzero) random variables Z_i, $i = 1, 2, \ldots$. Suppose that the random variables $(Z_1, Z_2, \ldots, Z_n)$ assume the values $(z_1, z_2, \ldots, z_n)$ and let X_n be the number of nonzeroes among these values. Moreover, suppose that the random variable Z_i assumes the values 0 and $q^{X_{i-1}}$, for $0 < q < 1$, with conditional probabilities

$$P(Z_i = 0 | X_{i-1} = j-1) = \theta q^{j-1}, \quad P(Z_i = q^{j-1} | X_{i-1} = j-1) = 1 - \theta q^{j-1},$$

for $j = 1, 2, \ldots, i$, $i = 1, 2, \ldots$, and $0 < \theta < 1$. Show that the sum $Y_n = \sum_{i=1}^{n} Z_i$ follows a q-binomial distribution of the second kind, with probability function

$$P(Y_n = [x]_q) = \begin{bmatrix} n \\ x \end{bmatrix}_q \theta^{n-x} \prod_{j=1}^{x} (1 - \theta q^{j-1}), \quad x = 0, 1, \ldots, n.$$

3.9 Let X_n be the number of failures in n independent Bernoulli trials, with probability of success at the jth geometric sequence of trials given by

$$p_j = 1 - \theta q^{j-1}, \quad j = 1, 2, \ldots, 0 < \theta < 1, \quad 0 < q < 1 \text{ or } 1 < q < \infty,$$

with $\theta q^{n-1} < 1$. Also, let $Y_n = n - X_n$.

(a) Show that the probability of the occurrence of at most r failures,

$$P(X_n \leq r) = \sum_{x=0}^{r} \begin{bmatrix} n \\ x \end{bmatrix}_q \theta^x \prod_{j=1}^{n-x} (1 - \theta q^{j-1}), \quad r = 0, 1, \ldots,$$

may be expressed by a q-integral as

$$P(X_n \leq r) = 1 - \frac{[n]_q!}{[r]_q![n-r-1]_q!} \int_0^\theta t^r \prod_{j=1}^{n-r-1} (1 - tq^j) d_q t.$$

(b) Also, show that the probability of the occurrence of at most r successes,

$$P(Y_n \leq r) = \sum_{y=0}^{r} \begin{bmatrix} n \\ y \end{bmatrix}_q \theta^{n-y} \prod_{j=1}^{y} (1 - \theta q^{j-1}), \quad r = 0, 1, \ldots,$$

may be expressed by a q-integral as

$$P(Y_n \leq r) = \frac{[n]_q!}{[r]_q![n-r-1]_q!} \int_0^\theta t^{n-r-1} \prod_{j=1}^{r} (1 - tq^j) d_q t.$$

3.10 *Moments of the absorption distribution.*

(a) Derive the q-factorial moments of the absorption distribution as

$$E([Y_n]_{m,q}) = [n]_{m,q} [r]_{m,q} (1 - q)^m, \quad m = 1, 2, \ldots, n,$$

and $E([Y_n]_{m,q}) = 0$, $m = n + 1, n + 2, \ldots$, for $0 < r < \infty$ and $0 < q < 1$, with $n \leq [r]$.

(b) Deduce the usual factorial moments as

$$E[(Y_n)_j] = j! \sum_{m=j}^{n} (-1)^{m-j} \begin{bmatrix} n \\ m \end{bmatrix}_q [r]_{m,q} (1 - q)^{2m-j} s_q(m, j),$$

for $j = 1, 2, \ldots, n$, where $s_q(m, j)$ is the q-Stirling number of the first kind.

3.11 *The size of the transitive closure in a random acyclic digraph.* Consider a random acyclic digraph $G_{n,q} = (V, E)$ of n vertices, $V = \{\upsilon_1, \upsilon_2, \ldots, \upsilon_n\}$, in which any edge, $(\upsilon_k, \upsilon_r) \in E$, occurs independently, with probability $p = 1 - q$, and let $\upsilon \in V$ be a fixed vertex. The subset of vertices

$$U = \{u \in V\text{: there exists a path from } u \text{ to } \upsilon\} \subseteq V$$

is called transitive closure of vertex $\upsilon \in V$. Let S_n be the size of the transitive closure U_1 of the first vertex $\upsilon_1 \in V$.

(a) Find the conditional probability

$$P(S_{i+1} = j + 1 | S_i = j), \ \ j = 1, 2, \ldots, i, \ \ i = 1, 2, \ldots, \ \ 0 < q < 1.$$

(b) Derive the probability function of the random variable S_n.

3.12 *Crossing a minefield.* Consider a queue of people attempting to cross one after the other a minefield containing m mines, which are not connected. If a person steps on a mine, he/she is killed from the explosion of the mine.

(a) Determine the conditional probability that a person attempting to cross the minefield is killed, given that $j - 1$ persons are killed, for $j = 1, 2, \ldots$.

(b) Deduce the probability function of the number X_n of people killed from a queue of n persons attempting to cross the minefield.

3.13 *Crossing a field with a random number of absorption points.* Consider a queue of particles that are required to cross a field containing a random number of absorption points (traps) acting independently. If a particle clashes (contacts) with any of the absorption points, it is absorbed (trapped) with probability $p = 1 - q$. An absorption point (trap) is ruined when it absorbs (traps) a particle. Let X_n be the number of absorbed particles from a queue of n particles and Y be the number of absorption points.

(a) Show that the conditional distribution of X_n, given that $Y = y$, is an absorption distribution, with probability function

$$P(X_n = x \,|\, Y = y) = \begin{bmatrix} n \\ x \end{bmatrix}_q q^{(n-x)(y-x)}(1 - q)^x [y]_{x,q}, \ \ x = 0, 1, \ldots, n.$$

(b) Assume that the number Y of absorption points follows a Heine distribution, with probability function

$$P(Y = y) = e_q(-\lambda)\frac{q^{\binom{y}{2}}\lambda^y}{[y]_q!}, \ \ y = 0, 1, \ldots, 0 < \lambda < \infty, \ \ 0 < q < 1,$$

where $e_q(t) = \prod_{i=1}^{\infty} (1 - (1 - q)q^{i-1}t)^{-1}$ is a q-exponential function. Show that the distribution of the number X_n of absorbed particles from a queue of n particles is a q-binomial distribution of the first kind, with probability function

$$P(X_n = x) = \begin{bmatrix} n \\ x \end{bmatrix}_q \frac{\theta^x q^{\binom{x}{2}}}{\prod_{i=1}^{n}(1+\theta q^{i-1})}, \quad x = 0, 1, \ldots, n,$$

where $\theta = \lambda(1-q)$.

(c) Suppose that the number Y of absorption points follows an Euler distribution, with probability function

$$P(Y = y) = E_q(-\lambda)\frac{\lambda^y}{[y]_q!}, \quad y = 0, \ 1, \ldots, 0 < \lambda < 1/(1-q), \ 0 < q < 1,$$

where $E_q(t) = \prod_{i=1}^{\infty}(1 + (1-q)q^{i-1}t)$ is a q-exponential function. Show that the distribution of the number X_n of absorbed particles from a queue of n particles is a q-binomial distribution of the second kind, with probability function

$$P(X_n = x) = \begin{bmatrix} n \\ x \end{bmatrix}_q \theta^x \prod_{j=1}^{n-x}(1-\theta q^{j-1}), \quad x = 0, 1, \ldots, n,$$

where $\theta = \lambda(1-q)$.

3.14 (*Continuation*). Let W_n be the number of nonabsorbed (surviving the crossing of the field) particles until the absorption of n particles.

(a) Show that the conditional distribution of W_n, given that $Y = y$, is an inverse absorption distribution, with probability function

$$P(W_n = w \mid Y = y) = \begin{bmatrix} n+w-1 \\ w \end{bmatrix}_q q^{(y-n+1)w}(1-q)^n[y]_{n,q},$$

for $w = 0, 1, \ldots$.

(b) Assume that the number Y of absorption points follows a Heine distribution, with probability function

$$P(Y = y) = e_q(-\lambda)\frac{q^{\binom{y}{2}}\lambda^y}{[y]_q!}, \quad y = 0, 1, \ldots, \quad 0 < \lambda < \infty, \quad 0 < q < 1,$$

where $e_q(t) = \prod_{i=1}^{\infty}(1-(1-q)q^{i-1}t)^{-1}$ is a q-exponential function. Show that the distribution of the number W_n of nonabsorbed particles until the absorption of n particles is a negative q-binomial distribution of the first kind, with probability function

$$P(W_n = w) = \begin{bmatrix} n+w-1 \\ w \end{bmatrix}_q \frac{\theta^n q^{\binom{n}{2}+w}}{\prod_{i=1}^{n+w}(1+\theta q^{i-1})}, \quad w = 0, 1, \ldots,$$

where $\theta = \lambda(1-q)$.

3.15 (*Continuation*).

(a) Assume that the number X_n of absorbed particles obeys a distribution with probability function $P(X_n = x)$, $x = 0, 1, \ldots, n$, which tends to a probability function $P(X = x)$, $x = 0, 1, \ldots$, as $n \to \infty$. Show that the number Y of absorption points obeys a distribution with probability function $P(Y = x) = P(X = x)$, $x = 0, 1, \ldots$.

(b) In particular, conclude that if X_n obeys a q-binomial distribution of the first or the second kind, then Y obeys a q-Poisson (Heine or Euler) distribution.

3.16 Let X be a nonnegative integer valued random variable obeying an Euler distribution, with parameters λ and q.

(a) Derive the probability generating function $P_X(t) = E(t^X)$ and conclude that $X = \sum_{j=1}^{\infty} U_j$, with $U_j, j = 1, 2 \ldots$, independent geometric random variables.

(b) Show that the mean and variance of X may be expressed as

$$E(X) = \sum_{j=1}^{\infty} \frac{\lambda(1-q)q^{j-1}}{1-\lambda(1-q)q^{j-1}}, \quad V(X) = \sum_{j=1}^{\infty} \frac{\lambda(1-q)q^{j-1}}{(1-\lambda(1-q)q^{j-1})^2}.$$

3.17 *Modality and failure rate of the q-Poisson distributions*. Consider the q-Poisson distributions, with probability function

$$f(x) = E_q(-\lambda)\frac{\lambda^x}{[x]_q!}, \quad x = 0, 1, \ldots,$$

where $0 < \lambda < 1/(1-q)$ and $0 < q < 1$ (Euler distribution) or $0 < \lambda < \infty$ and $1 < q < \infty$ (Heine distribution), with

$$E_q(t) = \prod_{i=1}^{\infty}(1 + t(1-q)q^{i-1}), \quad 0 < q < 1,$$

and

$$E_q(t) = e_{q^{-1}}(t) = \prod_{i=1}^{\infty}(1 - t(1-q^{-1})q^{-(i-1)})^{-1}, \quad 1 < q < \infty,$$

q-exponential functions.

(a) Show that the probability function $f(x)$, $x = 0, 1, \ldots$, is unimodal.

(b) Also, show that the failure rate $r(x) = f(x)/R(x)$, $x = 0, 1, \ldots$, where $R(x) = \sum_{k=x}^{\infty} f(k)$, $x = 0, 1, \ldots$, is monotonically increasing.

3.18 *Elementary derivation of the distribution of the Euler process*. Consider a stochastic model in which successes or failures (events A or A') may occur at continuous time (or space) points. Furthermore, consider a time interval $(0, t]$ and its partition in n subintervals

$$\left(\frac{[i-1]_q t}{[n]_q}, \frac{[i]_q t}{[n]_q}\right], \quad i = 1, 2, \ldots, n, \quad 0 < q < 1,$$

with lengths $\delta_{n,i}(t) = tq^{i-1}/[n]_q$, $i = 1, 2, \ldots, n$. Assume that in each subinterval either a success or a failure may occur. Also, suppose that the conditional probability of success at any subinterval, given that $j-1$ successes occur in the previous subintervals, is given by $p_{n,j}(t) = 1 - \lambda t q^{j-1}/[n]_q$, $j = 1, 2, \ldots, n$, with $0 < \lambda t < [n]_q$. Let $X_{t,n}$ be the number of failures that occur in the n subintervals of $(0, t]$. Derive the probability function $P(X_{t,n} = x)$, $x = 0, 1, \ldots, n$, and show that

$$P(X_t = x) = \lim_{n\to\infty} P(X_{t,n} = x) = E_q(-\lambda t)\frac{(\lambda t)^x}{[x]_q!}, \quad x = 0, 1, \ldots,$$

where $0 < \lambda t < 1/(1-q)$ and $0 < q < 1$.

3.19 Let X and Y be independent nonnegative integer valued random variables. Furthermore, assume that X follows a Heine (or an Euler) distribution, with parameters λ_1 and q, and Y follows an Euler (or a Heine) distribution, with parameters λ_2 and q. Show that the conditional distribution of X (or $n - X$) given that $X + Y = n$ is a q-binomial distribution of the first kind, with parameters n, $\theta = \lambda_1/\lambda_2$ and q.

3.20 (*Continuation*). Assume that the conditional distribution of X given that $X + Y = n$ is a q-binomial distribution of the first kind, with parameters n, θ, and q. Show that X follows a Heine (or an Euler) distribution, with parameters λ_1 and q, and Y follows an Euler (or a Heine) distribution, with parameters λ_2 and q, where $\lambda_1/\lambda_2 = \theta$.

3.21 *A characterization of the q-Poisson distributions*. Assume that a nonnegative integer valued random variables X obeys a power series distribution with probability function

$$P(X = x) = \frac{a_x\lambda^x}{g(\lambda)}, \quad x = 0, 1, \ldots, \quad 0 < \lambda < \rho,$$

and series function

$$g(\lambda) = \sum_{x=0}^{\infty} a_x\lambda^x, \quad 0 < \lambda < \rho,$$

where ρ is the radius of convergence of the power series. Show that the random variable X obeys a q-Poisson distribution if and only if

$$E([X]_{2,q}) = [E([X]_q)]^2,$$

for $0 < q < 1$ or $1 < q < \infty$ and for all $0 < \lambda < \rho$.

3.22 Consider a sequence of independent geometric sequences of trials and assume that the probability of success at the jth geometric sequences of trials is given by

$$p_j = \theta q^{j-1}, \quad j = 1, 2, \ldots, 0 < \theta \leq 1, \quad 0 < q < 1 \text{ or } 1 < q < \infty,$$

where for $0 < \theta \leq 1$ and $1 < q < \infty$ the number of trials is restricted by $i \leq [-r]$, the integral part of $-r$, with $r = \log\ \theta / \log\ q < 0$. Let T_k be the number of trials until the occurrence of the kth success.

(a) Derive the probability generating function $P_{T_k}(t) = E(t^{T_k})$.

(b) Find the jth ascending binomial moment $E\left[\binom{T_k+j-1}{j}\right]$, $j = 1, 2, \ldots$, by the aid of $P_{T_k}(t)$.

3.23 *Geometric sequence of trials with discrete q-uniform success probability.* Assume that balls (white or black) are randomly selected, one after the other, with probability θ for a white and $1 - \theta$ for a black ball. The selected balls are sequentially placed, one at a time, in the first (bottom) cell of the jth column of an $n \times n$ array of cells, for $j = 1, 2, \ldots, n$, as follows. A coin, with probability q of tails, is successively tossed until heads occurs. Each time tails occurs, the first ball of the sequence moves up a cell, with one exception: If tails occurs when the ball is in the nth (top) cell, then the ball moves back to the first cell. When heads occurs, the ball comes to rest. The sequential placement of balls, one at a time, in the first cell of the jth column and the step by step movement continues until a white ball fails to rest in the jth cell (diagonal) or until a black ball rests in any cell. Let W_n be the number of white balls that come to rest on the diagonal running from the first cell of the first column to the nth cell of the nth column.

(a) Derive the probability function of W_n as

$$P(W_n = w) = \begin{bmatrix} n+w-1 \\ w \end{bmatrix}_q \left(\frac{\theta}{[n]_q}\right)^w \prod_{j=1}^{n} \left(1 - \frac{\theta}{[n]_q} q^{j-1}\right), \quad w = 0, 1, \ldots .$$

(b) Show that the limiting distribution of W_n, as $n \to \infty$, is an Euler distribution with probability function

$$P(W = w) = E_q(-\theta) \frac{\theta^w}{[w]_q!}, \quad w = 0, 1, \ldots .$$

3.24 *A q-Foster process.* Consider a craftsman who specializes in a particular product and makes one item per unit time. Let X_n be the number of items in stock at the nth unit time, for $n = 0, 1, \ldots$. Assume that, at any unit time, the conditional probability that there are j items in stock, given that at the previous unit time there were i items in stock, is given by

$$p_{i,j} = P(X_{n+1} = j | X_n = i) = \frac{q^j}{[i+2]_q}, \quad j = 0, 1, \ldots, i+1,$$

and $p_{i,j} = 0$, $j > i + 1$, the probability function of a *discrete q-uniform distribution*, where $0 < q < 1$ or $1 < q < 2$, for all $n = 0, 1, \ldots$ Show that the probability function of the stationary distribution,

$$f(x) = P(X = x) = \lim_{n \to \infty} P(X_n = x), \quad x = 0, 1, \ldots,$$

satisfies the recurrence relation

$$f(x) = qf(x-1) - \frac{q^x}{[x]_q} f(x-2), \quad x = 2, 3, \ldots, \quad f(1) = qf(0),$$

where $0 < q < 1$ or $1 < q < 2$ and inductively conclude that

$$f(x) = e_q(-\lambda) \frac{q^{\binom{x}{2}} \lambda^x}{[x]_q!}, \quad x = 0, 1, \ldots,$$

for $0 < q < 1$ and $\lambda = q$ or

$$f(x) = E_{q^{-1}}(-\lambda) \frac{\lambda^x}{[x]_{q^{-1}}!}, \quad x = 0, 1, \ldots,$$

for $1 < q < 2$ and $\lambda = q$.

4

SUCCESS PROBABILITY VARYING WITH THE NUMBER OF TRIALS AND THE NUMBER OF SUCCESSES

4.1 q-PÓLYA DISTRIBUTION

Consider a random experiment with sample space $\Omega = \{f, s\}$, where the sample points (elementary events) f and s are characterized as *failure* and *success*, respectively. An experiment with such a sample space is called *Bernoulli trial*. Furthermore, consider a sequence of independent Bernoulli trials, with the conditional probability of success at the ith trial, given that $j-1$ successes occur in the $i-1$ previous trials,

$$P_{i,j}(\{s\}) = p_{i,j}, \quad j = 1, 2, \ldots, i, \quad i = 1, 2, \ldots,$$

varying with the number of trials and the number of successes.

Note that, in the case the probability $p_{i,j}$, for $j = 1, 2, \ldots, i$ and $i = 1, 2, \ldots$, is of a general functional form very little can be inferred from it about the distributions of the various random variables that may be defined in this model. The particular cases in which the conditional probability $p_{i,j}$ is a quotient or a product of a function of the number j, of successes, only and a function of the number i, of trials, only, are of interest and allow a thorough study of the distributions of the various random variables that may be defined.

Discrete q-Distributions, First Edition. Charalambos A. Charalambides.

In the first part of this chapter, we focus on the case in which the conditional probability $p_{i,j}$ is a quotient of a function a_j of j only and a function b_i of i only,

$$p_{i,j} = \frac{a_j}{b_i}, \quad j = 1, 2, \ldots, i, \quad i = 1, 2, \ldots,$$

with $0 < a_j \leq b_i$, for $j = 1, 2, \ldots, i$ and $i = 1, 2, \ldots$.

A q-Pólya urn model, which belongs to the preceding family of stochastic models, may be introduced, by first defining a q-analogue of the notion of a random drawing of a ball from an urn.

Consider an urn containing r white balls, $\{b_1, b_2, \ldots, b_r\}$, and s black balls, $\{b_{r+1}, b_{r+2}, \ldots, b_{r+s}\}$. A *random q-drawing* of a ball from the urn is carried out as follows. Assume that the balls in the urn are forced to pass through a random mechanism, one by one, in the order $(b_1, b_2, \ldots, b_{r+s})$ or in the reverse order $(b_{r+s}, b_{r+s-1}, \ldots, b_1)$. Also, assume that each passing ball may or may not be caught by the mechanism, with probabilities $p = 1 - q$ and q, respectively. The first caught ball is drawn out of the urn. In the case all $r + s$ balls pass through the mechanism and no ball is caught, the ball passing procedure is repeated, with the same order. Clearly, the probability that ball b_x is drawn from the urn is given by

$$\sum_{k=0}^{\infty}(1-q)q^{(x-1)+k(r+s)} = (1-q)q^{x-1}\sum_{k=0}^{\infty} q^{(r+s)k} = \frac{q^{x-1}}{[r+s]_q},$$

or by

$$\sum_{k=0}^{\infty}(1-q)q^{(r+s-x)+k(r+s)} = \frac{q^{r+s-x}}{[r+s]_q} = \frac{q^{-(x-1)}}{[r+s]_{q^{-1}}},$$

where $0 < q < 1$, according to whether the ball passing order is $(b_1, b_2, \ldots, b_{r+s})$ or $(b_{r+s}, b_{r+s-1}, \ldots, b_1)$. These probabilities may be expressed as

$$p_{r+s}(x; q) = P(X_{r+s} = x) = \frac{q^{x-1}}{[r+s]_q}, \quad x = 1, 2, \ldots, r+s,$$

where $0 < q < 1$ or $1 < q < \infty$. Note that this is the probability function of the discrete q-uniform distribution on the set $\{1, 2, \ldots, r+s\}$. Also, the probability $P_{r+s}(r; q)$, that a white ball is drawn from the urn is given by

$$P_{r+s}(r; q) = P(X_{r+s} \leq r) = \frac{[r]_q}{[r+s]_q} = \frac{(q^{-1})^s[r]_{q^{-1}}}{[r+s]_{q^{-1}}},$$

where $0 < q < 1$ or $1 < q < \infty$. It is worth noticing that the probability $Q_{r+s}(s; q)$ that a black ball is drawn from the urn is given by

$$Q_{r+s}(s; q) = P(r < X_{r+s} \leq r+s) = \frac{q^r[s]_q}{[r+s]_q} = \frac{[s]_{q^{-1}}}{[r+s]_{q^{-1}}},$$

where $0 < q < 1$ or $1 < q < \infty$, which conforms with the relation

$$P_{r+s}(r;q) + Q_{r+s}(s;q) = 1.$$

Finally, notice that a random q-drawing of a ball, for $q \to 1$ and since

$$\lim_{q\to 1} P_{r+s}(r;q) = \frac{r}{r+s}, \quad \lim_{q\to 1} Q_{r+s}(s;q) = \frac{s}{r+s},$$

reduces to the usual random drawing of a ball from the urn.

Furthermore, assume that random q-drawings of balls are sequentially carried out, one after the other, from an urn, initially containing r white and s black balls, according to the following scheme. After each q-drawing, the drawn ball is placed back in the urn together with k balls of the same color. Then, the conditional probability of drawing a white ball at the ith q-drawing, given that $j-1$ white balls are drawn in the previous $i-1$ q-drawings, is given by

$$p_{i,j} = \frac{1-q^{r+k(j-1)}}{1-q^{r+s+k(i-1)}} = \frac{[\alpha - j + 1]_{q^{-k}}}{[\alpha + \beta - i + 1]_{q^{-k}}}, \tag{4.1}$$

for $j = 1, 2, \ldots, i$ and $i = 1, 2, \ldots$, where $0 < q < 1$ or $1 < q < \infty$ and $\alpha = -r/k$ and $\beta = -s/k$, with r and s positive integers and k an integer. This model, which for $q \to 1$ and since

$$\lim_{q\to 1} p_{i,j} = \frac{r + k(j-1)}{r+s+k(i-1)} = \frac{\alpha - j + 1}{\alpha + \beta - i + 1},$$

for $j = 1, 2, \ldots, i$ and $i = 1, 2, \ldots$, reduces to the (classical) Pólya urn model, may be called *q-Pólya urn model.*

Characterizing the q-drawing of a white ball as success and the q-drawing of a black ball as failure, the q-Pólya urn model reduces to the stochastic model of a sequence of independent Bernoulli trials, with probability of success at a trial varying with the number of trials and the number of previous successes, according to (4.1).

The study of the distribution of the number of white balls drawn (successes) in a given number of q-drawings (trials) in a q-Pólya urn model is of theoretical and practical interest. In this respect, the following definition is introduced.

Definition 4.1. *Let X_n be the number of white balls drawn in n q-drawings in a q-Pólya urn model, with the conditional probability of drawing a white ball at the ith q-drawing, given that $j-1$ white balls are drawn in the previous $i-1$ q-drawings, given by (4.1). The distribution of the random variable X_n is called q-Pólya distribution, with parameters n, α, β, k, and q.*

The probability function, the q-factorial moments, and the usual factorial moments of the q-Pólya distribution are obtained in the following theorem.

Theorem 4.1. *The probability function of the q-Pólya distribution, with parameters n, α, β, k, and q, is given by*

$$P(X_n = x) = \begin{bmatrix} n \\ x \end{bmatrix}_{q^{-k}} q^{-k(n-x)(\alpha-x)} \frac{[\alpha]_{x,q^{-k}}[\beta]_{n-x,q^{-k}}}{[\alpha+\beta]_{n,q^{-k}}}$$
$$= q^{-k(n-x)(\alpha-x)} \begin{bmatrix} \alpha \\ x \end{bmatrix}_{q^{-k}} \begin{bmatrix} \beta \\ n-x \end{bmatrix}_{q^{-k}} \Big/ \begin{bmatrix} \alpha+\beta \\ n \end{bmatrix}_{q^{-k}}, \tag{4.2}$$

for $x = 0, 1, \ldots, n$, where $0 < q < 1$ or $1 < q < \infty$, and $\alpha = -r/k$, $\beta = -s/k$, with r and s positive integers and k an integer. Its q-factorial moments are given by

$$E([X_n]_{i,q^{-k}}) = \frac{[n]_{i,q^{-k}}[\alpha]_{i,q^{-k}}}{[\alpha+\beta]_{i,q^{-k}}}, \tag{4.3}$$

for $i = 1, 2, \ldots, n$ and $E([X_n]_{i,q^{-k}}) = 0$, for $i = n+1, n+2, \ldots$. Furthermore, its factorial moments are given by

$$E[(X_n)_j] = j! \sum_{i=j}^{n} (-1)^{i-j} \begin{bmatrix} n \\ i \end{bmatrix}_{q^{-k}} \frac{s_{q^{-k}}(i,j)(1-q^{-k})^{i-j}[\alpha]_{i,q^{-k}}}{[\alpha+\beta]_{i,q^{-k}}}, \tag{4.4}$$

for $j = 1, 2, \ldots, n$, where $s_q(i,j)$ is the q-Stirling number of the first kind, and $E[(X_n)_j] = 0$, for $j = n+1, n+2, \ldots$.

Proof. The probability function of X_n, since, by (4.1),

$$P(X_n = x | X_{n-1} = x-1) = p_{n,x} = \frac{[\alpha - x + 1]_{q^{-k}}}{[\alpha+\beta-n+1]_{q^{-k}}}$$

and

$$P(X_n = x | X_{n-1} = x) = 1 - p_{n,x+1} = \frac{q^{-k(\alpha-x)}[\beta-n+x+1]_{q^{-k}}}{[\alpha+\beta-n+1]_{q^{-k}}},$$

satisfies the recurrence relation

$$P(X_n = x) = \frac{q^{-k(\alpha-x)}[\beta-n+x+1]_{q^{-k}}}{[\alpha+\beta-n+1]_{q^{-k}}} P(X_{n-1} = x)$$
$$+ \frac{[\alpha-x+1]_{q^{-k}}}{[\alpha+\beta-n+1]_{q^{-k}}} P(X_{n-1} = x-1),$$

for $x = 1, 2, \ldots, n$ and $n = 1, 2, \ldots$, with initial conditions $P(X_0 = 0) = 1$ and

$$P(X_n = 0) = \frac{\prod_{i=1}^{n} q^{-k\alpha}[\beta-i+1]_{q^{-k}}}{\prod_{i=1}^{n} [\alpha+\beta-i+1]_{q^{-k}}} = \frac{q^{-kn\alpha}[\beta]_{n,q^{-k}}}{[\alpha+\beta]_{n,q^{-k}}}, \quad n > 0.$$

Clearly, the sequence

$$c_{n,x} = q^{k(n-x)(\alpha-x)} \frac{[\alpha+\beta]_{n,q^{-k}}}{[\alpha]_{x,q^{-k}}[\beta]_{n-x,q^{-k}}} P(X_n = x), \tag{4.5}$$

for $x = 1, 2, \ldots, n$, and $n = 1, 2, \ldots$, satisfies the recurrence relation

$$c_{n,x} = c_{n-1,x} + q^{-k(n-x)} c_{n-1,x-1}, \quad x = 1, 2, \ldots, n, \quad n = 1, 2, \ldots,$$

with initial conditions $c_{0,0} = 1$, $c_{n,0} = 1$, for $n > 0$, and $c_{0,x} = 0$, for $x > 0$. Since this recurrence relation, according to (1.1), uniquely determines the q-binomial coefficient,

$$c_{n,x} = \begin{bmatrix} n \\ x \end{bmatrix}_{q^{-k}}, \quad x = 0, 1, \ldots, n, \quad n = 0, 1, \ldots,$$

expression (4.2) is readily deduced from (4.5). Note that the q-Vandermonde's formula (1.6) guarantees that the probabilities (4.2) sum to unity.

The ith q-factorial moment of X_n, on using (4.2), is expressed as

$$\begin{aligned} E([X_n]_{i,q^{-k}}) &= \sum_{x=i}^{n} [x]_{i,q^{-k}} \begin{bmatrix} n \\ x \end{bmatrix}_{q^{-k}} q^{-k(n-x)(\alpha-x)} \frac{[\alpha]_{x,q^{-k}} [\beta]_{n-x,q^{-k}}}{[\alpha+\beta]_{n,q^{-k}}} \\ &= \frac{[n]_{i,q^{-k}} [\alpha]_{i,q^{-k}}}{[\alpha+\beta]_{n,q^{-k}}} \sum_{x=i}^{n} \begin{bmatrix} n-i \\ x-i \end{bmatrix}_{q^{-k}} q^{-k(n-x)(\alpha-x)} [\alpha - i]_{x-i,q^{-k}} [\beta]_{n-x,q^{-k}}, \end{aligned}$$

which, by using the q-Vandermonde's formula (1.6), yields the expression

$$E([X_n]_{i,q^{-k}}) = [n]_{i,q^{-k}} [\alpha]_{i,q^{-k}} \frac{[\alpha+\beta-i]_{n-i,q^{-k}}}{[\alpha+\beta]_{n,q^{-k}}},$$

for $i = 1, 2, \ldots, n$ and $E([X_n]_{i,q^{-k}}) = 0$, for $i = n+1, n+2, \ldots$. Then, since

$$[\alpha+\beta]_{n,q^{-k}} = [\alpha+\beta]_{i,q^{-k}} [\alpha+\beta-i]_{n-i,q^{-k}},$$

expression (4.3) is established. Furthermore, using the expression of the factorial moments in terms of the q-factorial moments (1.61), with q^{-k} instead of q, expression (4.4) is deduced. □

Remark 4.1. *The q-Pólya as stationary distribution in a birth and death process.* The q-Pólya distribution, according to Remark 2.1, may be considered as the stationary distribution of a birth and death process with birth and death rates proportional to

$$\begin{aligned} \lambda_j &= [n-j]_{q^{-k}} [\alpha - j]_{q^{-k}} q^{-kj}, \quad j = 0, 1, \ldots, n, \\ \mu_j &= [j]_{q^{-k}} [\beta - n + j]_{q^{-k}} q^{-k(\alpha+n-j)}, \quad j = 1, 2, \ldots, n, \end{aligned}$$

where $0 < q < 1$ or $1 < q < \infty$ and $\alpha = -r/k$ and $\beta = -s/k$, with r and s positive integers and k an integer. Indeed,

$$\begin{aligned} \prod_{j=1}^{x} \frac{\lambda_{j-1}}{\mu_j} &= \prod_{j=1}^{x} \frac{[n-j+1]_{q^{-k}} [\alpha - j + 1]_{q^{-k}} q^{-k(j-1)}}{[j]_{q^{-k}} [\beta - n + j]_{q^{-k}} q^{-k(\alpha+n-j)}} \\ &= \begin{bmatrix} n \\ x \end{bmatrix}_{q^{-k}} q^{k(\alpha+n)x - kx^2} \frac{[\alpha]_{x,q^{-k}}}{[\beta - (n-x)]_{x,q^{-k}}} \end{aligned}$$

and since

$$k(\alpha+n)x-x^2=-k(\alpha-x)(n-x)+kn\alpha, \quad [\beta-(n-x)]_{x,q^{-k}}=\frac{[\beta]_{n,q^{-k}}}{[\beta]_{n-x,q^{-k}}},$$

it follows that the probability function of the stationary distribution,

$$P(X=x)=P(X=0)\prod_{j=1}^{x}\frac{\lambda_{j-1}}{\mu_j}, \quad x=1,2,\ldots,$$

is given by

$$P(X=x)=c\begin{bmatrix} n \\ x \end{bmatrix}_{q^{-k}} q^{-k(\alpha-x)(n-x)}\frac{[\alpha]_{x,q^{-k}}[\beta]_{n-x,q^{-k}}}{[\beta]_{n,q^{-k}}}, \quad x=0,1,\ldots,n,$$

where, by the q-Vandermonde's formula (1.6), $c=[\beta]_{n,q^{-k}}/[\alpha+\beta]_{n,q^{-k}}$, and so the probability function (4.2) is deduced.

The q-Pólya distribution, for large $r+s$, can be approximated by a q-binomial distribution of the second kind. Specifically, the following limiting theorem is derived.

Theorem 4.2. *Consider the q-Pólya distribution with probability function (4.2). For* $0<q<1$, *assume that*

$$\lim_{r+s\to\infty}\frac{[s]_{q^{-1}}}{[r+s]_{q^{-1}}}=\lim_{r+s\to\infty}\frac{q^{-s}-1}{q^{-(r+s)}-1}=\theta \tag{4.6}$$

and in the case of a negative integer k assume, in addition, that $\theta<q^{-k(m-1)}$, *for some positive integer m. Then,*

$$\lim_{r+s\to\infty}P(X_n=x)=\begin{bmatrix} n \\ x \end{bmatrix}_{q^k}\theta^{n-x}\prod_{i=1}^{x}(1-\theta q^{k(i-1)}), \tag{4.7}$$

for $x=0,1,\ldots,n$, *where* $0<q<1$ *and* $0<\theta<1$, *in the case k is a positive integer, or* $0<\theta<q^{-k(m-1)}$, *for some positive integer* $m\geq n$, *in the case k is a negative integer.*

Also, for $1<q<\infty$, *assume that*

$$\lim_{r+s\to\infty}\frac{[r]_q}{[r+s]_q}=\lim_{r+s\to\infty}\frac{q^r-1}{q^{r+s}-1}=\lambda \tag{4.8}$$

and in the case of a negative integer k assume, in addition, that $\lambda<q^{k(m-1)}$, *for some positive integer m. Then,*

$$\lim_{r+s\to\infty}P(X_n=x)=\begin{bmatrix} n \\ x \end{bmatrix}_{q^{-k}}\lambda^x\prod_{i=1}^{n-x}(1-\lambda q^{-k(i-1)}), \tag{4.9}$$

for $x=0,1,\ldots,n$, *where* $1<q<\infty$ *and* $0<\lambda<1$, *in the case k is a positive integer, or* $0<\lambda<q^{-k(m-1)}$, *for some positive integer* $m\geq n$, *in the case k is a negative integer.*

Proof. For $0 < q < 1$, the probability function (4.2) may be written as

$$P(X_n = x) = \begin{bmatrix} n \\ x \end{bmatrix}_{q^{-k}} q^{-k(n-x)(\alpha-x)} \frac{[\alpha]_{x,q^{-k}} [\beta]_{n-x,q^{-k}}}{[\alpha+\beta]_{n,q^{-k}}}$$
$$= \begin{bmatrix} n \\ x \end{bmatrix}_{q^{-k}} q^{kx(n-x)+r(n-x)} \frac{\prod_{j=1}^{x}(1-q^{r+k(j-1)}) \prod_{j=1}^{n-x}(1-q^{s+k(j-1)})}{\prod_{i=1}^{n}(1-q^{r+s+k(i-1)})}$$
$$= \begin{bmatrix} n \\ x \end{bmatrix}_{q^{k}} \frac{\prod_{j=1}^{x}(q^{-r-k(j-1)}-1)q^{-s+k(j-1)} \prod_{j=1}^{n-x}(q^{-s-k(j-1)}-1)q^{k(j-1)}}{\prod_{i=1}^{n}(q^{-(r+s)-k(i-1)}-1)q^{k(i-1)}}.$$

Furthermore, by the assumption (4.6), it follows that

$$\lim_{r+s\to\infty} \frac{(q^{-r-k(j-1)}-1)q^{-s+k(j-1)}}{q^{-(r+s)}-1} = 1 - q^{k(j-1)} \lim_{r+s\to\infty} \frac{q^{-s}-1}{q^{-(r+s)}-1} - \lim_{r+s\to\infty} \frac{q^{k(j-1)}-1}{q^{-(r+s)}-1} = 1 - \theta q^{k(j-1)}$$

and

$$\lim_{r+s\to\infty} \frac{(q^{-s-k(j-1)}-1)q^{k(j-1)}}{q^{-(r+s)}-1} = \lim_{r+s\to\infty} \frac{q^{-s}-1}{q^{-(r+s)}-1} - \lim_{r+s\to\infty} \frac{q^{k(j-1)}-1}{q^{-(r+s)}-1} = \theta.$$

Also,

$$\lim_{r+s\to\infty} \frac{(q^{-(r+s)-k(i-1)}-1)q^{k(i-1)}}{q^{-(r+s)}-1} = 1 - \lim_{r+s\to\infty} \frac{q^{k(i-1)}-1}{q^{-(r+s)}-1} = 1.$$

Thus, dividing both the numerator and denominator of the last expression of the probability function (4.2) by $(q^{-(r+s)}-1)^n$ and then taking the limits as $r+s \to \infty$, the limiting expression (4.7) is readily deduced.

Also, for $1 < q < \infty$, the probability function (4.2) may be written as

$$P(X_n = x) = \begin{bmatrix} n \\ x \end{bmatrix}_{q^{-k}} q^{-k(n-x)(\alpha-x)} \frac{[\alpha]_{x,q^{-k}} [\beta]_{n-x,q^{-k}}}{[\alpha+\beta]_{n,q^{-k}}}$$
$$= \begin{bmatrix} n \\ x \end{bmatrix}_{q^{-k}} q^{kx(n-x)+r(n-x)} \frac{\prod_{j=1}^{x}(1-q^{r+k(j-1)}) \prod_{j=1}^{n-x}(1-q^{s+k(j-1)})}{\prod_{i=1}^{n}(1-q^{r+s+k(i-1)})}$$
$$= \begin{bmatrix} n \\ x \end{bmatrix}_{q^{-k}} \frac{\prod_{j=1}^{x}(1-q^{r+k(j-1)})q^{-k(j-1)} \prod_{j=1}^{n-x}(1-q^{s+k(j-1)})q^{r-k(j-1)}}{\prod_{i=1}^{n}(1-q^{r+s+k(i-1)})q^{-k(i-1)}}.$$

Furthermore, by the assumption (4.8), it follows that

$$\lim_{r+s\to\infty} \frac{(q^{r+k(j-1)}-1)q^{-k(j-1)}}{q^{r+s}-1} = 1 - \lim_{r+s\to\infty} \frac{q^{r}-1}{q^{r+s}-1} - \lim_{r+s\to\infty} \frac{q^{-k(j-1)}-1}{q^{r+s}-1} = \lambda$$

and

$$\lim_{r+s\to\infty} \frac{(q^{s+k(j-1)}-1)q^{r-k(j-1)}}{q^{r+s}-1} = 1 - q^{-k(j-1)} \lim_{r+s\to\infty} \frac{q^r-1}{q^{r+s}-1} - \lim_{r+s\to\infty} \frac{q^{-k(j-1)}-1}{q^{r+s}-1} = 1 - \lambda q^{-k(j-1)}.$$

Also,

$$\lim_{r+s\to\infty} \frac{(q^{r+s-k(i-1)}-1)q^{-k(i-1)}}{q^{r+s}-1} = 1 - \lim_{r+s\to\infty} \frac{q^{k(i-1)}-1}{q^{r+s}-1} = 1.$$

Thus, dividing both the numerator and denominator of the last expression of the probability function (4.2) by $(q^{r+s}-1)^n$ and then taking the limits as $r+s\to\infty$, the limiting expression (4.9) is obtained. □

Remark 4.2. *The limiting probability of drawing a white ball.*

For $0<q<1$, by the assumption (4.6), it follows that

$$\lim_{r+s\to\infty} \frac{q^{-s-k(i-j)}-1}{q^{-(r+s)}-1} = q^{-k(i-j)}\left(\lim_{r+s\to\infty} \frac{q^{-s}-1}{q^{-(r+s)}-1} - \lim_{r+s\to\infty} \frac{q^{k(i-j)}-1}{q^{-(r+s)}-1}\right) = \theta q^{-k(i-j)}$$

and

$$\lim_{r+s\to\infty} \frac{q^{-(r+s)-k(i-1)}-1}{q^{-(r+s)}-1} = q^{-k(i-1)}\left(1 - \lim_{r+s\to\infty} \frac{q^{k(i-1)}-1}{q^{-(r+s)}-1}\right) = q^{-k(i-1)}.$$

Therefore,

$$\lim_{r+s\to\infty} p_{i,j} = \lim_{r+s\to\infty} \frac{1-q^{r+k(j-1)}}{1-q^{r+s+k(i-1)}} = 1 - \lim_{r+s\to\infty} \frac{q^{-s-k(i-j)}-1}{q^{-(r+s)-k(i-1)}-1} = 1 - \theta q^{k(j-1)},$$

for $j=1,2,\ldots,\min\{i,m\}$ and $i=1,2,\ldots,$ where $0<q<1$ and $0<\theta<1$ or $0<\theta<q^{-k(m-1)}$ in the case of a negative integer k.

Also, for $1<q<\infty$, by the assumption (4.8), it follows that

$$\lim_{r+s\to\infty} \frac{q^{r+k(j-1)}-1}{q^{r+s}-1} = q^{k(j-1)}\left(\lim_{r+s\to\infty} \frac{q^r-1}{q^{r+s}-1} - \lim_{r+s\to\infty} \frac{q^{-k(j-1)}-1}{q^{r+s}-1}\right) = \lambda q^{k(j-1)}$$

and

$$\lim_{r+s\to\infty} \frac{q^{r+s+k(i-1)}-1}{q^{r+s}-1} = q^{k(i-1)}\left(1 - \lim_{r+s\to\infty} \frac{q^{-k(i-1)}-1}{q^{r+s}-1}\right) = q^{k(i-1)}.$$

Therefore,

$$\lim_{r+s\to\infty} p_{i,j} = \lim_{r+s\to\infty} \frac{q^{r+k(j-1)}-1}{q^{r+s+k(i-1)}-1} = \lambda q^{-k(i-j)},$$

for $j = 1, 2, \ldots, i$ and $i = 1, 2, \ldots,$ with $i - j \leq m$, where $0 < q^{-1} < 1$ and $0 < \lambda < 1$ or $0 < \lambda < q^{k(m-1)}$ in the case of a negative integer k.

Note that these limiting probabilities compared with the success probabilities (3.1), of the q-binomial distribution of the second kind, with q replaced by q^{-k}, imply the limiting probability functions (4.7) and (4.9).

Remark 4.3. *Alternative expressions for certain limiting distributions.* The additional assumptions that $\theta < q^{-k(m-1)}$ and $\lambda < q^{k(m-1)}$ in Theorem 4.2, in the case k is a negative integer, are required for the limiting probability functions (4.7) and (4.9) to be in the interval (0, 1). In this case, it is preferable to replace the parameter θ by $q^{-k\theta}$, with $\theta > 0$, and the parameter λ by $q^{k\lambda}$, with $\lambda > 0$, and take the integral parts of these new parameters as m. Then, the limiting probability functions (4.7) and (4.9), in the case k is a negative integer, are transformed to

$$\lim_{r+s\to\infty} P(X_n = x) = \begin{bmatrix} n \\ x \end{bmatrix}_{q^{-k}} q^{-k(n-x)(\theta-x)}(1-q^{-k})^x [\theta]_{x,q^{-k}}, \tag{4.10}$$

for $x = 0, 1, \ldots, n$, where $0 < q < 1$ and $\theta \geq n$, and

$$\lim_{r+s\to\infty} P(X_n = x) = \begin{bmatrix} n \\ x \end{bmatrix}_{q^k} q^{kx(\lambda-n+x)}(1-q^k)^{n-x} [\lambda]_{n-x,q^k}, \tag{4.11}$$

for $x = 0, 1, \ldots, n$, where $1 < q < \infty$ and $\lambda \geq n$, respectively. Note that, according to Remark 3.3, these limiting distributions are *absorption distributions.*

4.2 q-HYPERGEOMETRIC DISTRIBUTIONS

The q-Pólya urn model in the particular case $k = 0$ reduces to q-drawings with replacement and the distribution (4.2) reduces to the classical binomial distribution with success probability $[r]_q/[r+s]_q$. Also, for $k = -1$, the case which corresponds to q-drawings without replacement, the conditional probability (4.1) of drawing a white ball at the ith q-drawing, given that $j-1$ white balls are drawn in the previous $i-1$ q-drawings, reduces to

$$p_{i,j} = \frac{1-q^{r-j+1}}{1-q^{r+s-i+1}} = \frac{[r-j+1]_q}{[r+s-i+1]_q}, \tag{4.12}$$

for $j = 1, 2, \ldots, \min\{i, r\}$, $i = 1, 2, \ldots, r+s$, where $0 < q < 1$ or $1 < q < \infty$ and r and s are positive integers. This model, which for $q \to 1$ and since

$$\lim_{q\to 1} p_{i,j} = \frac{r-j+1}{r+s-i+1},$$

for $j = 1, 2, \ldots, \min\{i, r\}$ and $i = 1, 2, \ldots, r + s$, reduces to the (classical) hypergeometric urn model, may be called *q-hypergeometric urn model.*

Definition 4.2. *Let X_n be the number of white balls drawn in n q-drawings in a q-hypergeometric urn model, with the conditional probability of drawing a white ball at the ith q-drawing, given that $j - 1$ white balls are drawn in the previous $i - 1$ q-drawings, given by (4.12). The distribution of the random variable X_n is called q-hypergeometric distribution, with parameters n, r, s, and q.*

The probability function, the q-factorial moments, and the factorial moments of the q-hypergeometric distribution, with parameters n, r, s, and q, are obtained in the following corollary of Theorem 4.1.

Corollary 4.1. *The probability function of the q-hypergeometric distribution, with parameters n, r, s, and q, is given by*

$$P(X_n = x) = \begin{bmatrix} n \\ x \end{bmatrix}_q q^{(n-x)(r-x)} \frac{[r]_{x,q}[s]_{n-x,q}}{[r+s]_{n,q}}$$

$$= q^{(n-x)(r-x)} \begin{bmatrix} r \\ x \end{bmatrix}_q \begin{bmatrix} s \\ n-x \end{bmatrix}_q \Big/ \begin{bmatrix} r+s \\ n \end{bmatrix}_q, \tag{4.13}$$

for $x = 0, 1, \ldots, n$, where $0 < q < 1$ or $1 < q < \infty$, and r and s are positive integers. Its q-factorial moments are given by

$$E([X_n]_{i,q}) = \frac{[n]_{i,q}[r]_{i,q}}{[r+s]_{i,q}}, \tag{4.14}$$

for $j = 1, 2, \ldots, n$ and $E([X_n]_{i,q}) = 0$, for $j = n + 1, n + 2, \ldots$. Moreover, its factorial moments are given by

$$E[(X_n)_j] = j! \sum_{i=j}^{n} (-1)^{i-j} \begin{bmatrix} n \\ i \end{bmatrix}_q \frac{s_q(i,j)(1-q)^{i-j}[r]_{i,q}}{[r+s]_{i,q}}, \tag{4.15}$$

for $j = 1, 2, \ldots, n$, where $s_q(i, j)$ is the q-Stirling number of the first kind, and $E[(X_n)_j] = 0$, for $j = n + 1, n + 2, \ldots$.

Example 4.1. *An estimator of the number of errors in a manuscript.* Assume that a proofreader reads a manuscript, which contains an unknown number of errors, m, and when he/she finds an error corrects it and starts reading the manuscript from the beginning. Also, the proofreader starts reading the manuscript from the beginning when he/she reaches its end. A scan (reading) of the manuscript is successful if the proofreader finds (and corrects) an error and is a failure otherwise. Thus, a scan of the manuscript constitutes a Bernoulli trial. Assume that the probability of finding any particular error is $p = 1 - q$. Then, the conditional probability that a scan (trial) is successful, given that $j - 1$ scans (trials) were successful in the previous scans, is

$$p_j = 1 - q^{m-j+1}, \quad j = 1, 2, \ldots, m, \quad 0 < q < 1.$$

Also, assume that a known number of errors, r, are contained in the first part of the manuscript. Considering the m errors of the manuscript as m balls in an urn, with the r errors of the first part of the manuscript as white balls and the other $m-r$ errors as black balls, a successful scan of the manuscript by the proofreader corresponds to a q-drawing of a ball from the urn. In particular, a successful scan of the manuscript by the proofreader, in which he/she finds an error in the first part of it, corresponds to a q-drawing of a white ball. Then, the conditional probability of drawing a white ball at the ith q-drawing, given that $j-1$ white balls are drawn in the previous $i-1$ q-drawings, is given by

$$p_{i,j} = \frac{1-q^{r-j+1}}{1-q^{m-i+1}} = \frac{[r-j+1]_q}{[m-i+1]_q},$$

for $j = 1, 2, \ldots, \min\{i, r\}$ and $i = 1, 2, \ldots, m$, where $0 < q < 1$ and r and m are positive integers. Suppose that the proofreader, after a number of scans, finds n errors in the manuscript. Clearly, according to Corollary 4.1, the distribution of the X_n number of errors the proofreader finds in the first part of the manuscript is the q-hypergeometric with probability function (4.13), where $0 < q < 1$ and $s = m - r$.

An estimator of the unknown number of errors, m, in a manuscript, for given n, r, x, and q, may be obtained by considering the probability function (4.13) as a function of m,

$$p(m) = q^{(n-x)(r-x)} \begin{bmatrix} r \\ x \end{bmatrix}_q \begin{bmatrix} m-r \\ n-x \end{bmatrix}_q \Big/ \begin{bmatrix} m \\ n \end{bmatrix}_q, \quad m = 0, 1, \ldots,$$

and take as estimator of m the value $\hat{m}$ that maximizes $p(m)$. In order to find this maximum, consider the ratio

$$\frac{p(m)}{p(m-1)} = \frac{[m-n]_q/[m]_q}{[(m-r)-(n-x)]_q/[m-r]_q}.$$

Using the relations

$$\frac{[m-n]_q}{[m]_q} = \frac{[m]_q - q^{m-n}[n]_q}{[m]_q} = 1 - q^{m-n}\frac{[n]_q}{[m]_q}$$

and

$$\frac{[(m-r)-(n-x)]_q}{[m-r]_q} = 1 - q^{(m-r)-(n-x)}\frac{[n-x]_q}{[m-r]_q}$$
$$= 1 - q^{m-n}\frac{q^x[n-x]_q}{q^r[m-r]_q} = 1 - q^{m-n}\frac{[n]_q - [x]_q}{[m]_q - [r]_q},$$

it may be written as

$$\frac{p(m)}{p(m-1)} = \frac{1 - q^{m-n}([n]_q/[m]_q)}{1 - q^{m-n}([n]_q - [x]_q)/([m]_q - [r]_q)}.$$

Clearly, the ratio $p(m)/p(m-1)$ is greater than unity for

$$\frac{[n]_q}{[m]_q} < \frac{[n]_q - [x]_q}{[m]_q - [r]_q},$$

or, equivalently, for $[m]_q < [n]_q[r]_q/[x]_q$ and it is smaller than unity for $[m]_q > [n]_q[r]_q/[x]_q$. Therefore, the probability $p(m)$, for $m = 0, 1, \ldots,$ first increases and then decreases; it attains its maximum at the integer part $[\hat{m}]$ of $\hat{m}$, for which

$$[\hat{m}]_q = [n]_q[r]_q/[x]_q.$$

The number $[\hat{m}]$ is the maximum likelihood estimator of m.

The q-hypergeometric distribution may be obtained as the conditional distribution of a q-binomial distribution of the first kind, given its sum with another q-binomial distribution of the first kind independent of it, according to the following theorem.

Theorem 4.3. *Consider a sequence of independent Bernoulli trials and assume that the probability of success at the ith trial is given by*

$$p_i = \frac{\theta q^{i-1}}{1+\theta q^{i-1}}, \quad i = 1, 2, \ldots, \quad 0 < \theta < \infty, \quad 0 < q < 1 \ \textit{or} \ 1 < q < \infty.$$

Let X_r be the number of successes in the first r trials and Y_s be the number of successes in the next s trials. Then, the conditional probability function of X_r, given that $X_r + Y_s = n$, is the q-hypergeometric distribution with probability function (4.13).

Proof. The random variables X_r and Y_s are independent, with probability functions, according to Theorem 2.1, given by

$$P(X_r = x) = \begin{bmatrix} r \\ x \end{bmatrix}_q \frac{\theta^x q^{\binom{x}{2}}}{\prod_{i=1}^{r}(1+\theta q^{i-1})}, \qquad x = 0, 1, \ldots, r,$$

and

$$P(Y_s = y) = \begin{bmatrix} s \\ y \end{bmatrix}_q \frac{(\theta q^r)^y q^{\binom{y}{2}}}{\prod_{i=1}^{r}(1+\theta q^{r+i-1})}, \qquad y = 0, 1, \ldots, s.$$

Also, the probability function of the sum $X_r + Y_s$, since $X_r + Y_s = X_{r+s}$, is given by

$$P(X_r + Y_s = n) = \begin{bmatrix} r+s \\ n \end{bmatrix}_q \frac{\theta^n q^{\binom{n}{2}}}{\prod_{i=1}^{r+s}(1+\theta q^{i-1})}, \qquad n = 0, 1, \ldots, r+s.$$

Then, the conditional probability function of X_r, given that $X_r + Y_s = n$,

$$P(X_r = x | X_r + Y_s = n) = \frac{P(X_r = x)P(Y_s = n-x)}{P(X_r + Y_s = n)}, \qquad x = 0, 1, \ldots, n,$$

on using these expressions, is readily deduced as (4.13). □

Another interesting particular case of the q-Pólya urn model is obtained by setting $k = 1$. In this case, the conditional probability (4.1) of drawing a white ball at the ith q-drawing, given that $j - 1$ white balls are drawn in the previous $i - 1$ q-drawings, reduces to

$$p_{i,j} = \frac{1 - q^{r+j-1}}{1 - q^{r+s+i-1}} = \frac{[r+j-1]_q}{[r+s+i-1]_q}, \tag{4.16}$$

for $j = 1, 2, \ldots, i$ and $i = 1, 2, \ldots$, where $0 < q < 1$ or $1 < q < \infty$ and r and s are positive integers. This model, which for $q \to 1$ and since

$$\lim_{q \to 1} p_{i,j} = \frac{r+j-1}{r+s+i-1},$$

for $j = 1, 2, \ldots, i$ and $i = 1, 2, \ldots$, reduces to the (classical) negative hypergeometric urn model, may be called *negative q-hypergeometric urn model.*

Definition 4.3. *Let X_n be the number of white balls drawn in n q-drawings in a negative q-hypergeometric urn model, with the conditional probability of drawing a white ball at the ith q-drawing, given that $j - 1$ white balls are drawn in the previous $i - 1$ q-drawings, given by (4.16). The distribution of the random variable X_n is called negative q-hypergeometric distribution, with parameters n, r, s, and q.*

The probability function, the q-factorial moments, and the factorial moments of the negative q-hypergeometric distribution, with parameters n, r, s, and q, are deduced from Theorem 4.1, by using the relations

$$\begin{bmatrix} n \\ x \end{bmatrix}_{q^{-1}} = q^{-x(n-x)} \begin{bmatrix} n \\ x \end{bmatrix}_q, \quad [-m]_{k,q^{-1}} = (-1)^k q^k [m+k-1]_{k,q},$$

and

$$[-m]_{k,q^{-1}} = (-1)^k q^{mk+\binom{k}{2}} [m+k-1]_{k,q^{-1}}.$$

Corollary 4.2. *The probability function of the negative q-hypergeometric distribution, with parameters n, r, s, and q, is given by*

$$\begin{aligned} P(X_n = x) &= \begin{bmatrix} n \\ x \end{bmatrix}_q q^{r(n-x)} \frac{[r+x-1]_{x,q}[s+n-x-1]_{n-x,q}}{[r+s+n-1]_{n,q}} \\ &= q^{r(n-x)} \begin{bmatrix} r+x-1 \\ x \end{bmatrix}_q \begin{bmatrix} s+n-x-1 \\ n-x \end{bmatrix}_q \Big/ \begin{bmatrix} r+s+n-1 \\ n \end{bmatrix}_q, \end{aligned} \tag{4.17}$$

for $x = 0, 1, \ldots, n$, where $0 < q < 1$ or $1 < q < \infty$, and r and s are positive integers. Its q-factorial moments are given by

$$E([X_n]_{i,q^{-1}}) = \frac{[n]_{i,q^{-1}}[-r]_{i,q^{-1}}}{[-r-s]_{i,q^{-1}}} = \frac{[n]_{i,q^{-1}}[r+i-1]_{i,q^{-1}}}{q^{si}[r+s+i-1]_{i,q^{-1}}}, \tag{4.18}$$

for $i = 1, 2, \ldots, n$, *and* $E([X_n]_{i,q^{-1}}) = 0$, *for* $j = n+1, n+2, \ldots$. *Also, its factorial moments are given by*

$$E[(X_n)_j] = j! \sum_{i=j}^{n} (-1)^{i-j} \begin{bmatrix} n \\ i \end{bmatrix}_{q^{-1}} \frac{s_{q^{-1}}(i,j)(1-q^{-1})^{i-j}[-r]_{i,q^{-1}}}{[-r-s]_{i,q^{-1}}}$$

$$= j! \sum_{i=j}^{n} (-1)^{i-j} \begin{bmatrix} n \\ i \end{bmatrix}_{q^{-1}} \frac{s_{q^{-1}}(i,j)(1-q^{-1})^{i-j}[r+i-1]_{i,q^{-1}}}{q^{si}[r+s+i-1]_{i,q^{-1}}}, \tag{4.19}$$

for $j = 1, 2, \ldots, n$, *where* $s_q(i,j)$ *is the q-Stirling number of the first kind, and* $E[(X_n)_j] = 0$, *for* $j = n+1, n+2, \ldots$.

The negative q-hypergeometric distribution may be obtained as the conditional distribution of a negative q-binomial distribution of the second kind, given its sum with another negative q-binomial distribution of the second kind independent of it, according to the following theorem.

Theorem 4.4. *Consider a sequence of independent geometric sequences of trials and assume that the probability of success at the jth geometric sequence of trials is given by*

$$p_j = 1 - \theta q^{j-1}, \quad j = 1, 2, \ldots, \quad 0 < \theta < 1, \quad 0 < q < 1,$$

or

$$p_j = 1 - q^{m-j+1}, \quad j = 1, 2, \ldots, m, \quad m \text{ positive integer}, \quad 0 < q < 1.$$

Let W_r *be the number of failures until the occurrence of the rth success and* U_s *be the number of failures after the rth success and until the occurrence of the* $(r+s)$*th success, with* $r + s \leq m$ *in the second case. Then, the conditional probability function of* W_r, *given that* $W_r + U_s = n$, *is the negative q-hypergeometric distribution, with probability function (4.17).*

Proof. The random variables W_r and U_s are independent and, in the first case, their probability functions, according to Theorem 3.1, are given by

$$P(W_r = x) = \begin{bmatrix} r+x-1 \\ x \end{bmatrix}_q \theta^x \prod_{j=1}^{r} (1 - \theta q^{j-1}), \quad x = 0, 1, \ldots,$$

and

$$P(U_s = u) = \begin{bmatrix} s+u-1 \\ u \end{bmatrix}_q (\theta q^r)^u \prod_{j=1}^{s} (1 - \theta q^{r+j-1}), \quad u = 0, 1, \ldots.$$

Also, the probability function of the sum $W_r + U_s$, since $W_r + U_s = W_{r+s}$, is given by

$$P(W_r + U_s = n) = \begin{bmatrix} r+s+n-1 \\ n \end{bmatrix}_q \theta^n \prod_{j=1}^{r+s} (1 - \theta q^{j-1}), \quad n = 0, 1, \ldots.$$

Then, the conditional probability function of W_r, given that $W_r + U_s = n$,

$$P(W_r = x | W_r + U_s = n) = \frac{P(W_r = x)P(U_s = n - x)}{P(W_r + U_s = n)}, \quad x = 0, 1, \ldots, n,$$

on using these expressions, is readily deduced as (4.17).

In the second case, the probability functions of W_r and U_s, according to Theorem 3.1, are given by

$$\begin{aligned} P(W_r = x) &= \begin{bmatrix} r + x - 1 \\ x \end{bmatrix}_q q^{(m-r+1)x} \prod_{j=1}^{r} (1 - q^{m-j+1}) \\ &= \begin{bmatrix} r + x - 1 \\ x \end{bmatrix}_{q^{-1}} q^{mx} \prod_{j=1}^{r} (1 - q^{m-j+1}), \quad x = 0, 1, \ldots, \end{aligned}$$

and

$$P(U_s = u) = \begin{bmatrix} s + u - 1 \\ u \end{bmatrix}_{q^{-1}} q^{(m-r)u} \prod_{j=1}^{s} (1 - q^{m-r-j+1}), \quad u = 0, 1, \ldots .$$

Also, the probability function of the sum $W_r + U_s$, since $W_r + U_s = W_{r+s}$, is given by

$$P(W_r + U_s = n) = \begin{bmatrix} r + s + n - 1 \\ n \end{bmatrix}_{q^{-1}} q^{mn} \prod_{j=1}^{r+s} (1 - q^{m-j+1}), \quad n = 0, 1, \ldots .$$

Then, the conditional probability function of W_r, given that $W_r + U_s = n$, on using these expressions, is deduced as

$$P(W_r = x | W_r + U_s = n) = \frac{q^{-r(n-x)} \begin{bmatrix} r + x - 1 \\ x \end{bmatrix}_{q^{-1}} \begin{bmatrix} s + n - x - 1 \\ n - x \end{bmatrix}_{q^{-1}}}{\begin{bmatrix} r + s + n - 1 \\ n \end{bmatrix}_{q^{-1}}},$$

for $x = 0, 1, \ldots, n$, which is (4.17), with q replaced by q^{-1} and $1 < q^{-1} < \infty$. □

4.3 INVERSE q-PÓLYA DISTRIBUTION

Consider again the q-Pólya urn model. Specifically, assume that random q-drawings of balls are sequentially carried out, one after the other, from an urn, initially containing r white and s black balls, according to the following scheme. After each q-drawing, the drawn ball is placed back in the urn together with k balls of the same color. Assume that the conditional probability of drawing a white ball at the ith q-drawing, given that $j - 1$ white balls are drawn in the previous $i - 1$ q-drawings, is given by (4.1). In this section, the interest is turned to the study of the number of black balls drawn until the nth white ball is drawn. For this reason, the following definition is introduced.

Definition 4.4. *Let W_n be the number of black balls drawn until the nth white ball is drawn in a q-Pólya urn model, with the conditional probability of drawing a white ball at the ith q-drawing, given that $j-1$ white balls are drawn in the previous $i-1$ q-drawings, given by (4.1). The distribution of the random variable W_n is called inverse q-Pólya distribution, with parameters n, α, β, k, and q.*

The probability function, the q-factorial moments, and the usual factorial moments of the inverse q-Pólya distribution are obtained in the following theorem.

Theorem 4.5. *The probability function of the inverse q-Pólya distribution, with parameters n, α, β, k, and q, is given by*

$$P(W_n = w) = \begin{bmatrix} n+w-1 \\ w \end{bmatrix}_{q^{-k}} q^{-wk(\alpha-n+1)} \frac{[\alpha]_{n,q^{-k}}[\beta]_{w,q^{-k}}}{[\alpha+\beta]_{n+w,q^{-k}}}, \tag{4.20}$$

for $w = 0, 1, \ldots,$ where $0 < q < 1$ or $1 < q < \infty$, $\alpha = -r/k$ and $\beta = -s/k$, with r and s positive integers and k an integer. Its q-factorial moments are given by

$$E([W_n]_{i,q^{-k}}) = \frac{[n-i+1]_{i,q^{-k}}[\beta]_{i,q^{-k}}}{q^{ik(\alpha-n+1)}[\alpha+i]_{i,q^{-k}}}, \tag{4.21}$$

for $i = 1, 2, \ldots,$ provided $\alpha + i \neq 0$. Furthermore, its factorial moments are given by

$$E[(W_n)_j] = j! \sum_{i=j}^{\infty} (-1)^{i-j} \begin{bmatrix} n+i-1 \\ i \end{bmatrix}_{q^{-k}} \frac{s_{q^{-k}}(i,j)(1-q^{-k})^{i-j}[\beta]_{i,q^{-k}}}{q^{ik(\alpha-n+1)}[\alpha+i]_{i,q^{-k}}}, \tag{4.22}$$

for $j = 1, 2, \ldots,$ provided $\alpha + j \neq 0$, where $s_q(i,j)$ is the q-Stirling number of the first kind.

Proof. The probability function of the inverse q-Pólya distribution is closely connected to the probability function of q-Pólya distribution. Precisely,

$$P(W_n = w) = P(X_{n+w-1} = n-1)p_{n+w,n},$$

where $P(X_{n+w-1} = n-1)$ is the probability of drawing $n-1$ white balls in $n+w-1$ q-drawings and $p_{n+w,n} = [\alpha-n+1]_{q^{-k}}/[\alpha+\beta-n-w+1]_{q^{-k}}$ is the conditional probability of drawing a white ball at the $(n+w)$th q-drawing, given that $n-1$ white balls are drawn in the previous $n+w-1$ q-drawings. Thus, using (4.2), expression (4.20) is deduced. Note that the negative q-Vandermonde formula (1.12) guarantees that the probabilities (4.20) sum to unity. The ith q-factorial moment of W_n, on using (4.20), is expressed as

$$E([W_n]_{i,q^{-k}}) = \sum_{w=i}^{\infty} [w]_{i,q^{-k}} \begin{bmatrix} n+w-1 \\ w \end{bmatrix}_{q^{-k}} q^{-wk(\alpha-n+1)} \frac{[\alpha]_{n,q^{-k}}[\beta]_{w,q^{-k}}}{[\alpha+\beta]_{n+w,q^{-k}}}$$

$$= \frac{[n-i+1]_{i,q^{-k}}[\alpha]_{n,q^{-k}}[\beta]_{i,q^{-k}}}{q^{ik(\alpha-n+1)}}$$

$$\times \sum_{w=i}^{\infty} \begin{bmatrix} n+w-1 \\ w-i \end{bmatrix}_{q^{-k}} q^{-(w-i)k(\alpha-n+1)} \frac{[\beta-i]_{w-i,q^{-k}}}{[\alpha+\beta]_{n+w,q^{-k}}},$$

which, by using the negative q-Vandermonde formula (1.12), yields the expression

$$E([W_n]_{i,q^{-k}}) = \frac{[n-i+1]_{i,q^{-k}}[\alpha]_{n,q^{-k}}[\beta]_{i,q^{-k}}}{q^{ik(\alpha-n+1)}[\alpha+i]_{n+i,q^{-k}}},$$

for $i = 1, 2, \ldots$, provided $\alpha + i \neq 0$. Then, since

$$[\alpha+i]_{n+i,q^{-k}} = [\alpha+i]_{i,q^{-k}}[\alpha]_{n,q^{-k}},$$

expression (4.21) is readily deduced. Furthermore, using the expression of the factorial moments in terms of the q-factorial moments (1.61), with q^{-k} instead of q, expression (4.22) is deduced. □

Remark 4.4. *The inverse q-Pólya as stationary distribution in a birth and death process.* The inverse q-Pólya distribution, according to Remark 2.1, may be considered as the stationary distribution of a birth and death process with birth and death rates proportional to

$$\lambda_j = [n+j]_{q^{\,k}}[\beta-j]_{q^{-k}}q^{k(\beta-1)}, \quad j = 0, 1, \ldots,$$

and

$$\mu_j = [j]_{q^{-k}}[\alpha+\beta-n-j+1]_{q^{-k}}q^{k(\alpha+\beta-n)}, \quad j = 1, 2, \ldots,$$

where $0 < q < 1$ or $1 < q < \infty$, $\alpha = -r/k$ and $\beta = -s/k$, with r and s positive integers and k an integer. Indeed,

$$\prod_{j=1}^{w} \frac{\lambda_{j-1}}{\mu_j} = \prod_{j=1}^{w} \frac{[n+j-1]_{q^{-k}}[\beta-j+1]_{q^{-k}}q^{k(\beta-1)}}{[j]_{q^{-k}}[\alpha+\beta-n-j+1]_{q^{-k}}q^{k(\alpha+\beta-n)}}$$

$$= \begin{bmatrix} n+w-1 \\ w \end{bmatrix}_{q^{-k}} q^{-wk(\alpha-n+1)} \frac{[\beta]_{w,q^{-k}}}{[\alpha+\beta-n]_{w,q^{-k}}}$$

and since

$$[\alpha+\beta-n]_{w,q^{-k}} = \frac{[\alpha+\beta]_{n+w,q^{-k}}}{[\alpha+\beta]_{n,q^{-k}}},$$

it follows that the probability function of the stationary distribution,

$$P(W = w) = P(W = 0)\prod_{j=1}^{w} \frac{\lambda_{j-1}}{\mu_j}, \qquad w = 1, 2, \ldots,$$

is given by

$$P(W = w) = c\begin{bmatrix} n+w-1 \\ w \end{bmatrix}_{q^{-k}} q^{-wk(\alpha-n+1)} \frac{[\beta]_{w,q^{-k}}}{[\alpha+\beta]_{n+w,q^{-k}}},$$

for $w = 0, 1, \ldots,$ where the constant, by the negative q-Vandermonde's formula (1.12), equals $c = [\alpha]_{n,q^{-k}}$, and so the probability function (4.20) is deduced.

The inverse q-Pólya distribution, for large $r + s$, can be approximated by a negative q-binomial distribution of the second kind. Specifically, the following limiting theorem is derived.

Theorem 4.6. *Consider the inverse q-Pólya distribution with probability function (4.20).*

For $0 < q < 1$, assume that the limiting expression (4.6) holds true. Then,

$$\lim_{r+s\to\infty} P(W_n = w) = \begin{bmatrix} n+w-1 \\ w \end{bmatrix}_{q^k} \theta^w \prod_{i=1}^{n}(1 - \theta q^{k(i-1)}), \tag{4.23}$$

for $w = 0, 1, \ldots,$ where $0 < q < 1$ and $0 < \theta < 1$, in the case k is a positive integer, or $0 < \theta < q^{-k(m-1)}$, for some positive integer $m \geq n$, in the case k is a negative integer.

Also, for $1 < q < \infty$, assume that the limiting expression (4.8) holds true. Then,

$$\lim_{r+s\to\infty} P(W_n = w) = \begin{bmatrix} n+w-1 \\ w \end{bmatrix}_{q^{-k}} q^{-kw}\lambda^n \prod_{i=1}^{w}(1 - \lambda q^{-k(i-1)}), \tag{4.24}$$

for $w = 0, 1, \ldots,$ where $1 < q < \infty$ and $0 < \lambda < 1$, in the case k is a positive integer, or $\lambda = q^{km}$, for some positive integer m, in the case k is a negative integer.

Proof. The probability function (4.20) may be written as

$$\begin{aligned} P(W_n = w) &= \begin{bmatrix} n+w-1 \\ w \end{bmatrix}_{q^{-k}} q^{-kw(\alpha-n+1)} \frac{[\alpha]_{n,q^{-k}}[\beta]_{w,q^{-k}}}{[\alpha+\beta]_{n+w,q^{-k}}} \\ &= \begin{bmatrix} n+w-1 \\ w \end{bmatrix}_{q^{-k}} q^{kw(n-1)+rw} \frac{\prod_{j=1}^{n}(1 - q^{r+k(j-1)})\prod_{j=1}^{w}(1 - q^{s+k(j-1)})}{\prod_{i=1}^{n+w}(1 - q^{r+s+k(i-1)})} \\ &= \begin{bmatrix} n+w-1 \\ w \end{bmatrix}_{q^k} \frac{\prod_{j=1}^{n}(q^{-r-k(j-1)} - 1)q^{-s+k(j-1)}\prod_{j=1}^{w}(q^{-s-k(j-1)} - 1)q^{k(j-1)}}{\prod_{i=1}^{n+w}(q^{-(r+s)-k(i-1)} - 1)q^{k(i-1)}}. \end{aligned}$$

and, alternatively, as

$$P(W_n = w) = \begin{bmatrix} n+w-1 \\ w \end{bmatrix}_{q^{-k}} q^{-kw(\alpha-n+1)} \frac{[\alpha]_{n,q^{-k}}[\beta]_{w,q^{-k}}}{[\alpha+\beta]_{n+w,q^{-k}}}$$

$$= \begin{bmatrix} n+w-1 \\ w \end{bmatrix}_{q^{-k}} q^{kw(n-1)+rw} \frac{\prod_{j=1}^{n}(1-q^{r+k(j-1)})\prod_{j=1}^{w}(1-q^{s+k(j-1)})}{\prod_{i=1}^{n+w}(1-q^{r+s+k(i-1)})}$$

$$= \begin{bmatrix} n+w-1 \\ w \end{bmatrix}_{q^{-k}} q^{-kw} \frac{\prod_{j=1}^{n}(1-q^{r+k(j-1)})q^{-k(j-1)}\prod_{j=1}^{w}(1-q^{s+k(j-1)})q^{r-k(j-1)}}{\prod_{i=1}^{n+w}(1-q^{r+s+k(i-1)})q^{-k(i-1)}}.$$

Then, proceeding as in the derivation of Theorem 4.2, the required limiting expressions (4.23) and (4.24) are readily deduced. □

Remark 4.5. *Alternative expressions for certain limiting distributions.* The additional assumptions that $\theta < q^{-k(m-1)}$ and $\lambda < q^{k(m-1)}$ in Theorem 4.6, in the case k is a negative integer, are required for the limiting probability functions (4.23) and (4.24) to be in the interval (0, 1). In this case, it is preferable to replace the parameter θ by $q^{-k\theta}$, with $\theta > 0$, and the parameter λ by $q^{k\lambda}$, with $\lambda > 0$, and take the integral parts of these new parameters as m. Then, the limiting probability functions (4.23) and (4.24), in the case k is a negative integer, are transformed to

$$\lim_{r+s\to\infty} P(W_n = w) = \begin{bmatrix} n+w-1 \\ w \end{bmatrix}_{q^{-k}} q^{-kw(\theta-n+1)}(1-q^{-k})^n[\theta]_{n,q^{-k}}, \qquad (4.25)$$

for $w = 0, 1, \ldots$, where $0 < q < 1$ and $\theta \geq n$, and

$$\lim_{r+s\to\infty} P(W_n = w) = \begin{bmatrix} n+w-1 \\ w \end{bmatrix}_{q^k} q^{kn(m-w)}(1-q^k)^w[m]_{w,q^k}, \qquad (4.26)$$

for $w = 0, 1, \ldots, m$, where $1 < q < \infty$ and $\lambda = m$ is a positive integer, respectively.

4.4 INVERSE q-HYPERGEOMETRIC DISTRIBUTIONS

The q-Pólya urn model in the particular case $k = -1$, which corresponds to q-drawings without replacement, reduces to the *q-hypergeometric urn model.* In this model, the conditional probability of drawing a white ball at the ith q-drawing, given that $j-1$ white balls are drawn in the previous $i-1$ q-drawings, is given by (4.12). In this section the interest is turned to the study of the number of black balls drawn until the nth white ball is drawn. For this reason, the following definition is introduced.

Definition 4.5. *Let W_n be the number of black balls drawn until the nth white ball is drawn in a q-hypergeometric urn model, with the conditional probability of drawing a white ball at the ith q-drawing, given that $j-1$ white balls are drawn in the previous $i-1$ q-drawings, given by (4.12). The distribution of the random variable W_n is called inverse q-hypergeometric distribution, with parameters n, r, s, and q.*

The probability function, the q-factorial moments, and the usual factorial moments of the inverse q-hypergeometric distribution, with parameters n, r, s, and q, in the following corollary of Theorem 4.5.

Corollary 4.3. *The probability function of the inverse q-hypergeometric distribution, with parameters n, r, s, and q, is given by*

$$P(W_n = w) = \begin{bmatrix} n+w-1 \\ w \end{bmatrix}_q q^{-w(r-n+1)} \frac{[r]_{n,q}[s]_{w,q}}{[r+s]_{n+w,q}}, \tag{4.27}$$

for $w = 0, 1, \ldots, s$, *where* $0 < q < 1$ *or* $1 < q < \infty$ *and r and s positive integers. Its q-factorial moments are given by*

$$E([W_n]_{i,q}) = \frac{[n+i-1]_{i,q}[s]_{i,q}}{q^{-i(r-n+1)}[r+i]_{i,q}}, \quad i = 1, 2, \ldots, s. \tag{4.28}$$

Furthermore, its factorial moments are given by

$$E[(W_n)_j] = j! \sum_{i=j}^{s} \begin{bmatrix} n+i-1 \\ i \end{bmatrix}_q \frac{s_q(i,j)(1-q)^{i-j}[s]_{i,q^{-k}}}{q^{-i(r-n+1)}[r+i]_{i,q}}, \tag{4.29}$$

for $j = 1, 2, \ldots, s$, *where* $s_q(i,j)$ *is the q-Stirling number of the first kind, and* $E[(W_n)_j] = 0$, *for* $j = s+1, s+2, \ldots$.

4.5 GENERALIZED q-FACTORIAL COEFFICIENT DISTRIBUTIONS

Another interesting particular case, in which the conditional probability $p_{i,j}$, of success at the ith trial, given that $j-1$ successes occur in the $i-1$ previous trials, is a product of a function c_i of i only and a function q_j of j only,

$$p_{i,j} = c_i q_j, \quad j = 1, 2, \ldots, i, \quad i = 1, 2, \ldots,$$

with $0 < c_i \leq 1$, for $i = 1, 2, \ldots$, and $0 \leq q_j \leq 1$, for $j = 1, 2, \ldots$, is examined in the remainder of the chapter. It should be noted that in this case the distribution of the number of successes up to a specific number of trials (or the number of trials up to a specific number of successes), is a mixture distribution. The mixed distribution is that of the number of successes up to a specific number of trials (or the number of trials up to a specific number of successes), with the conditional probability of success at any trial, given that $j-1$ successes occur in the previous trails, given by $p_j = 1 - q_j$, for $j = 1, 2, \ldots$. The mixing distribution is that of the number of successes up to a specific number of trials (or the number of trials up to a specific number of successes), with the probability of success at the ith trial given by c_i, for $i = 1, 2, \ldots$; for details see Exercises 4.2 and 4.3.

Let us first consider the case of a sequence of independent Bernoulli trials and assume that the conditional probability of success at the ith trial, given that $j-1$ successes occur in the previous $i-1$ trials, is given by

$$p_{i,j} = q^{-a(i-1)+b(j-1)+c}, \quad j = 1, 2, \ldots, i, \quad i = 1, 2, \ldots,$$

where the parameters a, b, c, and q are such that $0 \leq p_{i,j} \leq 1$ for all $j = 1, 2, \ldots, i$ and $i = 1, 2, \ldots$. This success probability varies geometrically in both the number of trials and the number of successes with rates q^{-a} and q^b, respectively. Excluding the particular case $a = 0$, which was studied in Chapter 3, let us assume that $a \neq 0$ and set $s = b/a$ and $r = c/a$. Then, the study, with theoretical and practical interest, focuses on the particular case in which the conditional probability $p_{i,j}$ is given by

$$p_{i,j} = q^{a\{r-(i-1)+s(j-1)\}}, \quad j = 1, 2, \ldots, i, \quad i = 1, 2, \ldots, \tag{4.30}$$

where $0 < q^a < 1$, $0 \leq r < \infty$, and $0 \leq s < \infty$, with the restriction $i \leq [r]$, or by $1 < q^a < \infty$, $-\infty < r < 0$, and $-\infty < s < 0$.

The probability function, the q-factorial moments, and the (usual) factorial moments of the number of successes up to a specific number of trials are derived in the following theorem.

Theorem 4.7. *Let X_n be the number of successes in a sequence of n independent Bernoulli trials, with the conditional probability of success at the ith trial, given that $j-1$ successes occur in the $i-1$ previous trials, given by (4.30). The probability function of X_n is given by*

$$P(X_n = x) = q^{a\left\{rn-\binom{n}{2}+s\binom{x}{2}\right\}} \frac{(1-q^a)^n}{(1-q^{as})^x} C_{q^a}(n, x; s, r), \tag{4.31}$$

for $x = 0, 1, \ldots, n$, where $0 < q^a < 1$, $0 \leq r < \infty$, and $0 \leq s < \infty$, with the restriction $n \leq [r]$, and by

$$P(X_n = x) = q^{a\left\{rn-\binom{n}{2}+s\binom{x}{2}\right\}} \frac{(1-q^{-a})^n}{(1-q^{as})^x} |C_{q^a}(n, x; s, r)|, \tag{4.32}$$

for $x = 0, 1, \ldots, n$, with $1 < q^a < \infty$, $-\infty < r < 0$, and $-\infty < s < 0$, where $C_{q^a}(n, k; s, r)$ is the noncentral generalized q-factorial coefficient. Its q-factorial moments are given by

$$E([X_n]_{m,q^{-b}}) = [m]_{q^{-b}}! q^{(b-a)\binom{m}{2}+cm} \begin{bmatrix} n \\ m \end{bmatrix}_{q^{-a}}, \tag{4.33}$$

for $m = 1, 2, \ldots, n$ and $E([X_n]_{m,q^{-b}}) = 0$, for $m = n+1, n+2, \ldots$. Moreover, its factorial moments are given by

$$E[(X_n)_j] = j! \sum_{m=j}^{n} (-1)^{m-j} \begin{bmatrix} n \\ m \end{bmatrix}_{q^{-a}} q^{(b-a)\binom{m}{2}+cm} s_{q^{-b}}(m, j)(1-q^{-b})^{m-j}, \tag{4.34}$$

for $j = 1, 2, \ldots, n$ *and* $E[(X_n)_j] = 0$, *for* $j = n+1, n+2, \ldots,$ *where* $s_q(m,j)$ *is the* q*-Stirling number of the first kind. In particular, its mean and variance are given by*

$$E(X_n) = \sum_{m=1}^{n} q^{-a\binom{m}{2}+cm} \begin{bmatrix} n \\ m \end{bmatrix}_{q^{-a}} (q^b - 1)^{m-1}[m-1]_{q^b}!, \tag{4.35}$$

$$\begin{aligned} V(X_n) = 2\sum_{m=2}^{n} q^{-a\binom{m}{2}+cm} \begin{bmatrix} n \\ m \end{bmatrix}_{q^{-a}} (q^b - 1)^{m-2}[m-1]_{q^b}! h_{m-1,q^b}(1) \\ + E(X_n) - [E(X_n)]^2, \end{aligned} \tag{4.36}$$

where $h_{m,q}(k) = \sum_{j=1}^{m} q^j/[j]_q^k$, $k \geq 1$, $m = 1, 2, \ldots,$ *is the incomplete* q*-zeta function.*

Proof. Using the conditional probabilities (4.30), it follows that the probability function of X_n satisfies the recurrence relation

$$P(X_n = x) = (1 - q^{a\{r-(n-1)+sx\}})P(X_{n-1} = x) + q^{a\{r-(n-1)+s(x-1)\}}P(X_{n-1} = x-1),$$

for $x = 1, 2, \ldots, n$ and $n = 1, 2, \ldots,$ with initial conditions

$$P(X_0 = 0) = 1, \quad P(X_0 = x) = 0, x > 0,$$

and

$$P(X_n = 0) = \prod_{i=1}^{n}(1 - q^{a(r-i+1)}) = (1 - q^a)^n [r]_{n,q^a}, \quad n > 0.$$

Multiplying both sides by

$$\begin{aligned} & q^{-a\left\{rn-\binom{n}{2}+s\binom{x}{2}\right\}} \frac{(1-q^{as})^x}{(1-q^a)^n} \\ & = q^{-a\left\{r(n-1)-\binom{n-1}{2}+s\binom{x-1}{2}\right\}-as(x-1)-a(r-n+1)} \frac{(1-q^{as})(1-q^{as})^{x-1}}{(1-q^a)(1-q^a)^{n-1}}, \end{aligned}$$

we deduce for the sequence

$$c_{n,x} = q^{-a\left\{rn-\binom{n}{2}+s\binom{x}{2}\right\}} \frac{(1-q^{as})^x}{(1-q^a)^n} P(X_n = x), \tag{4.37}$$

for $x = 0, 1, \ldots, n$ and $n = 0, 1, \ldots,$ the recurrence relation

$$c_{n,x} = ([sx]_{q^a} - [n-r-1]_{q^a})c_{n-1,x} + [s]_{q^a} c_{n-1,x-1},$$

for $x = 1, 2, \ldots, n$ and $n = 1, 2, \ldots,$ with initial conditions

$$c_{0,0} = 1, \quad c_{0,x} = 0, x > 0, \quad c_{n,0} = q^{-a\left\{rn-\binom{n}{2}\right\}}[r]_{n,q}, n > 0.$$

Comparing the last recurrence relation with the recurrence relation (1.49), of the generalized q-factorial coefficients, we conclude that

$$c_{n,x} = C_{q^a}(n, x; s, r), \quad x = 0, 1, \ldots, n, \quad n = 0, 1, \ldots .$$

Thus, for $0 < q^a < 1$, $0 \le r < \infty$, and $0 \le s < \infty$, by (4.37), (4.31) is established. Also, for $1 < q^a < \infty$, $-\infty < r < 0$, and $-\infty < s < 0$, $C_{q^a}(n,x;s,r)$ has the sign of $(-1)^n$ and specifically

$$C_{q^a}(n,x;s,r) = [-1]_{q^a}^n |C_{q^a}(n,x;s,r)| = (-1)^n q^{-an} |C_{q^a}(n,x;s,r)|.$$

Consequently, for $1 < q^a < \infty, -\infty < r < 0$, and $-\infty < s < 0$, expression (4.31) may be rewritten in the form (4.32). Furthermore, note that, by (1.52),

$$C_{q^a}(n,x;s,r) = \frac{(1-q^b)^x}{(1-q^a)^n} \sum_{j=x}^{n} (-1)^{j-x} q^{a\binom{n-j}{2}-c(n-j)} \begin{bmatrix} n \\ j \end{bmatrix}_{q^a} \begin{bmatrix} j \\ x \end{bmatrix}_{q^b}, \tag{4.38}$$

and so (4.31) may be written, equivalently, as

$$P(X_n = x) = q^{b\binom{x}{2}} \sum_{j=x}^{n} (-1)^{j-x} q^{cj-a\binom{i}{2}-aj(n-j)} \begin{bmatrix} n \\ j \end{bmatrix}_{q^a} \begin{bmatrix} j \\ x \end{bmatrix}_{q^b}, \tag{4.39}$$

for $x = 0, 1, \ldots, n$.

The mth q-factorial moment $E([X_n]_{m,q^{-b}})$, $m = 1, 2, \ldots$, using (4.39), and interchanging the order of summation, is obtained as

$$\begin{aligned} E([X_n]_{m,q^{-b}}) &= \sum_{x=m}^{n} [x]_{m,q^{-b}} q^{b\binom{x}{2}} \sum_{j=x}^{n} (-1)^{j-x} q^{cj-a\binom{j}{2}-aj(n-j)} \begin{bmatrix} n \\ j \end{bmatrix}_{q^a} \begin{bmatrix} j \\ x \end{bmatrix}_{q^b} \\ &= [m]_{q^{-b}}! \sum_{j=m}^{n} (-1)^{j-m} q^{cj-a\binom{j}{2}} \begin{bmatrix} n \\ j \end{bmatrix}_{q^{-a}} \sum_{x=m}^{j} (-1)^{x-m} q^{b\binom{x}{2}} \begin{bmatrix} x \\ m \end{bmatrix}_{q^{-b}} \begin{bmatrix} j \\ x \end{bmatrix}_{q^b} \\ &= [m]_{q^{-b}}! \sum_{j=m}^{n} (-1)^{j-m} q^{cj-a\binom{j}{2}+b\binom{m}{2}} \begin{bmatrix} n \\ j \end{bmatrix}_{q^{-a}} \sum_{x=m}^{j} (-1)^{x-m} q^{b\binom{x-m}{2}} \begin{bmatrix} x \\ m \end{bmatrix}_{q^b} \begin{bmatrix} j \\ x \end{bmatrix}_{q^b} \end{aligned}$$

and since, by (1.16),

$$\sum_{x=m}^{j} (-1)^{x-m} q^{b\binom{x-m}{2}} \begin{bmatrix} x \\ m \end{bmatrix}_{q^b} \begin{bmatrix} j \\ x \end{bmatrix}_{q^b} = \delta_{j,m},$$

expression (4.33) is readily deduced. Furthermore, applying (1.61) with q^{-b} instead of q, expression (4.34) is obtained. In particular, on using the expressions

$$s_{q^{-b}}(m,1) = (-1)^{m-1}[m-1]_{q^{-b}}! = (-1)^{m-1} q^{-b\binom{m-1}{2}} [m-1]_{q^b}!$$

and

$$\begin{aligned} s_{q^{-b}}(m,2) &= (-1)^{m-2}[m-1]_{q^{-b}}! \zeta_{m-1,q^{-b}}(1) \\ &= (-1)^{m-2} q^{-b\binom{m-1}{2}-b} [m-1]_{q^b}! h_{m-1,q^b}(1), \end{aligned}$$

the mean and variance of X_n are deduced from (4.34) as (4.35) and (4.36). □

The mixture distribution, with (distribution to be) mixed a q-Stirling distribution of the second kind and mixing a q-Stirling distribution of the first kind, is a generalized q-factorial coefficient distribution, according to the following theorem.

Theorem 4.8. *The mixture distribution, with mixed the q-Stirling distribution of the second kind*

$$P(Z_m = x) = q^{b\binom{x}{2}+c_2x}(1-q^b)^{m-x}S_{q^b}(m,x;r_2),\quad x=0,1,\ldots,m, \tag{4.40}$$

with $0 \le r_2 < \infty$ *and* $0 < q^b < 1$ *or* $-\infty < r_2 < 0,\ 1 < q^b < \infty,$ *and* $r_2 + m < 0,$ *where* $r_2 = c_2/b$, *and mixing the q-Stirling distribution of the first kind*

$$P(Y_n = m) = q^{-a\binom{n}{2}+c_1n}(1-q^a)^{n-m}s_{q^a}(n,m;-r_1),\quad m=0,1,\ldots,n, \tag{4.41}$$

with $0 \le r_1 < \infty,\ 0 < q^a < 1,$ *and* $n \le [r_1]$ *and*

$$P(Y_n = m) = q^{-a\binom{n}{2}+c_1n}(1-q^{-a})^{n-m}|s_{q^a}(n,m;-r_1)|,\quad m=0,1,\ldots,n, \tag{4.42}$$

with $-\infty < r_1 < 0$ *and* $1 < q^a < \infty$, *where* $r_1 = c_1/a$ *and* $c_1 + c_2 = c$, *is the generalized q-factorial coefficient distribution (4.31) or (4.32).*

Proof. The mixture distribution, with (distribution to be) mixed the q-Stirling distribution of the second kind (4.40), for $0 \le r_2 < \infty$ and $0 < q^b < 1$, and mixing the q-Stirling distribution of the first kind (4.41), with $0 \le r_1 < \infty,\ 0 < q^a < 1$, is expressed as

$$\begin{aligned}P(X_n = x) &= \sum_{m=x}^{n} P(Y_n = m)P(Z_m = x)\\ &= q^{-a\binom{n}{2}+b\binom{x}{2}+c_1n+c_2x}\frac{(1-q^a)^n}{(1-q^b)^x}\sum_{m=x}^{n} s_{q^a}(n,m;-r_1)S_{q^b}(m,x;r_2)[s]_{q^a}^m,\end{aligned}$$

where $s = b/a > 0$, since $0 < q^b = (q^a)^s < 1$ and $0 < q^a < 1$. Also, from (1.50), by replacing q by q^a, r by $-r_1$ and ρ by r_2 and using the relation

$$r_1 + sr_2 = \frac{ar_1 + br_2}{a} = \frac{c_1 + c_2}{a} = \frac{c}{a} = r,$$

we get the expression

$$\sum_{m=x}^{n} s_{q^a}(n,m;-r_1)S_{q^b}(m,x;r_2)[s]_{q^a}^m = q^{c_2(n-x)}C_{q^a}(n,x;s,r).$$

Therefore, using the last relation, the expression of the probability function of the mixture distribution reduces to (4.31).

Similarly, with mixed the q-Stirling distribution of the second kind (4.40), for $-\infty < r_2 < 0,\ 1 < q^b < \infty$, and mixing the q-Stirling distribution of the first kind

(4.42), with $-\infty < r_1 < 0$ and $1 < q^a < \infty$, the mixture distribution is reduced to (4.32). □

The probability function of the number of trials until the occurrence of the kth success, T_k, is readily deduced from (4.36) by using the relation

$$P(T_k = n) = P(X_{n-1} = k-1)p_{n,k}, \quad n = k, k+1, \ldots .$$

Specifically, the following corollary is deduced.

Corollary 4.4. *The probability function of the number of trials until the occurrence of the kth success, T_k, is given by*

$$P(T_k = n) = q^{a\left\{rn-\binom{n}{2}+s\binom{x}{2}\right\}} \frac{(1-q^{-a})^{n-1}}{(1-q^{as})^{k-1}} |C_{q^a}(n-1, k-1; s, r)|, \tag{4.43}$$

for $n = k, k+1, \ldots,$ with $1 < q^a < \infty$, $-\infty < r < 0$, and $-\infty < s < 0$, where $C_{q^a}(n, k; s, r)$ is the noncentral generalized q-factorial coefficient.

The mixture distribution, with mixed a waiting time q-Stirling distribution of the first kind and mixing a waiting time q-Stirling distribution of the second kind is a generalized waiting time q-factorial coefficient distribution, according to the following theorem.

Theorem 4.9. *The mixture distribution, with mixed the waiting time q-Stirling distribution of the first kind,*

$$P(W_m = n) = q^{-a\binom{n}{2}+c_1 n}(1-q^{-a})^{n-m}|s_{q^a}(n-1, m-1; -r_1)|, \tag{4.44}$$

for $n = m, m+1, \ldots,$ with $-\infty < r_1 < 0$ and $1 < q^a < \infty$, where $r_1 = c_1/a$, and mixing the q-Stirling distribution of the second kind,

$$P(U_k = m) = q^{b\binom{k}{2}+c_2 k}(1-q^b)^{m-k}S_{q^b}(m-1, k-1; r_2), \tag{4.45}$$

for $m = k, k+1, \ldots,$ with $0 \le r_2 < \infty$ and $0 < q^b < 1$, where $r_2 = c_2/b$ and $c_1 + c_2 = c$, is the generalized waiting time q-factorial coefficient distribution, with probability function (4.43).

Proof. The mixture distribution, with mixed the waiting time q-Stirling distribution of the first kind (4.44) and mixing the waiting time q-Stirling distribution of the second kind (4.45), is expressed as

$$\begin{aligned} P(T_k = n) &= \sum_{m=k}^{n} P(U_k = m)P(W_m = n) \\ &= q^{-a\binom{n}{2}+b\binom{x}{2}+c_1 n+c_2 k} \frac{(1-q^a)^{n-1}}{(1-q^b)^{k-1}} \\ &\quad \times \sum_{m=k}^{n} |s_{q^a}(n-1, m-1; -r_1)|S_{q^b}(m-1, k-1; r_2)[-s]_{q^{-a}}^{m-1}, \end{aligned}$$

where $s = b/a < 0$, since $0 < q^b = (q^a)^s < 1$ and $1 < q^a < \infty$. Also, from (1.50), by replacing q by q^a, r by $-r_1$ and ρ by r_2 and using the relations

$$r_1 + sr_2 = \frac{ar_1 + br_2}{a} = \frac{c_1 + c_2}{a} = \frac{c}{a} = r, \quad [-s]_{q^{-a}}^{m-1} = [-1]_{q^a}^{-(m-1)}[s]_{q^a}^{m-1},$$

$$|s_{q^a}(n-1, m-1; -r_1)| = [-1]_{q^a}^{-(n-m)} s_{q^a}(n-1, m-1; -r_1)$$

and

$$|C_{q^a}(n-1, k-1; s, r)| = [-1]_{q^a}^{-(n-1)} C_{q^a}(n-1, k-1; s, r),$$

we get the expression

$$\sum_{m=k}^{n} |s_{q^a}(n-1, m-1; -r_1)| S_{q^b}(m-1, k-1; r_2)[s]_{q^a}^{m-1}$$
$$= q^{c_2(n-k)} |C_{q^a}(n-1, k-1; s, r)|.$$

Therefore, using the last relation, the expression of the probability function of the mixture distribution reduces to (4.43). □

Let us now turn the study to the particular case $b = a$, in which the probability of success at the ith trial, given that $j-1$ successes occur in the previous $i-1$ trials,

$$p_{i,j} = q^{a(r-i+j)}, \quad j = 1, 2, \ldots, i, \quad i = 1, 2, \ldots, \tag{4.46}$$

for $0 < q^a < 1$ and $0 \leq r < \infty$, with the restriction $i \leq [r]$, is geometrically increasing with the number of trials and decreasing with the number of successes, with the same rate q^a. In the other interesting particular case $b = -a$, the probability of success at the ith trial, given that $j-1$ successes occur in the previous $i-1$ trials,

$$p_{i,j} = q^{-a\{(-r)+i+j-2\}}, \quad j = 1, 2, \ldots, i, \quad i = 1, 2, \ldots, \tag{4.47}$$

for $1 < q^a < \infty$ and $-\infty < r < 0$, is geometrically decreasing in both the number of trials and the number of successes with the same rate q^{-a}.

The distribution of the number of successes in a given number of trials is given in the following corollary of Theorem 4.7.

Corollary 4.5. *The probability function of the number of successes in a sequence of n independent Bernoulli trials, X_n, with the conditional probability of success at the ith trial, given that $j-1$ successes occur in the $i-1$ previous trials, given by (4.46), is*

$$P(X_n = x) = \begin{bmatrix} n \\ x \end{bmatrix}_{q^a} q^{ax(r-n+x)} (1 - q^a)^{n-x} [r]_{n-x, q^a}, \tag{4.48}$$

for $x = 0, 1, \ldots, n$, with $0 < q^a < 1$ and $0 \leq r < \infty$.

Also, the probability function of the number of successes in a sequence of n independent Bernoulli trials, X_n, with the conditional probability of success at the ith trial, given that $j-1$ successes occur in the $i-1$ previous trials, given by (4.47), is

$$P(X_n = x) = \begin{bmatrix} n \\ x \end{bmatrix}_{q^{-a}} q^{-ax(x-r-1)}(1-q^{-a})^{n-x}[n-r+1]_{n-x,q^{-a}}, \qquad (4.49)$$

for $x = 0, 1, \ldots, n$, with $1 < q^a < \infty$ and $-\infty < r < 0$.

Proof. The noncentral generalized q-factorial coefficient (4.38), in the particular case $b = a$, on using the relation

$$\begin{bmatrix} n \\ j \end{bmatrix}_{q^a} \begin{bmatrix} j \\ x \end{bmatrix}_{q^a} = \begin{bmatrix} n \\ x \end{bmatrix}_{q^a} \begin{bmatrix} n-x \\ j-x \end{bmatrix}_{q^a},$$

may be expressed as

$$\begin{aligned} C_{q^a}(n,x;1,r) &= \begin{bmatrix} n \\ x \end{bmatrix}_{q^a} \frac{1}{(1-q^a)^{n-x}} \sum_{j=x}^{n} (-1)^{j-x} q^{a\binom{n-j}{2}-ar(n-j)} \begin{bmatrix} n-x \\ j-x \end{bmatrix}_{q^a} \\ &= \begin{bmatrix} n \\ x \end{bmatrix}_{q^a} \frac{(-1)^{n-x}}{(1-q^a)^{n-x}} \sum_{k=0}^{n-x} q^{a\binom{k}{2}} \begin{bmatrix} n-x \\ k \end{bmatrix}_{q^a} (-q^{-ar})^k \\ &= \begin{bmatrix} n \\ x \end{bmatrix}_{q^a} \frac{\prod_{i=1}^{n-x}\left(1-q^{-a(r-i+1)}\right)}{(q^a-1)^{n-x}}. \end{aligned}$$

Therefore,

$$C_{q^a}(n,x;1,r) = \begin{bmatrix} n \\ x \end{bmatrix}_{q^a} q^{a\binom{n-x}{2}-ar(n-x)} [r]_{n-x,q^a}$$

and introducing it into expression (4.31), we deduce (4.48).

Similarly, we get

$$\begin{aligned} |C_{q^a}(n,x;-1,r)| &= \begin{bmatrix} n \\ x \end{bmatrix}_{q^{-a}} \frac{q^{a\binom{n}{2}-arn}}{(1-q^{-a})^{n-x}} \sum_{j=x}^{n} (-1)^{j-x} q^{-a\binom{j}{2}+arj} \begin{bmatrix} n-x \\ j-x \end{bmatrix}_{q^{-a}} \\ &= \begin{bmatrix} n \\ x \end{bmatrix}_{q^{-a}} \frac{q^{a\binom{n-x}{2}-a(r-x)(n-x)}}{(1-q^{-a})^{n-x}} \sum_{k=0}^{n-x} q^{-a\binom{k}{2}} \begin{bmatrix} n-x \\ k \end{bmatrix}_{q^{-a}} (-q^{-a(x-r)})^k \\ &= \begin{bmatrix} n \\ x \end{bmatrix}_{q^{-a}} \frac{q^{a\binom{n-x}{2}-a(r-x)(n-x)}}{(1-q^{-a})^{n-x}} \prod_{i=1}^{n-x} \left(1-q^{-a(x-r+i-1)}\right) \\ &= \begin{bmatrix} n \\ x \end{bmatrix}_{q^{-a}} q^{a\binom{n-x}{2}-a(r-x)(n-x)} [n-r-1]_{n-x,q^{-a}} \end{aligned}$$

and from (4.32), we find (4.49). □

Remark 4.6. *The absorption distribution revisited.* The probability function of the number of failures in $n \leq [r]$ trials, Y_n, since $P(Y_n = y) = P(X_n = n - y)$, is readily deduced from (4.48) as

$$P(Y_n = y) = \begin{bmatrix} n \\ y \end{bmatrix}_{q^a} q^{a(n-y)(r-y)}(1 - q^a)^y [r]_{y,q^a}, \quad y = 0, 1, \ldots, n,$$

for $0 < q^a < 1$ and $0 < r < \infty$. Notice that, for r a positive integer, this is exactly the probability function (3.14) of the number of successes in n independent Bernoulli trials, with $p_j = 1 - q^{a(r-j+1)}, j = 1, 2, \ldots, r, 0 < q^a < 1$, the probability of success at any trial given that $j - 1$ successes occur in the previous trials. This is not a coincidence; it can be explained as follows. The assumption that the probability of success at the ith trial, given that $j - 1$ successes occur in the $i - 1$ previous trials, is given by

$$p_{i,j} = q^{a(r-i+j)}, \quad j = 1, 2, \ldots, i, \quad i = 1, 2, \ldots, [r],$$

for $0 < q^a < 1$ and $0 < r < \infty$, is equivalent to the assumption that the probability of failure at the ith trial, given that $i - j$ failures occur in the $i - 1$ previous trials, which, in turn, is equivalent to the assumption that the probability of failure at the ith trial, given that $j - 1$ failures occur in the $i - 1$ previous trials, is given by

$$q_{i,j} = 1 - q^{a(r-j+1)}, \quad j = 1, 2, \ldots, [r], \quad i = 1, 2, \ldots,$$

for $0 < q^a < 1$ and $0 < r < \infty$. Interchanging the notions of success and failure, this is exactly the assumption in the absorption model, with $0 < q^a < 1$ and r a positive integer.

Let us now consider the case of a sequence of independent Bernoulli trials and assume that the conditional probability of success at the ith trial, given that $j - 1$ successes occur in the previous $i - 1$ trials, is given by

$$p_{i,j} = 1 - q^{-a(i-1)+b(j-1)+c}, \quad j = 1, 2, \ldots, i, \quad i = 1, 2, \ldots, \tag{4.50}$$

where the parameters a, b, c, and q are such that $0 \leq p_{i,j} \leq 1$ for all $j = 1, 2, \ldots, i$ and $i = 1, 2, \ldots,$ which is closely connected to the case discussed in this section. Specifically, this assumption is equivalent to the assumption that the conditional probability of failure at the ith trial, given that $i - j$ failures occur in the $i - 1$ previous trials, is given by

$$q_{i,i-j} = 1 - p_{i,j} = q^{-a(i-1)+b(j-1)+c}, \quad j = 1, 2, \ldots, i, \quad i = 1, 2, \ldots,$$

which, by replacing $i - j$ by $j - 1$, is equivalent to the assumption that the probability of failure at the ith trial, given that $j - 1$ failures occur in the $i - 1$ previous trials, is given by

$$q_{i,j} = 1 - p_{i,i-j} = q^{-(a-b)(i-1)+b(j-1)+c}, \quad j = 1, 2, \ldots, i, \quad i = 1, 2, \ldots .$$

Excluding the particular case $a = b$, which was studied in Chapter 3, let us assume that $a \neq b$ and set $s = b/(a-b)$ and $r = c/(a-b)$. Then, the study, with theoretical and practical interest, focuses on the particular case in which the conditional probability $q_{i,j}$ is given by

$$q_{i,j} = q^{(a-b)\{r-(i-1)+s(j-1)\}}, \quad j = 1, 2, \ldots, i, \quad i = 1, 2, \ldots, \tag{4.51}$$

where $0 < q^{a-b} < 1$, $0 \leq r < \infty$, and $0 \leq s < \infty$, with the restriction $i \leq [r]$, or by $1 < q^{a-b} < \infty$, $-\infty < r < 0$, and $-\infty < s < 0$. Consequently, if X_n is the number of successes in n Bernoulli trials, with success probability given by (4.50), and Y_n is the number of failures in n Bernoulli trials, with failure probability given by (4.51), then

$$P(X_n = x) = P(Y_n = n - x), \quad x = 0, 1, \ldots, n.$$

The probability function, the q-factorial moments, and the mean and variance of the number of successes up to a given number of trials, on using this relation, are deduced in the following corollary of Theorem 4.7.

Corollary 4.6. *Let X_n be the number of successes in a sequence of n independent Bernoulli trials, with the conditional probability of success at the ith trial, given that $j-1$ successes occur in the $i-1$ previous trials, given by (4.50). The probability function of X_n is given by*

$$P(X_n = x) = q^{(a-b)\left\{rn-\binom{n}{2}+s\binom{x}{2}\right\}} \frac{(1-q^{a-b})^n}{(1-q^{(a-b)s})^x} C_{q^{a-b}}(n, n-x; s, r), \tag{4.52}$$

for $x = 0, 1, \ldots, n$, where $0 < q^{a-b} < 1$, $0 \leq r < \infty$, and $0 \leq s < \infty$, with the restriction $n \leq [r]$, and by

$$P(X_n = x) = q^{(a-b)\left\{rn-\binom{n}{2}+s\binom{x}{2}\right\}} \frac{(1-q^{-(a-b)})^n}{(1-q^{(a-b)s})^x} |C_{q^{a-b}}(n, n-x; s, r)|, \tag{4.53}$$

for $x = 0, 1, \ldots, n$, with $1 < q^{a-b} < \infty$, $-\infty < r < 0$, and $-\infty < s < 0$, where $C_{q^{a-b}}(n, k; s, r)$ is the noncentral generalized q-factorial coefficient. Its mean and variance are given by

$$E(X_n) = n - \sum_{m=1}^{n} q^{(b-a)\binom{m}{2}+cm} \begin{bmatrix} n \\ m \end{bmatrix}_{q^{b-a}} (q^b - 1)^{m-1} [m-1]_{q^b}!, \tag{4.54}$$

$$V(X_n) = 2\sum_{m=2}^{n} q^{(b-a)\binom{m}{2}+cm} \begin{bmatrix} n \\ m \end{bmatrix}_{q^{b-a}} (q^b - 1)^{m-2} [m-1]_{q^b}! h_{m-1,q^b}(1) + E(X_n) - [E(X_n)]^2, \tag{4.55}$$

where $h_{m,q}(k) = \sum_{j=1}^{m} q^j/[j]_q^k$, $k \geq 1$, $m = 1, 2, \ldots$, is the incomplete q-zeta function.

4.6 REFERENCE NOTES

A stochastic model of a sequence of independent Bernoulli trials, with success probability varying geometrically both with the number of trials and the number of successes in two specific cases, is considered. In the first stochastic model, the success probability is assumed to be a quotient of a function of the number of successes only and a function of the number of trials only. Charalambides (2012a) introduced the q-Pólya urn model, which belongs to this family of stochastic models, and studied in detail the q-Pólya and the inverse q-Pólya distributions. In particular, the q-hypergeometric and inverse q-hypergeometric were discussed. Kupershmidt (2000) introduced a q-hypergeometric distribution and a q-contagious distribution (q-Pólya distribution) and represented the corresponding random variable as a sum of two-valued dependent random variables. Exercise 4.4 and the second part of Exercise 4.8 are extracted from this paper. The coin-tossing game, presented in Exercises 4.10 and 4.11, was discussed by Moritz and Williams (1988) and further studied by Treadway and Rawlings (1994).

Kemp (2005) starting from a confluent form of the Chu–Vandermonde sum as a probability generating function obtained two q-confluent hypergeometric distributions with infinite support and one with finite support. She also deduced these distributions as steady-state birth and death Markov chains.

In the second stochastic model, which is considered in this chapter, the success probability is assumed to be a product of a function of the number of successes only and a function of trials only. The distribution of the number of successes in a given number of trials and the distribution of the number trials until the occurrence of a given number of successes are expressed in terms of the generalized Lah numbers in Exercises 4.1–4.3. An equivalent expression of the distribution of the number of successes in a given number of trials was derived by Bickel et al. (2001) in terms of symmetric polynomials. Furthermore, Crippa and Simon (1997) discussed the case of the probability of success varying geometrically both with the number of trials and the number of successes, with rate q. Charalambides (2004), inspired by the q-distributions obtained by Crippa and Simon, introduced the noncentral generalized q-factorial coefficients and used them to express these distributions. Also, Louchard and Prodinger (2008), using generating functions, derived these distributions together with some new results.

4.7 EXERCISES

4.1 Consider a sequence of independent Bernoulli trials and assume that the probability of success at the ith trial, given that $j-1$ successes occur in the $i-1$ previous trials, is given by

$$P_{i,j}(\{s\}) = p_{i,j} = \frac{1-b_j}{1+a_i}, \quad i = 1, 2, \dots, \quad j = 1, 2, \dots,$$

where $a_i \geq 0$, $i = 1, 2, \dots$, and $0 \leq b_j \leq 1$, $j = 1, 2, \dots$.

(a) Let X_n be the number of successes in n trials. Show that its probability function is given by

$$P(X_n = k) = \frac{\prod_{j=1}^{k}(1-b_j)}{\prod_{i=1}^{n}(1+a_i)}\ C(n,k;-\boldsymbol{a},\boldsymbol{b}), \quad k=0,1,\ldots,n,$$

where $C(n,k;-\boldsymbol{a},\boldsymbol{b})$ is the generalized Lah number.

(b) Let T_k be the number of trials until the occurrence of the kth success. Show that its probability function is given by

$$P(T_k = n) = \frac{\prod_{j=1}^{k}(1-b_j)}{\prod_{i=1}^{n}(1+a_i)}\ C(n-1,k-1;-\boldsymbol{a},\boldsymbol{b}), \quad n=k,k+1,\ldots .$$

4.2 (*Continuation*). Show that the family of distributions with probability function

$$P(X_n = k) = \frac{\prod_{j=1}^{k}(1-b_j)}{\prod_{i=1}^{n}(1+a_i)}\ C(n,k;-\boldsymbol{a},\boldsymbol{b}), \quad k=0,1,\ldots,n,$$

may be represented as a mixture distribution,

$$P(X_n = k) = \sum_{m=k}^{n} P(Y_n = m)P(Z_m = k), \ \ k=0,1,\ldots,n, \ \ (n=1,2,\ldots),$$

with mixed distribution belonging to the family of distributions with probability function

$$P(Z_m = k) = \prod_{j=1}^{k}(1-b_j)S(m,k;\boldsymbol{b}), \quad k=0,1,\ldots,m, \quad (m=1,2,\ldots),$$

where $S(m,k;\boldsymbol{b})$ is the generalized Stirling number of the second kind, and mixing distribution belonging to the family of distributions with probability function

$$P(Y_n = m) = \frac{|s(n,m;\boldsymbol{a})|}{\prod_{i=1}^{n}(1+a_i)}, \quad m=0,1,\ldots,n, \quad (n=1,2,\ldots),$$

where $|s(n,m;\boldsymbol{a})|$ is the generalized signless Stirling number of the first kind.

4.3 (*Continuation*). Show that the family of distributions with probability function

$$P(T_k = n) = \frac{\prod_{j=1}^{k}(1-b_j)}{\prod_{i=1}^{n}(1+a_i)}\ C(n-1,k-1;-\boldsymbol{a},\boldsymbol{b}), \ \ n=k,k+1,\ldots,$$

may be represented as a mixture distribution,

$$P(T_k = n) = \sum_{m=k}^{n} P(U_k = m)P(W_m = k), \ \ n=k,k+1,\ldots, \ \ (k=1,2,\ldots),$$

with mixed distribution belonging to the family of distributions with probability function

$$P(W_m = n) = \frac{|s(n-1, m-1; \boldsymbol{a})|}{\prod_{i=1}^{n}(1+a_i)}, \quad n = m, m+1, \ldots, \quad (m = 1, 2, \ldots),$$

and mixing distribution belonging to the family of distributions with probability function

$$P(U_k = m) = \prod_{j=1}^{k}(1 - b_j)S(m-1, k-1; \boldsymbol{b}), \quad m = k, k+1, \ldots, \quad (k = 1, 2, \ldots)$$

4.4 *q-Pólya distribution as sum of two-point distributions.* Consider a sequence of two-point (zero and nonzero) random variables Z_i, $i = 1, 2, \ldots$. Suppose that the random vector $(Z_1, Z_2, \ldots, Z_n)$ assumes the values $(z_1, z_2, \ldots, z_n)$ and let X_n be the number of nonzeroes among these values. Furthermore, suppose that the random variable Z_i assumes the values 0 and $q^{X_{i-1}}$, for $0 < q < 1$ or $1 < q < \infty$, with conditional probabilities

$$P(Z_i = 0 | X_{i-1} = j-1) = 1 - \frac{[\alpha - j + 1]_{q^{-k}}}{[\alpha + \beta - i + 1]_{q^{-k}}} = \frac{q^{-k(\alpha - j + 1)}[\beta - i + j]_{q^{-k}}}{[\alpha + \beta - i + 1]_{q^{-k}}},$$

$$P(Z_i = q^{j-1} | X_{i-1} = j-1) = \frac{[\alpha - j + 1]_{q^{-k}}}{[\alpha + \beta - i + 1]_{q^{-k}}},$$

for $j = 1, 2, \ldots, i$ and $i = 1, 2, \ldots$, where $0 < q < 1$ or $1 < q < \infty$, and $\alpha = -r/k$, $\beta = -s/k$, with r and s positive integers and k an integer. Show that the sum $Y_n = \sum_{i=1}^{n} Z_i$ follows a q-Pólya distribution, with probability function

$$P(Y_n = [x]_q) = P(X_n = x) = \begin{bmatrix} n \\ x \end{bmatrix}_q q^{-k(n-x)(\alpha - x)} \frac{[\alpha]_{x,q^{-k}}[\beta]_{n-x,q^{-k}}}{[\alpha + \beta]_{n,q^{-k}}},$$

for $x = 0, 1, \ldots, n$.

4.5 *A discrete q-uniform distribution.* Assume that random q-drawings of balls are sequentially carried out, one after the other, from an urn, initially containing one white and one black ball, according to the following scheme. After each q-drawing, the drawn ball is placed back in the urn together with another ball of the same color. Then, the conditional probability of drawing a white ball at the ith q-drawing, given that $j-1$ white balls are drawn in the previous $i-1$ q-drawings, is given by

$$p_{i,j} = \frac{1 - q^j}{1 - q^{i+1}} = \frac{[j]_q}{[i+1]_q}, \quad j = 1, 2, \ldots, i, \quad i = 1, 2, \ldots,$$

where $0 < q < 1$ or $1 < q < \infty$.

(a) Show that the probability function of the number of black balls drawn in n q-drawings, Y_n, is given by

$$P(Y_n = y) = \frac{q^y}{[n+1]_q}, \quad y = 0, 1, \ldots, n.$$

The distribution of Y_n may be called *discrete q-uniform distribution*, with parameters n and q, since its probability function, for $q \to 1$, converges to the probability function of the discrete uniform distribution, with parameter n.

(b) Derive the q-factorial moments $E([Y_n]_{m,q})$, for $m = 1, 2, \ldots$, and in particular deduce the q-mean $E([Y_n]_q)$ and the q-variance $V([Y_n]_q)$.

(c) Also, deduce the factorial moments $E[(Y_n)_j]$, for $j = 1, 2, \ldots$.

4.6 (*Continuation*). Consider a sequence of two-point (zero and nonzero) random variables Z_i, $i = 1, 2, \ldots$. Assume that the random vector $(Z_1, Z_2, \ldots, Z_n)$ take the values $(z_1, z_2, \ldots, z_n)$ and let Y_n be the number of zeroes among these values. Furthermore, suppose that the random variable Z_i assumes the values 0 and $q^{(i-1)-Y_{i-1}}$, for $0 < q < 1$ or $1 < q < \infty$, with conditional probabilities

$$P(Z_i = 0 | Y_{i-1} = i - j) = 1 - \frac{[j]_q}{[i+1]_q} = \frac{q^j[i-j+1]_q}{[i+1]_q},$$

$$P(Z_i = q^{j-1} | Y_{i-1} = i - j) = \frac{[j]_q}{[i+1]_q},$$

for $j = 1, 2, \ldots, i$ and $i = 1, 2, \ldots$, where $0 < q < 1$ or $1 < q < \infty$. Show that the random variable Y_n follows a discrete q-uniform distribution, with probability function

$$P(Y_n = y) = \frac{q^y}{[n+1]_q}, \quad y = 0, 1, \ldots, n.$$

where $0 < q < 1$ or $1 < q < \infty$.

4.7 (*Continuation*). Consider a sequence of independent Bernoulli trials and assume that the conditional probability of success at any trial, given that $j - 1$ successes occur at the previous trials, is given by

$$p_j = 1 - \theta q^{j-1}, \quad j = 1, 2, \ldots, \quad 0 < \theta < 1, \quad 0 < q < 1 \quad \text{or} \quad 1 < q < \infty,$$

where, for $0 < \theta < 1$ and $1 < q < \infty$, the number j is restricted by $\theta q^{j-1} < 1$. Let W be the number of failures until the occurrence of the first success and U be the number of failures after the occurrence of the first success and until the occurrence of the second success. Show that the conditional distribution of W, given that $W + U = n$, is the discrete q-uniform distribution, with parameters n and q^{-1}.

4.8 (*Continuation*). Let $X_{1,n}, X_{2,n}, \ldots, X_{m,n}$ be independent and identically distributed discrete random variables, with common distribution a discrete q-uniform distribution on the set $\{0, 1, \ldots, n\}$. Also, let $Y_{m,n} = \min\{X_{1,n}, X_{2,n}, \ldots, X_{m,n}\}$ and $Z_{m,n} = \max\{X_{1,n}, X_{2,n}, \ldots, X_{m,n}\}$.

(a) Find the joint and the marginal probability functions of $Y_{m,n}$ and $Z_{m,n}$ and

(b) Determine the probability function of the range $R_{m,n} = Z_{m,n} - Y_{m,n}$.

4.9 Assume that n players $\{1, 2, \ldots, n\}$ toss a coin, one after the other, in the order $(1, 2, \ldots, n)$, and the first one to toss heads wins. In the case in which all n players toss the coin and no one gets heads, the play continues with another round of sequential coin tosses. If q is the probability of tossing tails and $p = 1 - q$ is the probability of tossing heads, at any trial, show that the probability for player x to win the game is given by

$$P(X_n = x) = \frac{q^{x-1}}{[n]_q}, \quad x = 1, 2, \ldots, n.$$

Note that this is the probability function of a discrete q-uniform distribution on the set $\{1, 2, \ldots, n\}$.

4.10 *A coin-tossing game*. Assume that n players $\{1, 2, \ldots, n\}$ toss a coin, one after the other, in the order $(1, 2, \ldots, n)$. A player, upon tossing heads, goes out of the game and passes the coin to the next player still in the game. The remaining players continue to toss the coin until all go out. Let $\sigma_n = (i_1, i_2, \ldots, i_n) \in S_n$ be the order in which the players go out of the game, where S_n is the set of the permutations of $\{1, 2, \ldots, n\}$. If q is the probability of tossing tails and $p = 1 - q$ is the probability of tossing heads, at any trial, show that

$$P(\sigma_n) = \frac{q^{|\sigma_n|}}{[n]_q!}, \quad \sigma_n \in S_n,$$

where the norm $|\sigma_n|$ is the number of tails in the shortest sequence of coin tosses for which the game ends in the order σ_n.

4.11 (*Continuation*). Consider the set of permutations $\sigma_n = (i_1, i_2, \ldots, i_n)$ of $\{1, 2, \ldots, n\}$ with norm k, $S_{n,k} = \{\sigma_n \in S_n : |\sigma_n| = k\}$, and let $a(n, k)$ denote its cardinality. Also, let Y_n be the number of tails in the shortest sequence of coin tosses with which the game ends.

(a) Show that the probability function of the random variable Y_n is given by

$$P(Y_n = k) = \frac{a(n, k)q^k}{[n]_q!}, \quad k = 0, 1, \ldots, m_n, \quad m_n = \binom{n}{2}.$$

(b) Deduce the probability generating function of Y_n as

$$E(t^{Y_n}) = [n]_{qt}!/[n]_q!.$$

(c) Show that the numbers $a(n,k)$, $k = 0, 1, \ldots, m_n$, $n = 1, 2, \ldots$, satisfy the horizontal recurrence relation

$$a(n,k) = \sum_{j=k-n+1}^{k} a(n-1,j), \quad k = 0, 1, \ldots, m_n, \quad n = 1, 2, \ldots,$$

with initial conditions $a(n, 0) = 1$ and $a(n, k) = 0$, for $k > m_n$.

4.12 Suppose that m random q-drawings of balls are sequentially carried out, one after the other, without replacement, from an urn containing r white and s black balls. The m drawn balls are placed into an empty urn. Furthermore, assume that n random q-drawings of balls are sequentially carried out, one after the other, without replacement, from the second urn. Let Y_n be the number of white balls drawn from the second urn. Show that the distribution of the random variable Y_n is a q-hypergeometric, independent of m.

4.13 Assume that random q-drawings of balls are sequentially carried out one after the other, without replacement, from an urn initially containing r white and s black balls. Let U be the number of white balls drawn until the first black ball is drawn.

(a) Show that the probability function of the random variable U_n is given by

$$P(U = u) = q^{r-u} \begin{bmatrix} r+s-u-1 \\ r-u \end{bmatrix}_q \Big/ \begin{bmatrix} r+s \\ r \end{bmatrix}_q, \quad u = 0, 1, \ldots, r,$$

where $0 < q < 1$ or $1 < q < \infty$.

(b) Derive the q-factorial moments $E([U]_{i,q^{-1}})$, for $i = 1, 2, \ldots$.

(c) Also, deduce the factorial moments $E[(U)_j]$, for $j = 1, 2, \ldots$.

4.14 *Another inverse q-Pólya distribution.* Assume that random q-drawings of balls are sequentially carried out, one after the other, from an urn, initially containing r white and s black balls, according to the following scheme. After each q-drawing, the drawn ball is placed back in the urn together with k balls of the same color. Then, the conditional probability of drawing a white ball at the ith q-drawing, given that $j-1$ white balls are drawn in the previous $i-1$ q-drawings, is given by (4.1). Let U_n be the number of white balls drawn until the nth black ball is drawn.

(a) Show that the probability function of the random variable U_n is given by

$$P(U_n = u) = \begin{bmatrix} n+u-1 \\ u \end{bmatrix}_{q^{-k}} q^{-kn(\alpha-u)} \frac{[\alpha]_{u,q^{-k}}[\beta]_{n,q^{-k}}}{[\alpha+\beta]_{n+u,q^{-k}}},$$

for $u = 0, 1, \ldots$, where $0 < q < 1$ or $1 < q < \infty$ and $\alpha = -r/k$, $\beta = -s/k$, with r and s positive integers and k an integer. Notice that this probability function, on introducing the parameter $p = q^{-1}$ and interchanging the roles of α and β, reduces to probability function (4.20) of the inverse q-Pólya distribution.

(b) Derive the q-factorial moments $E([U_n]_{i,q^k})$, for $i = 1, 2, \ldots$.

(c) Also, deduce the factorial moments $E[(U_n)_j]$, for $j = 1, 2, \ldots$.

4.15 Suppose that random q-drawings of balls are sequentially carried out, one after the other, from an urn, initially containing r white and s black balls, according to the following scheme. After each q-drawing, the drawn ball is placed back in the urn together with one white ball. Determine the probability functions of

(a) the number X_n of black balls drawn in n q-drawings and

(b) the number T_k of q-drawings until the kth black ball is drawn.

4.16 Assume that random q-drawings of balls are sequentially carried out, one after the other, from an urn, initially containing r white and s black balls, according to the following scheme. After each q-drawing, if the drawn ball is white, it is placed back in the urn, while if the drawn ball is black, a white ball is placed in the urn. Determine the probability functions of

(a) the number X_n of black balls drawn in n q-drawings and

(b) the number T_k of q-drawings until the kth black ball is drawn.

4.17 *A q-Ehrenfest model.* Assume that random q-drawings of balls are sequentially carried out, one after the other, from an urn, containing a total of m white and black balls, according to the following scheme. After each q-drawing, the drawn ball is replaced in the urn by a ball of the opposite color. Let X_n be the number of white balls in the urn after the nth q-drawing, for $n = 1, 2, \ldots$. Show that the stationary distribution of the number of white balls in the urn is a q-binomial distribution of the first kind, with parameters $\theta = 1$ and q.

4.18 (*Continuation*). Suppose that random q-drawings of balls are sequentially carried out, one after the other, from an urn, containing a total of m white and black balls, according to the following scheme. After each q-drawing, the drawn ball is randomly replaced in the urn by a white or black ball, with probabilities $\theta/(1+\theta)$ and $1/(1+\theta)$, respectively, where $0 < \theta < \infty$. Let X_n be the number of white balls in the urn after the nth q-drawing, for $n = 1, 2, \ldots$. Show that the stationary distribution of the number of white balls in the urn is a q-binomial distribution of the first kind, with parameters θ and q.

4.19 Consider a sequence of independent Bernoulli trials and assume that the conditional probability of success at the ith trial, given that $j-1$ successes occur in the previous $i-1$ trials, is given by

$$p_{i,j} = 1 - q^{-a(i-1)+b(j-1)+c}, \quad j = 1, 2, \ldots, i, \quad i = 1, 2, \ldots,$$

where the parameters a, b, c, and q are such that $0 \le p_{i,j} \le 1$, for all $j = 1, 2, \ldots, i$ and $i = 1, 2, \ldots$. Let X_n be the number of successes up to the nth trial.

(a) Derive the combinatorial identity

$$\begin{bmatrix} x \\ m \end{bmatrix}_q = \sum_{k=0}^{m} (-1)^k q^{k(n-m)-\binom{k}{2}} \begin{bmatrix} n-k \\ m-k \end{bmatrix}_q \begin{bmatrix} n-x \\ k \end{bmatrix}_{q^{-1}},$$

for m and n positive integers and x a real number and

(b) using it, show that

$$E\left(\begin{bmatrix} X_n \\ m \end{bmatrix}_{q^b}\right) = \sum_{k=0}^{m} (-1)^k q^{(b-a)\binom{k}{2}+bk(n-m)+ck} \begin{bmatrix} n-k \\ m-k \end{bmatrix}_{q^b} \begin{bmatrix} n \\ k \end{bmatrix}_{q^{b-a}}.$$

5

LIMITING DISTRIBUTIONS

5.1 INTRODUCTION

The study of limiting distributions is of great theoretical, as well as of practical, importance. There are two types of limiting theorems: the local and the global. The *local limit theorems* are concerned with the limit of a sequence of probability mass or density functions. Particular cases of such theorems, which were examined in the previous chapters, constitute the theorems on the limit of the sequence of probability (mass) functions

(a) of the q-binomial distribution of the first kind, as the number of trials tends to infinity, and the negative q-binomial distribution of the first kind, as the number of failures tends to infinity, to the probability function of the Heine distribution (Theorem 2.4),

(b) of the q-binomial distribution of the second kind, as the number of trials tends to infinity, and the negative q-binomial distribution of the second kind, as the number of successes tends to infinity, to the probability function of the Euler distribution (Theorem 3.4),

(c) of the q-Pólya and inverse q-Pólya distributions, as the total number of balls tends to infinity, to the probability functions of the q-binomial and the negative q-binomial distributions of the second kind (Theorems 4.2 and 4.6).

Discrete q-Distributions, First Edition. Charalambos A. Charalambides.

The *global (or integral) limit theorems* examine the limit of a sequence of distribution functions. If the limiting distribution is a degenerate (one-point) distribution, we refer to these theorems as *laws of large numbers*, while if the limiting distribution is the standard normal distribution, we refer to them as *central limit theorems*.

In this chapter, after introducing the mode of *stochastic convergence* (or *convergence in probability*) and the mode of *convergence in distribution* (or *convergence in law*), we present the *Chebyshev's law of large numbers*. In the particular case of a sequence of independent Bernoulli trials, with the probability of success varying with the number of trials, which is studied in Chapter 2, the *Poisson's law of large numbers* is concluded. In the other particular case of a sequence of independent geometric sequences of (Bernoulli) trials, with the probability of success varying with the number of geometric sequences of trials, which is studied in Chapter 3, another particular case of Chebyshev's law of large numbers is deduced. The central limit theorems for independent and not necessarily identically distributed random variables are presented next. Specifically, the *Lyapunov* and the *Lindeberg–Feller central limit theorems* are given, without proof, and their use in investigating the limiting q-distributions are discussed. This chapter is concluded with some local limit theorems, which examine the converges of the probability (mass) functions of particular discrete q-deformed distributions to the density function of a *Stieltjes–Wigert distribution*.

5.2 STOCHASTIC AND IN DISTRIBUTION CONVERGENCE

Consider a sequence of random variables X_n, $n = 1, 2, \ldots$, and a random variable X, defined on the same sample space. The notion of stochastic convergence (or convergence in probability) is introduced in the next definition.

Definition 5.1. *The sequence of random variables X_n, $n = 1, 2, \ldots$, converges stochastically (or in probability) to the random variable X, for $n \to \infty$, if for every $\epsilon > 0$, the relation*

$$\lim_{n\to\infty} P(|X_n - X| < \epsilon) = 1$$

or, equivalently, the relation

$$\lim_{n\to\infty} P(|X_n - X| \geq \epsilon) = 0$$

holds true.

The notion of stochastic convergence is closely connected to the notion of convergence in distribution (or in law), which is introduced in the following definition.

Definition 5.2. *The sequence of random variables X_n, $n = 1, 2, \ldots$, converges in distribution (or in law) to the random variable X, for $n \to \infty$, if the sequence of* *distribution functions* $F_n(x) = P(X_n \leq x)$, $n = 1, 2, \ldots$, *converges to the distribution function* $F(x) = P(X \leq x)$,

$$\lim_{n \to \infty} F_n(x) = F(x),$$

for every continuity point x *of* $F(x)$. *The distribution function* $F(x)$ *is called the limiting distribution function.*

The presentation of the connection of the stochastic convergence and the convergence in distribution of a sequence of random variables requires the degenerate (one-point) distribution. Specifically, if the probability mass is concentrated at only one point x_0,

$$f(x_0) = P(X = x_0) = 1,$$

X is a *one-point random variable*, that is a constant, and has a *degenerate distribution*. Its distribution function is readily deduced as

$$F(x) = \begin{cases} 0, & -\infty < x < x_0, \\ 1, & x_0 \leq x < \infty. \end{cases}$$

Theorem 5.1. *The sequence of random variables* X_n, $n = 1, 2, \ldots$, *converges stochastically to a constant* c, *if and only if the sequence of distribution functions* $F_n(x) = P(X_n \leq x)$, $n = 1, 2, \ldots$, *converges to the distribution function,*

$$F(x) = \begin{cases} 0, & -\infty < x < c, \\ 1, & c \leq x < \infty, \end{cases}$$

of a degenerate distribution.

Proof. Assume that the sequence of random variables X_n, $n = 1, 2, \ldots$, converges stochastically to a constant c and consider an arbitrary $\epsilon > 0$. Then,

$$\lim_{n \to \infty} F_n(c - \epsilon) = \lim_{n \to \infty} P(X_n \leq c - \epsilon) = \lim_{n \to \infty} P(|X_n - c| \geq \epsilon) = 0$$

and

$$1 - \lim_{n \to \infty} F_n(c + \epsilon) = \lim_{n \to \infty} P(X_n > c + \epsilon) = \lim_{n \to \infty} P(|X_n - c| > \epsilon) = 0,$$

since $\lim_{n \to \infty} P(|X_n - c| > \epsilon) \leq \lim_{n \to \infty} P(|X_n - c| \geq \epsilon) = 0$. Setting $x = c - \epsilon$ in the first and $x = c + \epsilon$ in the second limiting expression, we deduce the required limiting distribution function of the sequence $F_n(x)$, $n = 1, 2, \ldots$.

Assume now that $\lim_{n \to \infty} F_n(x) = F(x)$, for every continuity point $x \neq c$ of $F(x)$,

$$\lim_{n \to \infty} F_n(x) = \begin{cases} 0, & -\infty < x < c, \\ 1, & c < x < \infty. \end{cases}$$

Then, for an arbitrary $\epsilon > 0$, it follows that

$$\lim_{n \to \infty} P(X_n \leq c - \epsilon) = \lim_{n \to \infty} F_n(c - \epsilon) = 0$$

and

$$\lim_{n\to\infty} P(X_n \geq c+\epsilon) = 1 - \lim_{n\to\infty} F_n(c+\epsilon-) = 0.$$

Therefore,

$$\lim_{n\to\infty} P(|X_n - c| \geq \epsilon) = 0$$

and so the sequence of random variables X_n, $n = 1, 2, \ldots$, converges stochastically to the constant c. □

It is worth noticing that the existence of a global (integral) limit theorem does not guarantee the existence of a corresponding local limit theorem. Precisely, let $f_n(x)$ and $F_n(x)$, $n = 1, 2, \ldots$, be the probability density and the distribution functions of a sequence of random variables X_n, $n = 1, 2, \ldots$. Then, the convergence of the sequence of distribution functions, $\lim_{n\to\infty} F_n(x) = F(x)$, at every continuity point x of $F(x)$, does not necessarily imply the convergence of the sequence of the probability density functions $f_n(x)$, $n = 1, 2, \ldots$, to a probability density function, as the following example illustrates.

Example 5.1. Consider a sequence of two-point random variables X_n, $n = 1, 2, \ldots$, with probability function

$$f_n(x) = P(X_n = x) = \begin{cases} 1/2, & x = x_0 - 1/n,\ x_0 + 1/n, \\ 0, & x \neq x_0 - 1/n,\ x_0 + 1/n, \end{cases}$$

where x_0 is a fixed real number. The distribution function of X_n, $n = 1, 2, \ldots$, is readily deduced as

$$F_n(x) = P(X_n \leq x) = \begin{cases} 0, & -\infty < x < x_0 - 1/n, \\ 1/2, & x_0 - 1/n \leq x < x_0 + 1/n, \\ 1, & x_0 + 1/n \leq x < \infty. \end{cases}$$

Clearly, $\lim_{n\to\infty} F_n(x) = F(x)$, for every continuity point $x \neq x_0$ of

$$F(x) = \begin{cases} 0, & -\infty < x < x_0, \\ 1, & x_0 \leq x < \infty, \end{cases}$$

while the sequence of probability functions $f_n(x)$, $n = 1, 2, \ldots$, does not converge to a probability function. Specifically, we get $\lim_{n\to\infty} f_n(x) = 0$, for every real number x.

5.3 LAWS OF LARGE NUMBERS

Consider a sample (sequence, collection) of n independent random variables X_i, $i = 1, 2, \ldots, n$, and let us denote by $S_n = \sum_{i=1}^{n} X_i$ the sample sum and by $\overline{X}_n = S_n/n$ the sample mean. In statistical inference, the statistics S_n and $\overline{X}_n$ are taken as estimators of unknown parameters. The consistency of an estimator is a (limiting) criterion for evaluating the performance of an estimator. Specifically, the statistic $\overline{X}_n$ is a *consistent estimator* of the parameter θ, if, for $n \to \infty$, the sequence $\overline{X}_n$, $n = 1, 2, \dots$, converges stochastically to θ. The question of convergence may be generalized by considering a sequence of parameters, θ_i, $i = 1, 2, \dots$, and examined the conditions under which the sequence $\overline{X}_n - \overline{\theta}_n$, $n = 1, 2, \dots$, converges stochastically to zero, where $\overline{\theta}_n = \sum_{i=1}^{n} \theta_i / n$. The *Chebyshev's weak law of large numbers* gives a sufficient condition for such a convergence; its derivation is based on the *Chebyshev's inequality*, which is given in the following lemma.

Lemma 5.1. *Let X be a random variable with mean $\mu = E(X)$ and variance $\sigma^2 = V(X)$. Then, for every positive number c, the inequality*

$$P(|X - \mu| \geq c) \leq \frac{\sigma^2}{c^2} \tag{5.1}$$

holds true.

Theorem 5.2. *Let X_i, $i = 1, 2, \dots$, be a sequence of pairwise uncorrelated random variables, with $E(X_i) = \mu_i$ and $V(X_i) = \sigma_i^2$, for $i = 1, 2, \dots$, and assume that*

$$\lim_{n\to\infty} \frac{1}{n^2} \sum_{i=1}^{n} \sigma_i^2 = 0. \tag{5.2}$$

Then, the sequence of differences $\overline{X}_n - \overline{\mu}_n$, $n = 1, 2, \dots$, where

$$\overline{X}_n = \frac{1}{n} \sum_{i=1}^{n} X_i, \quad \overline{\mu}_n = \frac{1}{n} \sum_{i=1}^{n} \mu_i, \quad n = 1, 2, \dots,$$

converges stochastically to zero.

Proof. The mean of $\overline{X}_n$ is readily deduced as

$$E(\overline{X}_n) = \frac{1}{n} E\left(\sum_{i=1}^{n} X_i\right) = \frac{1}{n} \sum_{i=1}^{n} E(X_i) = \frac{1}{n} \sum_{i=1}^{n} \mu_i = \overline{\mu}_n.$$

Furthermore, since the random variables X_i, $i = 1, 2, \dots, n$, are pairwise uncorrelated, the variance of $\overline{X}_n$ is obtained as

$$V(\overline{X}_n) = \frac{1}{n^2} V\left(\sum_{i=1}^{n} X_i\right) = \frac{1}{n^2} \sum_{i=1}^{n} V(X_i) = \frac{1}{n^2} \sum_{i=1}^{n} \sigma_i^2.$$

Thus, applying Chebyshev's inequality (5.1), on the random variable $\overline{X}_n$, we get for every $\epsilon > 0$, the inequality

$$P(|\overline{X}_n - \overline{\mu}_n| \geq \epsilon) \leq \frac{V(\overline{X}_n)}{\epsilon^2} = \frac{1}{\epsilon^2} \cdot \frac{1}{n^2} \sum_{i=1}^{n} \sigma_i^2,$$

where the right-hand side, by assumption (5.2), converges to zero for $n \to \infty$. Therefore,

$$\lim_{n\to\infty} P(|\overline{X}_n - \overline{\mu}_n| \geq \epsilon) \leq 0,$$

and since the probability is always nonnegative, it follows that

$$\lim_{n\to\infty} P(|\overline{X}_n - \overline{\mu}_n| \geq \epsilon) = 0$$

and according to Definition 5.1, the sequence of differences $\overline{X}_n - \overline{\mu}_n$, $n = 1, 2, \ldots,$ converges stochastically to zero. □

The *Poisson's law of large numbers*, which constitutes a particular case of the Chebyshev's weak law of large numbers, is concerned with the number Z_i of successes at the ith trial in a sequence of independent Bernoulli trials, with the probability of success at the ith trial varying with the number of trials, $P_i(\{s\}) = p_i$, $i = 1, 2, \ldots$. In this case, $Z_1, Z_2, \ldots, Z_n$ are independent zero-one Bernoulli random variables, with probability function

$$P(Z_i = 0) = 1 - p_i, \quad P(Z_i = 1) = p_i, \quad i = 1, 2, \ldots,$$

and expected value and variance

$$\mu_i = E(Z_i) = p_i, \quad \sigma_i^2 = V(Z_i) = p_i(1 - p_i), \quad i = 1, 2, \ldots .$$

Note that the variance of Z_i as a function of p_i assumes its maximum value, $\max_{0<p_i<1}\{\sigma_i^2\} = 1/4$, at $p_i = 1/2$. Hence,

$$0 \leq \frac{1}{n^2} \sum_{i=1}^{n} \sigma_i^2 \leq \frac{1}{4n}$$

and condition (5.2) is fulfilled,

$$\lim_{n\to\infty} \frac{1}{n^2} \sum_{i=1}^{n} \sigma_i^2 = 0.$$

Therefore, the Poisson's law of large numbers may be expressed as a corollary of Theorem 5.2.

Corollary 5.1. *Let Z_i be the number of successes at the ith trial, in a sequence of independent Bernoulli trials, with the probability of success at the ith trial varying with the number of trials, $P_i(\{s\}) = p_i$, $i = 1, 2, \ldots$. Then, the sequence of differences $\overline{X}_n - \overline{p}_n$, $n = 1, 2, \ldots,$ where*

$$\overline{X}_n = \frac{1}{n} \sum_{i=1}^{n} Z_i, \quad \overline{p}_n = \frac{1}{n} \sum_{i=1}^{n} p_i, \quad n = 1, 2, \ldots,$$

converges stochastically to zero.

Two interesting consistent estimators of unknown parameters are presented in the following examples.

Example 5.2. Consider a sequence of independent Bernoulli trials and assume that the probability of success at the ith trial is given by

$$p_i = \theta q^{i-1}, \quad i = 1, 2, \dots , \quad 0 < \theta \leq 1, \quad 0 < q < 1.$$

Let Z_i be the number of successes at the ith trial and let $X_n = \sum_{i=1}^{n} Z_i$ be the number of successes in n trials. The expected value of X_n is readily deduced as

$$E(X_n) = \sum_{i=1}^{n} E(Z_i) = \sum_{i=1}^{n} \theta q^{i-1} = \theta [n]_q.$$

Note that, the probability function and the binomial moments of X_n are given in Exercise 2.19. Therefore, according to Corollary 5.1, the sequence of the differences

$$\frac{X_n}{n} - \frac{\theta[n]_q}{n}, \quad n = 1, 2, \dots ,$$

converges stochastically to zero. This convergence implies, for large n, the approximate equality relation $\theta \simeq X_n/[n]_q$. Thus, for a known q, the statistic $X_n/[n]_q$ may be considered as an approximately consistent estimator of the parameter θ. Notice that, for $q = 1$, the statistic X_n/n, which gives the frequency of successes, is a consistent estimator of the (constant) probability of success θ; this is a well-known result.

Example 5.3. *Estimating the rate of increase of a geometrically increasing population.* Consider a population of athletes with its size at the ith epoch given by

$$\alpha_i = \theta q^{-(i-1)}, \quad i = 1, 2, \dots , \quad 0 < \theta < \infty, \quad 0 < q < 1.$$

This is an increasing population with rate $q^{-1} > 1$. Furthermore, consider the record indicator random variables Z_i, $i = 1, 2, \dots$, defined by $Z_i = 1$, if a record occurs at the ith epoch and $Z_i = 0$, if no record occurs at the ith epoch. The probability function of Z_i was derived in Example 2.6 as

$$P(Z_i = 1) = \frac{1}{[i]_q}, \quad P(Z_i = 0) = 1 - \frac{1}{[i]_q}, \quad i = 1, 2, \dots .$$

Let $X_n = \sum_{i=1}^{n} Z_i$ be the number of records up to time (epoch) n. The expected value of X_n is given by

$$E(X_n) = \sum_{i=1}^{n} E(Z_i) = \sum_{i=1}^{n} \frac{1}{[i]_q} = (1 - q)n + \sum_{i=1}^{n} \frac{q^i}{[i]_q}.$$

Note that $\lim_{n \to \infty} \sum_{i=1}^{n} q^i/[i]_q = -l_q(1 - q)$, the q-logarithmic function (1.25). Thus, according to Corollary 5.1, the sequence of the differences

$$\frac{X_n}{n} - \left((1 - q) - \frac{1}{n} l_q(1 - q)\right), \quad n = 1, 2, \dots ,$$

converges stochastically to zero. This convergence implies, for large n, the approximate equality relation $1 - q \simeq X_n/n$. Therefore, the statistic $1 - X_n/n$ may be considered as an approximately consistent estimator of the inverse rate of increase q.

Let us now turn to the discussion of a sequence of independent geometric sequences of (Bernoulli) trials, with probability of success at the jth geometric sequence of trials varying with the number of geometric sequences, $Q_j(\{s\}) = p_j$, $j = 1, 2, \ldots$. Let U_j be the number of failures at the jth geometric sequence of trials, $j = 1, 2, \ldots, n$ Then, the random variables $U_1, U_2, \ldots, U_n$ are independent with probability function

$$P(U_j = u) = p_j q_j^u, \quad u = 0, 1, \ldots, \quad j = 1, 2, \ldots, n,$$

and expected value and variance

$$\mu_j = E(U_j) = \frac{q_j}{p_j}, \quad \sigma_j^2 = V(U_j) = \frac{q_j}{p_j^2}, \quad j = 1, 2, \ldots, n.$$

Furthermore, let us denote by $W_n = \sum_{j=1}^n U_j$ the number of failures until the occurrence of the nth success. Also, consider the variance $V(U_j)$ as a function of p_j,

$$g(p_j) = (1 - p_j)/p_j^2, \quad 0 < p_j < 1.$$

Note that this function is not bounded from above and precisely $\lim_{p_j \downarrow 0} g(p_j) = \infty$. This disadvantage can be rectified by slightly restricting the domain of its definition. Specifically, consider a fix point $0 < \theta < 1$ and let

$$g(p_j) = (1 - p_j)/p_j^2, \quad \theta \leq p_j < 1.$$

Furthermore, the second-order equation

$$\{(a^2 - 1)/4\}t^2 + t - 1 = 0, \quad 1 < a < \infty,$$

has two roots: $t_0 = 2/(a + 1) > 0$ and $t_1 = -2/(a - 1) < 0$. For $t_0 \leq t < \infty$, it follows that $\{(a^2 - 1)/4\}t^2 + t - 1 \geq 0$ or equivalently that

$$g(t) = (1 - t)/t^2 \leq (1 - t_0)/t_0^2.$$

Hence, $\sigma_j^2 = V(U_j) \leq (1 - \theta)/\theta^2$, for $\theta \leq p_j < 1$, with $0 < \theta < 1$. Therefore,

$$0 \leq \frac{1}{n^2} \sum_{j=1}^n \sigma_j^2 \leq \frac{1 - \theta}{n\theta^2}, \quad 0 < \theta < 1,$$

and

$$\lim_{n \to \infty} \frac{1}{n^2} \sum_{j=1}^n \sigma_j^2 = 0.$$

Consequently, the following corollary of Theorem 5.2 is shown.

Corollary 5.2. *Let U_j be the number of failures at the jth geometric sequence of trials, with probability of success at the jth geometric sequence of trials varying with the number of geometric sequences, $Q_j(\{s\}) = p_j$, $j = 1, 2, \ldots, n$. Suppose that*

$$\theta \leq p_j < 1, \quad j = 1, 2, \ldots, n, \tag{5.3}$$

for a fixed θ, with $0 < \theta < 1$. Then, the sequence of differences $\overline{W}_n - E(\overline{W}_n)$, $n = 1, 2, \ldots$, where

$$\overline{W}_n = \frac{1}{n}\sum_{j=1}^{n} U_j, \quad E(\overline{W}_n) = \frac{1}{n}\sum_{j=1}^{n} \frac{1-p_j}{p_j}, \quad n = 1, 2, \ldots,$$

converges stochastically to zero.

Example 5.4. Consider a sequence of independent geometric sequences of trials and assume that the probability of success at the jth sequence of trials is given by

$$p_j = 1 - q^j, \quad j = 1, 2, \ldots, \quad 0 < q < 1.$$

Let U_j be the number of failures at the jth geometric sequence of trials and let $W_n = \sum_{j=1}^{n} U_j$ be the number of failures until the occurrence of the nth success. The expected value of the random variable W_n is readily deduced as

$$E(W_n) = \sum_{j=1}^{n} E(U_j) = \frac{1}{1-q}\sum_{j=1}^{n} \frac{q^j}{[j]_q}.$$

Note that $\lim_{n\to\infty} \sum_{j=1}^{n} q^j/[j]_q = -l_q(1-q)$, the q-logarithmic function (1.25). Thus, according to Corollary 5.2, with $\theta = 1 - q$, the sequence of the differences

$$\frac{W_n}{n} - \frac{-l_q(1-q)}{n(1-q)}, \quad n = 1, 2, \ldots,$$

converges stochastically to zero. Note that for small q, the q-logarithm of $1 - q$ can be approximated by $-l_q(1-q) \simeq q$. Therefore, the statistic W_n may be considered as an approximately consistent estimator of the parametric function $q/(1-q)$.

5.4 CENTRAL LIMIT THEOREMS

Consider a sequence of independent random variables X_i, $i = 1, 2, \ldots$, with $E(X_i) = \mu_i$ and $V(X_i) = \sigma_i^2$, $i = 1, 2, \ldots$. Furthermore, let

$$\overline{X}_n = \frac{1}{n}\sum_{i=1}^{n} X_i, \quad \overline{\mu}_n = \frac{1}{n}\sum_{i=1}^{n} \mu_i, \quad \overline{\sigma}_n = \left(\frac{1}{n}\sum_{i=1}^{n} \sigma_i^2\right)^{1/2}, \quad n = 1, 2, \ldots.$$

The law of large numbers, presented in Section 5.3, investigates the convergence of the random variable $\overline{X}_n - \overline{\mu}_n$, as $n \to \infty$, to a degenerate (one-point) random variable (constant equal to zero). But this gives us no idea as to how the distribution of $\overline{X}_n$ can be approximated for large n. In this section, we examine the convergence of the distribution of the standardized random variable $Z_n = \sqrt{n}(\overline{X}_n - \overline{\mu}_n)/\overline{\sigma}_n$, as $n \to \infty$, to the standard normal distribution. A sufficient condition for Z_n to have a limiting standard normal distribution is given by the following *Lyapunov central limit theorem.*

Theorem 5.3. *Let X_i, $i = 1, 2, \ldots$, be a sequence of independent random variables, with expected value $E(X_i) = \mu_i$, variance $V(X_i) = \sigma_i^2$, and absolute central moment of third-order $E(|X_i - \mu_i|^3) = \beta_i$, $i = 1, 2, \ldots$. Furthermore, let*

$$B_n = \left(\sum_{i=1}^{n} \beta_i\right)^{1/3}, \quad C_n = \left(\sum_{i=1}^{n} \sigma_i^2\right)^{1/2}$$

and assume that

$$\lim_{n\to\infty} B_n/C_n = 0. \tag{5.4}$$

Then, the distribution function $F_n(z) = P(Z_n \leq z)$ of the standardized random variable

$$Z_n = \frac{\sqrt{n}(\overline{X}_n - \overline{\mu}_n)}{\overline{\sigma}_n}$$

converges to the distribution function $\Phi(z)$, $-\infty < z < \infty$, of the standard normal distribution.

A necessary and sufficient condition for the standardized random variable $Z_n = \sqrt{n}(\overline{X}_n - \overline{\mu}_n)/\overline{\sigma}_n$ to have a limiting standard normal distribution is given by the following *Lindeberg–Feller central limit theorem.*

Theorem 5.4. *Let X_i, $i = 1, 2, \ldots$, be a sequence of independent random variables, with distribution function $G_i(x) = P(X_i \leq x)$, expected value $E(X_i) = \mu_i$ and variance $V(X_i) = \sigma_i^2$, $i = 1, 2, \ldots$. Also, let*

$$Z_n = \frac{\sqrt{n}(\overline{X}_n - \overline{\mu}_n)}{\overline{\sigma}_n}, \quad C_n = \left(\sum_{i=1}^{n} \sigma_i^2\right)^{1/2},$$

and $F_n(z) = P(Z_n \leq z)$. Then, the relations

$$\lim_{n\to\infty} \max_{1\leq i\leq n} \frac{\sigma_i}{C_n} = 0 \qquad \text{and} \qquad \lim_{n\to\infty} F_n(z) = \Phi(z), \quad -\infty < z < \infty,$$

hold if and only if, for every $\epsilon > 0$,

$$\lim_{n\to\infty} \frac{1}{C_n^2} \sum_{i=1}^{n} \int_{|x-\mu_i|>\epsilon C_n} (x - \mu_i)^2 dG_i(x) = 0. \tag{5.5}$$

The following useful corollary of Lindeberg–Feller central limit theorem may be deduced.

Corollary 5.3. *Let X_i, $i = 1, 2, \ldots$, be a sequence of independent and uniformly bounded random variables,*

$$P(|X_i| \leq a) = 1, \quad a > 0 \text{ constant}, \tag{5.6}$$

with expected value $E(X_i) = \mu_i$ and variance $V(X_i) = \sigma_i^2$, $i = 1, 2, \ldots$. Then, the distribution function $F_n(z) = P(Z_n \leq z)$ of the standardized random variable

$$Z_n = \frac{\sqrt{n}(\overline{X}_n - \overline{\mu}_n)}{\overline{\sigma}_n}$$

converges to the distribution function $\Phi(z)$, $-\infty < z < \infty$, of the standard normal distribution, if and only if

$$\lim_{n\to\infty} \sum_{i=1}^{n} \sigma_i^2 = \infty. \tag{5.7}$$

Proof. Assume that condition (5.7) is satisfied. Clearly, assumption (5.6) implies that the random variables $X_i - \mu_i$, $i = 1, 2, \ldots$, are uniformly bounded. Therefore, for every $\epsilon > 0$, we can find an n_0 such that for $n > n_0$, it follows that

$$P(|X_i - \mu_i| < \epsilon C_n, \text{ for } i = 1, 2, \ldots, n) = 1$$

and so condition (5.5) is fulfilled.

Furthermore, assume that the limiting relation

$$\lim_{n\to\infty} F_n(z) = \Phi(z), \quad -\infty < z < \infty,$$

holds and condition (5.7) does not. Then, there exists a finite number C, such that $\lim_{n\to\infty} C_n^2 = C^2$. The last limiting relation, together with the convergence of the standardized random variable $Z_n = \sqrt{n}(\overline{X}_n - \overline{\mu}_n)/\overline{\sigma}_n$, to the standard normal distribution, implies that the sum $S = \sum_{i=1}^{\infty}(X_i - \mu_i)$ has the normal distribution, with $E(S) = 0$ and $V(S) = C^2$. Now, consider the independent random variables

$$U = X_1 - \mu_1, \quad W = \sum_{i=2}^{\infty}(X_i - \mu_i),$$

with sum $U + W = S$ obeying a normal distribution. By the well-known Cramér theorem, both variables U and W have the normal distribution. However, the hypothesis that the random variables X_i, $i = 1, 2, \ldots$, are uniformly bounded, implies that U does not have the normal distribution, which is a contradiction. Hence, the proof of the theorem is completed. □

Corollary 5.4. *Let Z_i be the number of successes at the ith trial, in a sequence of independent Bernoulli trials, with the probability of success at the ith trial varying* *with the number of trials,* $P_i(\{s\}) = p_i$, $i = 1, 2, \ldots$. *Then, the distribution function* $F_n(z) = P(Z_n \leq z)$ *of the standardized random variable*

$$\frac{\sqrt{n}(\overline{X}_n - \overline{p}_n)}{\overline{\sigma}_n},$$

where

$$\overline{p}_n = \frac{1}{n}\sum_{i=1}^{n} p_i, \quad \overline{\sigma}_n = \left(\frac{1}{n}\sum_{i=1}^{n} p_i(1-p_i)\right)^{1/2},$$

converges to the distribution function $\Phi(z)$, $-\infty < z < \infty$, *of the standard normal distribution, if and only if the series of variances is divergent,* $\sum_{i=1}^{\infty} p_i(1-p_i) = \infty$.

Example 5.5. Consider a sequence of independent Bernoulli trials and assume that the probability of success at the ith trial is given by

$$p_i = \frac{\theta q^{i-1}}{1+\theta q^{i-1}}, \quad i = 1, 2, \ldots, \quad 0 < \theta < \infty, \quad 0 < q < 1 \quad \text{or} \quad 1 < q < \infty,$$

and let Z_i be the number of successes at the ith trial, $i = 1, 2, \ldots$. The number $X_n = \sum_{i=1}^{n} Z_i$ of successes in n trials, according to Theorem 2.1, obeys a q-binomial distribution of the first kind, with mean and variance

$$E(X_n) = \sum_{i=1}^{n} \frac{\theta q^{i-1}}{1+\theta q^{i-1}}, \quad V(X_n) = \sum_{i=1}^{n} \frac{\theta q^{i-1}}{(1+\theta q^{i-1})^2}.$$

Since the limit of the sum of the variances $\sigma_i^2 = p_i(1-p_i)$, $i = 1, 2, \ldots, n$, as the number of trials increases indefinitely, is finite,

$$\lim_{n\to\infty} V(X_n) = \sum_{i=1}^{\infty} \frac{\theta q^{i-1}}{(1+\theta q^{i-1})^2} < \infty,$$

the standardized random variable $(\overline{X}_n - \mu_{\overline{X}_n})/\sigma_{\overline{X}_n}$, according to Corollary 5.4, does not converge in distribution to the standard normal random variable.

Example 5.6. *Asymptotic normality of the number of records in Olympic games.* Consider a population of athletes with its size at the ith epoch given by

$$\alpha_i = \theta q^{-(i-1)}, \quad i = 1, 2, \ldots, \quad 0 < \theta < \infty, \quad 0 < q < 1.$$

This is an increasing population with rate $q^{-1} > 1$. Furthermore, consider the record indicator random variables Z_i, $i = 1, 2, \ldots$, defined by $Z_i = 1$, if a record occurs at the ith epoch and $Z_i = 0$, if no record occurs at the ith epoch. The probability function of Z_i was derived in Example 2.6 as

$$P(Z_i = 1) = \frac{1}{[i]_q}, \quad P(Z_i = 0) = 1 - \frac{1}{[i]_q}, \quad i = 1, 2, \ldots.$$

Let $X_n = \sum_{i=1}^{n} Z_i$ be the number of records up to time (epoch) n. The expected value and variance of X_n are given by

$$E(X_n) = \sum_{i=1}^{n} E(Z_i) = \sum_{i=1}^{n} \frac{1}{[i]_q}, \quad V(X_n) = \sum_{i=1}^{n} V(Z_i) = \sum_{i=1}^{n} \frac{1}{[i]_q} - \sum_{i=1}^{n} \frac{1}{[i]_q^2}.$$

Since

$$\sum_{i=1}^{n} \frac{1}{[i]_q} = (1-q)n + \sum_{i=1}^{n} \frac{q^i}{[i]_q}, \quad \sum_{i=1}^{n} \frac{1}{[i]_q^2} = (1-q)\sum_{i=1}^{n} \frac{1}{[i]_q} + \sum_{i=1}^{n} \frac{q^i}{[i]_q^2},$$

these distributional measures may be written as

$$E(X_n) = (1-q)n + \sum_{i=1}^{n} \frac{q^i}{[i]_q}, \quad V(X_n) = q(1-q)n + q\sum_{i=1}^{n} \frac{q^i}{[i]_q} - \sum_{i=1}^{n} \frac{q^i}{[i]_q^2}.$$

Clearly,

$$\lim_{n\to\infty} \{E(X_n) - (1-q)n\} = -l_q(1-q)$$

and

$$\lim_{n\to\infty} \{V(X_n) - q(1-q)n\} = -ql_q(1-q) - h_q(2), \quad \lim_{n\to\infty} V(X_n) = \infty,$$

where $l_q(1-t)$ is the q-logarithmic function (1.25) and $h_q(2) = \sum_{i=1}^{\infty} q^i/[i]_q^2$ is a q-zeta function. Therefore, by Corollary 5.4,

$$\lim_{n\to\infty} P\left(\frac{X_n - (1-q)n}{\sqrt{q(1-q)n}} \leq z \right) = \Phi(z),$$

where $\Phi(z)$ denotes the distribution function of the standard normal distribution.

5.5 STIELTJES–WIGERT DISTRIBUTION AS LIMITING DISTRIBUTION

In applied statistical inference and in a variety of situations it is observed that, while the random variable under consideration does not follow a normal distribution, its logarithm is a random variable obeying a normal distribution. This remark leads to the introduction of the following distribution. If the logarithm, $Y = \log X$, of a continuous random variable obeys the normal distribution, with mean $E(Y) = \mu$ and variance $V(Y) = \sigma^2$, then X has the *lognormal distribution*. Its density function is given by

$$f(x) = \frac{1}{\sigma x\sqrt{2\pi}} e^{-\frac{1}{2\sigma^2}(\log x-\mu)^2}, \quad 0 < x < \infty.$$

The nth (power) moment $E(X^n) = E(e^{nY})$, on using the moment generating function of the normal distribution, $E(e^{tY}) = e^{\mu t+\sigma^2 t^2/2}$, is deduced as

$$E(X^n) = e^{n\mu+n^2\sigma^2/2}, \quad n = 1, 2, \dots .$$

In probability theory and more generally in measure theory, the moment problem asks whether a sequence of (power) moments μ'_n, $n = 0, 1, \dots$, with $\mu'_0 = 1$, uniquely determines a probability distribution F and more generally a positive measure μ. For example, the normal distribution is uniquely determined by its moment sequence, while the lognormal distribution is not. Furthermore, there exists a sequence of normalized orthogonal polynomials, with respect to the positive measure μ, which are expressed by the moment sequence μ'_n, $n = 0, 1, \dots$. It is well known that the *Hermite polynomials*,

$$H_n(x) = n! \sum_{k=0}^{[n/2]} \frac{(-1)^{n+k} x^{n-2k}}{k!(n-2k)!2^k}, \quad n = 0, 1, \dots ,$$

are orthogonal with respect to the normal distribution. Wigert (1923), generalizing a problem previously examined by Stieltjes, considered the particular case $\mu = 0$, $\sigma^2 = \log q^{-1}$ of the lognormal distribution and its *size-biased (weighted) distribution* with density function, $h(x) = xf(x)/E(X)$,

$$h(x) = \frac{q^{1/2}}{\sqrt{2\pi \log q^{-1}}} e^{-\frac{(\log x)^2}{2\log q^{-1}}}, \quad 0 < x < \infty, \quad 0 < q < 1,$$

and moments $\mu'_n = q^{-n(1+n/2)}$, $n = 0, 1, \dots$, and succeeded in deducing the orthogonal polynomials, with respect to it, as

$$S_n(x; q) = \frac{(-1)^n q^{1/4}}{\sqrt{q^{-n}(1-q)^n [n]_q!}} \sum_{k=0}^{n} \begin{bmatrix} n \\ k \end{bmatrix}_q (-1)^k q^{k(k+1/2)} x^k, \quad n = 0, 1, \dots .$$

These polynomials are nowadays known as *Stieltjes–Wigert polynomials*. Furthermore, consider the shape transformation $W = X/\sqrt{q}$ of the size-biased lognormal random variable, the distribution of which appears as limiting distribution of several discrete q-distributions. Its density function $g(w) = \sqrt{q} f(w\sqrt{q})$, since

$$(\log(w\sqrt{q}))^2 = (\log w)^2 - \log q^{-1} \log w + \frac{1}{4}(\log q^{-1})^2$$

and

$$-\frac{(\log(w\sqrt{q}))^2}{2\log q^{-1}} = -\frac{(\log w)^2}{2\log q^{-1}} + \log(\sqrt{w}) + \log q^{1/8},$$

is obtained as

$$g(w) = \frac{q^{5/8}\sqrt{w}}{\sqrt{2\pi \log q^{-1}}} e^{-\frac{(\log w)^2}{2\log q^{-1}}}, \quad 0 < w < \infty, \tag{5.8}$$

with moments $E(W^n) = q^{-n(n+3)/2}$, $n = 0, 1, \ldots$. In particular,

$$\mu_W = E(W) = q^{-2}, \quad \sigma_W = q^{-2}\sqrt{q^{-1} - 1}. \tag{5.9}$$

The distribution with density function (5.8) may be called *Stieltjes–Wigert distribution*.

As was already noted, in the introduction of this chapter, the probability (mass) functions of several discrete q-deformed distributions converge to the density function of a *Stieltjes–Wigert distribution*. This convergence is facilitated by an approximation of the q-factorial of a positive integer, which is present in the expression of these probability (mass) functions. In the next theorem, an asymptotic expansion of a q-factorial of a positive integer is derived.

Theorem 5.5. *Let n be a positive integer and q a real number, with* $0 < q < 1$ *or* $1 < q < \infty$. *Then, the following asymptotic expansion of the q-factorial of n, for large n, holds true*

$$[n]_q! \simeq \begin{cases} q^{\binom{n}{2} - \frac{n}{2}}[n]_{q^{-1}}^{n+1/2} e_q(-q^{-1}[n]_{q^{-1}})\sqrt{\frac{2\pi(q^{-1}-1)}{\log q^{-1}}}, & 0 < q < 1, \\ q^{\frac{n}{2}}[n]_q^{n+1/2} E_q(-q[n]_q)\sqrt{\frac{2\pi(q-1)}{\log q}}, & 1 < q < \infty, \end{cases} \tag{5.10}$$

where $E_q(t) = \prod_{i=1}^{\infty}(1 + t(1-q)q^{i-1})$ *and* $e_q(t) = \prod_{i=1}^{\infty}(1 - t(1-q)q^{i-1})^{-1}$ *are the q-exponential functions of (1.23) and (1.24), respectively.*

Sketch of Proof. Let us first consider the case $0 < q < 1$. The q-exponential function $E_q(t)$ may be expanded into a power series as

$$E_q(t) = \sum_{k=0}^{\infty} q^{\binom{k}{2}} \frac{t^k}{[k]_q!}, \quad -\infty < t < \infty.$$

Also, it can be expressed as

$$E_q(t) = e^{g(t)}, \quad -\infty < t < \infty,$$

where

$$g(t) = \sum_{i=1}^{\infty} \log(1 + t(1-q)q^{i-1}), \quad -\infty < t < \infty.$$

An asymptotic expansion of the q-factorial of n, may be obtained by expressing it via Cauchy's integral formula that gives the coefficients of a power series

$$\frac{q^{\binom{n}{2}}}{[n]_q!} = \frac{1}{2\pi i}\int_{|t|=r} \frac{E_q(t)}{t^{n+1}}dt = \frac{1}{2\pi i}\int_{|t|=r} \frac{e^{g(t)}}{t^{n+1}}dt,$$

where the contour of integration is taken to be the circle of radius r. It turns out that it is convenient to switch to polar coordinates, by setting $x = re^{i\theta}$. Then,

$$\frac{q^{\binom{n}{2}}}{[n]_q!} = \frac{e^{g(r)}}{r^n 2\pi} \int_{-\pi}^{\pi} \exp\left[g(re^{i\theta}) - g(r) - in\theta\right] d\theta.$$

This integral is estimated by using the saddle point method, with the radius r being the solution of the equation $rg'(r) = n$. Expanding the exponent $G(\theta) = g(re^{i\theta}) - g(r) - in\theta$ into a Maclaurin series about $\theta = 0$, we get

$$G(\theta) = -\phi^2 + \phi^2 \sum_{k=1}^{\infty} \alpha_k(r) \frac{(\psi\phi)^k}{k!},$$

where

$$\phi = \frac{1}{\sqrt{2}}[rg'(r) + r^2 g''(r)]^{1/2}\theta, \quad \psi = [g'(r)]^{-1/2}$$

and

$$\alpha_k(r) = \frac{[1 + rg''(r)/g'(r)]^{-k/2-1}}{(k+1)(k+2)g'(r)(r/2)^{k/2+1}} \left[\frac{d^{k+2} g(re^{i\theta})}{d\theta^{k+2}}\right]_{\theta=0}.$$

The absence of a linear term in θ indicates a saddle point. The function $|e^{G(\theta)}|$ is unimodal with its peak at $\theta = 0$.

An estimation of $[n]_q!$ is obtained by isolating a small portion of the contour (corresponding to x near the real axis). Thus, consider the integrals

$$I_1 = \frac{1}{\sqrt{2\pi}} \int_{-\delta}^{\delta} e^{G(\theta)} d\theta, \quad I_2 = \frac{1}{\sqrt{2\pi}} \int_{\delta}^{2\pi-\delta} e^{G(\theta)} d\theta,$$

and choose $\delta = n^{-3/8}$ so that, for $n \to \infty$, the limiting relations $n\delta^2 \to \infty$ and $n\delta^3 \to 0$ hold true. This choice of δ implies that $e^{G(\delta)}$ is exponentially small, being dominated by a term of the form $e^{-O(n^{1/4})}$. As $|e^{G(\theta)}|$ decreases in $[\delta, 2\pi - \delta]$,

$$|e^{G(\theta)}| \le |e^{G(\delta)}|, \quad \delta \le \theta \le 2\pi - \delta.$$

Furthermore, $G(\delta) \sim -n\delta^2$ and so $|I_2| = O(e^{-\frac{1}{2}n^{1/4}})$. Therefore, the choice of δ is large enough so that the first integral I_1 captures most of the contribution, while the second integral I_2 is exponentially small. Evaluating the integral I_1, the following approximate expansion of the q-factorial of n is obtained:

$$[n]_q! \simeq \sqrt{2\pi} q^{\binom{n}{2}} e^{-g(r)} [rg'(r) + r^2 g''(r)]^{1/2} r^n, \tag{5.11}$$

where, according to the saddle point method principles, the radius r is the real solution of the equation $rg'(r) = n$. Applying the standard approximation of a sum by an integral to the series

$$rg'(r) = \sum_{i=1}^{\infty} \frac{r(1-q)q^{i-1}}{1 + r(1-q)q^{i-1}},$$

the following double inequality is deduced

$$-\frac{\log(1+r(1-q))}{\log q} \leq rg'(r) \leq -\frac{\log(1+r(1-q))}{\log q} + \frac{r(1-q)}{1+r(1-q)}.$$

Clearly, for $r = q^{-1}[n]_{q^{-1}}$, this double inequality reduces to

$$n \leq rg'(r) \leq n + (1-q^n). \tag{5.12}$$

Consequently, $r = q^{-1}[n]_{q^{-1}}$ is a solution of the equation $rg'(r) = n$. Similarly, applying the standard approximation of a sum by an integral to the series

$$r^2 g''(r) = -\sum_{j=1}^{\infty} \frac{(1-q)^2 r^2 q^{2(j-1)}}{(1+((1-q)r)q^{j-1})^2},$$

the following double inequality is obtained

$$\frac{\log(1+(1-q)r)}{\log q} - \frac{(1-q)r}{(1+(1-q)r)\log q} + \frac{(1-q)^2 r^2}{(1+(1-q)r)^2} \leq r^2 g''(r) \leq \frac{\log(1+(1-q)r)}{\log q} - \frac{(1-q)r}{(1+(1-q)r)\log q}.$$

Since $r = q^{-1}[n]_{q^{-1}}$, this double inequality reduces to

$$-n - \frac{1-q^n}{\log q} + (1-q^n)^2 \leq r^2 g''(r) \leq -n - \frac{1-q^n}{\log q}. \tag{5.13}$$

Combining the double inequalities (5.12) and (5.13), we deduce the double inequality

$$-\frac{1-q^n}{\log q} + (1-q^n)^2 \leq rg'(r) + r^2 g''(r) \leq -\frac{1-q^n}{\log q} + (1-q^n),$$

which implies the approximation

$$rg'(r) + r^2 g''(r) \simeq \frac{q^{-1}-1}{\log q^{-1}} q^n [n]_{q^{-1}}.$$

Introducing the last approximate value, together with

$$e^{-g(r)} = \frac{1}{E_q(r)} = e_q(-r), \quad r = q^{-1}[n]_{q^{-1}},$$

into (5.11), we deduce the first part of (5.10).

The case $1 < q < \infty$, by replacing q by q^{-1}, with $0 < q^{-1} < 1$, is transformed to the preceding case. Then, using the relations

$$[n]_{q^{-1}}! = q^{-\binom{n}{2}} [n]_q!, \qquad e_{q^{-1}}(t) = E_q(t),$$

the second part (5.10) is readily obtained. □

An approximation of the q-deformed binomial distribution of the first kind (cf. Remark 1.8) is presented in the following theorem.

Theorem 5.6. *Consider the q-deformed binomial distribution of the first kind, with probability function*

$$f_{[X_n]_{q^{-1}}}([x]_{q^{-1}}) = \begin{bmatrix} n \\ x \end{bmatrix}_q \frac{\theta^x q^{\binom{x}{2}}}{\prod_{i=1}^{n}(1+\theta q^{i-1})}, \quad x = 0, 1, \ldots, n, \tag{5.14}$$

for $0 < \theta < \infty$ *and* $0 < q < 1$. *Assume that the parameter* θ *depends on n,* θ_n, *and that* $\theta_n \to \infty$ *and* $\theta_n q^n \to 0$ *as* $n \to \infty$. *Furthermore, let*

$$[X_n]_{q^{-1}} = \theta_n q(1-q)^{-1} W_n. \tag{5.15}$$

Then, the limit of the probability function of W_n, *as* $n \to \infty$, *is the density function of the Stieltjes–Wigert distribution,*

$$\lim_{n\to\infty} f_{W_n}(w) = \frac{q^{5/8}\sqrt{w}}{\sqrt{2\pi \log q^{-1}}} e^{-\frac{(\log w)^2}{2\log q^{-1}}}, \quad 0 < w < \infty, \quad 0 < q < 1. \tag{5.16}$$

Proof. The probability function of the q-deformed binomial distribution of the first kind, on using the approximation

$$\begin{bmatrix} n \\ x \end{bmatrix}_q \simeq \frac{(1-q)^{-x}}{[x]_q!}, \quad x = 0, 1, \ldots, \text{ for large } n,$$

together with (5.10), is expressed as

$$f_{[X_n]_{q^{-1}}}([x]_{q^{-1}}) \simeq K \frac{(\theta_n^{-1}q^{-1}(1-q)q^{-x})^{1/2}}{(\theta_n^{-1}q^{-1}(1-q)[x]_{q^{-1}})^{x+1/2}} \cdot \frac{E_q(q^{-1}[x]_{q^{-1}})}{\prod_{i=1}^{\infty}(1+\theta_n q^{i-1})},$$

where K is a normalizing constant. Furthermore, expression (5.15), written in terms of the corresponding values of the random variables,

$$[x]_{q^{-1}} = \theta_n q(1-q)^{-1} w,$$

leads to the following approximations, for large n,

$$q^{-x} - 1 \simeq \theta_n w, \quad q^{-x} \simeq \theta_n w, \quad x \simeq \frac{\log(\theta_n w)}{\log q^{-1}} = \frac{\log \theta_n + \log w}{\log q^{-1}}.$$

Using these exact and approximate values, we get

$$\begin{aligned}(\theta_n^{-1}q^{-1}(1-q)[x]_{q^{-1}})^{x+1/2} &= w^{x+1/2} = \sqrt{w}\exp\{x\log w\} \\ &\simeq \sqrt{w}\exp\left\{\frac{(\log w)^2}{\log q^{-1}} + \frac{\log\theta_n \log w}{\log q^{-1}}\right\}, \\ (\theta_n^{-1}q^{-1}(1-q)q^{-x})^{1/2} &\simeq \sqrt{w}(q^{-1}-1)^{1/2}.\end{aligned}$$

Consequently, the probability function of W_n may be written as

$$f_{W_n}(w) \simeq K e^{-\frac{(\log w)^2}{\log q^{-1}}} e^{-\frac{\log \theta_n \log w}{\log q^{-1}}} \frac{\prod_{i=1}^{\infty}(1+\theta_n w q^{i-1})}{\prod_{i=1}^{\infty}(1+\theta_n q^{i-1})}. \tag{5.17}$$

Moreover, the approximation of the quotient of products

$$\frac{\prod_{i=1}^{\infty}(1+\theta_n w q^{i-1})}{\prod_{i=1}^{\infty}(1+\theta_n q^{i-1})} = \exp\{G(w;\theta_n) - G(1;\theta_n)\}, \tag{5.18}$$

with $G(w;\theta_n) = \sum_{i=1}^{\infty} \log(1+\theta_n w q^{i-1})$, may be carried out by approximating the sum $G(w;\theta_n)$. Since the function $h(t) = \log(1+\theta_n w q^{t-1})$, for $t \in [1,\infty)$, has derivatives of all orders, on applying the Euler–Maclaurin summation formula, $G(w;\theta_n)$ is approximated by

$$\begin{aligned} G(w;\theta_n) =& \int_1^{\infty} \log(1+\theta_n w q^{t-1})dt + \frac{1}{2}\log(1+\theta_n w) \\ &+ \frac{\log q}{12} \cdot \frac{\theta_n w}{1+\theta_n w} + O(\theta_n^{-1}) \\ =& \frac{1}{\log q} Li_2(-\theta_n w) + \frac{1}{2}\log(1+\theta_n w) \\ &+ \frac{\log q}{12} \cdot \frac{\theta_n w}{1+\theta_n w} + O(\theta_n^{-1}), \end{aligned}$$

where $Li_2(u) = \sum_{j=1}^{\infty} u^j/j^2$ is the dilogarithm function. Using Landen's identity

$$Li_2(-u) = -Li_2\left(\frac{u}{1+u}\right) - \frac{1}{2}(\log(1+u))^2,$$

we get

$$\begin{aligned} G(w;\theta_n) \simeq& \frac{(\log(\theta_n w))^2}{2\log q^{-1}} + \frac{1}{2}\log(1+\theta_n w) + \frac{1}{\log q^{-1}} Li_2\left(\frac{\theta_n w}{1+\theta_n w}\right) \\ &+ \frac{\log q}{12} \cdot \frac{\theta_n w}{1+\theta_n w} + O(\theta_n^{-1}) \end{aligned}$$

and, in particular,

$$\begin{aligned} G(1;\theta_n) \simeq& \frac{(\log \theta_n)^2}{2\log q^{-1}} + \frac{1}{2}\log(1+\theta_n) + \frac{1}{\log q^{-1}} Li_2\left(\frac{\theta_n}{1+\theta_n}\right) \\ &+ \frac{\log q}{12} \cdot \frac{\theta_n}{1+\theta_n} + O(\theta_n^{-1}). \end{aligned}$$

Since

$$(\log(\theta_n w))^2 = (\log \theta_n)^2 + 2\log \theta_n \log w + (\log w)^2,$$

we deduce the approximation

$$\begin{aligned} G(w;\theta_n) - G(1;\theta_n) \simeq & \frac{(\log w)^2}{2\log q^{-1}} + \frac{\log \theta_n \log w}{\log q^{-1}} + \frac{1}{2}\log\left(\frac{1+\theta_n w}{1+\theta_n}\right) \\ & + \frac{1}{\log q^{-1}}\left(Li_2\left(\frac{\theta_n w}{1+\theta_n w}\right) - Li_2\left(\frac{\theta_n}{1+\theta_n}\right)\right) \\ & + \frac{\log q}{12}\left(\frac{\theta_n w}{1+\theta_n w} - \frac{\theta_n}{1+\theta_n}\right) + O(\theta_n^{-1}) \end{aligned}$$

and so

$$\lim_{n\to\infty}\left\{G(w;\theta_n) - G(1;\theta_n) - \frac{\log\theta_n \log w}{\log q^{-1}}\right\} \simeq \frac{(\log w)^2}{2\log q^{-1}} + \frac{1}{2}\log w.$$

Consequently, by (5.18), expression (5.17) reduces to

$$\lim_{n\to\infty} f_{W_n}(w) \simeq K\sqrt{w}e^{-\frac{(\log w)^2}{2\log q^{-1}}}, \quad 0 < w < \infty, \quad 0 < q < 1,$$

where the normalizing constant, by (5.8), is given by $K = q^{5/8}/\sqrt{2\pi \log q^{-1}}$. □

Remark 5.1. *The transformation of the q-deformed random variable.* The mean and variance of the q-random variable $[X_n]_{q^{-1}}$ are given in Exercise 2.7 as

$$E([X_n]_{q^{-1}}) = \frac{[n]_q\theta}{1+\theta q^{n-1}},$$

and

$$\begin{aligned} V([X_n]_{q^{-1}}) &= \frac{[n]_q\theta}{(1+\theta q^{n-1})(1+\theta q^{n-2})} + \frac{[n]_q^2\theta^2(1-q)}{q(1+\theta q^{n-1})^2(1+\theta q^{n-2})} \\ &= \left(\frac{[n]_q\theta}{1+\theta q^{n-1}}\right)^2\left(\frac{1+\theta q^{n-1}}{[n]_q\theta\left(1+\theta q^{n-2}\right)} + \frac{1-q}{q(1+\theta q^{n-2})}\right). \end{aligned}$$

These expressions, on using the assumption that the parameter θ depends on n, θ_n, and that $\theta_n \to \infty$, as $n \to \infty$, can be approximated by

$$E([X_n]_{q^{-1}}) \simeq \theta_n(1-q)^{-1}, \quad V([X_n]_{q^{-1}}) \simeq (\theta_n(1-q)^{-1})^2(q^{-1}-1).$$

Consequently, the standardized q-random variable $[X_n]_{q^{-1}}$

$$Z_n = \frac{[X_n]_{q^{-1}} - E([X_n]_{q^{-1}})}{\sqrt{V([X_n]_{q^{-1}})}}$$

is approximated by

$$Z_n \simeq U_n = \frac{[X_n]_{q^{-1}} - \theta_n(1-q)^{-1}}{\theta_n(1-q)^{-1}\sqrt{q^{-1}-1}},$$

implying that

$$[X_n]_{q^{-1}} = \theta_n q(1-q)^{-1}(U_n q^{-1}\sqrt{q^{-1}-1} + q^{-1}).$$

In summary, the transformation (5.15) is formed by approximating the mean and variance of $[X_n]_{q^{-1}}$, deducing the corresponding approximation of the standardized q-random variable $[X_n]_{q^{-1}}$, solving it for $[X_n]_{q^{-1}}$ and setting

$$W_n = U_n q^{-1}\sqrt{q^{-1}-1} + q^{-1}.$$

Furthermore, according to Theorem 5.6, the limiting distribution of W_n is the Stieltjes–Wigert, with density function (5.8). Therefore, the mean and standard deviation of the limiting (in distribution) random variable $W = Uq^{-1}\sqrt{q^{-1}-1} + q^{-1}$ are given by (5.9), which imply

$$E(U) = \sqrt{q^{-1}-1}, \quad V(U) = q^{-2}.$$

Notice that, the limiting (in distribution) random variable U, of the standardized random variable Z_n, is not any more a standardized random variable.

5.6 REFERENCE NOTES

A law of large numbers and a central limit for the number of records in a geometrically increasing population were examined in Charalambides (2007), on which Examples 5.3 and 5.6 are based.

Kyriakoussis and Vamvakari (2013) derived an asymptotic formula of a q-factorial of a positive integer and used it to show that the limiting distribution of a q-deformed binomial distribution of the first kind is a Stieltjes–Wigert distribution; Theorems 5.5 and 5.6 and Exercises 5.8 and 5.9 are based on this paper. Christiansen (2003) studied in length the interesting moment problem associated with Stieltjes–Wigert polynomials.

5.7 EXERCISES

5.1 Let X_i, $i = 1, 2, \ldots$, be a sequence of random variables, with $E(X_i) = \mu_i$ and $V(X_i) = \sigma_i^2 \leq c < \infty$, for $i = 1, 2, \ldots$. If the covariance is negative, $C(X_i, X_j) < 0$, for $i, j = 1, 2, \ldots$, with $i \neq j$, show that the sequence of differences $\overline{X}_n - \overline{\mu}_n$, $n = 1, 2, \ldots$, where

$$\overline{X}_n = \frac{1}{n}\sum_{i=1}^{n} X_i, \quad \overline{\mu}_n = \frac{1}{n}\sum_{i=1}^{n} \mu_i, \quad n = 1, 2, \ldots,$$

converges stochastically to zero.

5.2 (*Continuation*). If the covariance satisfies the weaker condition

$$C(X_i, X_j) < \frac{1}{|i-j|}, \quad i,j = 1,2, \ldots , \quad i \neq j,$$

show that the sequence of differences $\overline{X}_n - \overline{\mu}_n$, $n = 1, 2, \ldots$, converges stochastically to zero.

5.3 (*Continuation*). If the random variable X_i depends only on the random variables X_{i-1} and X_{i+1}, $i = 2, 3, \ldots$, show that the sequence of differences $\overline{X}_n - \overline{\mu}_n$, $n = 1, 2, \ldots$, converges stochastically to zero.

5.4 Consider a sequence of independent Bernoulli trials and assume that the conditional probability of success at any trial, given that $j-1$ successes occur in the previous trials, is given by

$$p_j = 1 - q^{r+j-1}, \quad j = 1,2, \ldots , \quad 0 < q < 1, \; r \quad \text{positive integer.}$$

Let W_n be the number of failures until the occurrence of the nth success. Show that the sequence of differences

$$\frac{W_n}{n} - \frac{-l_q(1-q) - h_{r-1,q}}{(1-q)n}, \quad n = 1,2, \ldots ,$$

with $l_q(1-t) = -\sum_{i=1}^{\infty} t^i/[i]_q$ the q-logarithmic function and $h_{n,q} = \sum_{i=1}^{n} q^i/[i]_q$ the incomplete q-harmonic function, converges stochastically to zero.

5.5 Consider a set of n urns $\{u_1, u_2, \ldots , u_n\}$, with urn u_i containing one white and $i-1$ black balls, $i = 1, 2, \ldots , n$. Assume that random q-drawings of one ball from each urn are carried out. Let S_n be the number of white balls drawn from the n urns.

(a) Show that the sequence

$$\frac{S_n}{n} - (1-q) + \frac{1}{n} l_q(1-q), \quad n = 1,2, \ldots ,$$

with $l_q(1-t) = -\sum_{i=1}^{\infty} t^i/[i]_q$ the q-logarithmic function, converges stochastically to zero.

(b) Derive the limiting expression

$$\lim_{n\to\infty} P\left(\frac{S_n - (1-q)n}{\sqrt{q(1-q)n}} \leq z \right) = \Phi(z),$$

where $\Phi(z)$ is the distribution function of the standard normal distribution.

5.6 Assume that random q-drawings of balls are sequentially carried out, one after the other, from an urn, initially containing r white and s black balls, according to the following scheme. After each q-drawing, the drawn ball is placed back in the urn together with one white ball. Let Z_i be the number of black balls drawn at the ith q-drawing, $i = 1, 2, \ldots, n$, and $X_n = \sum_{i=1}^{n} Z_i$ be the (total) number of black balls drawn in n q-drawings. Show that

$$\lim_{n\to\infty} P\left(\frac{X_n - (1-q^s)n}{\sqrt{q^s(1-q^s)n}} \leq z\right) = \Phi(z),$$

where $\Phi(z)$ is the distribution function of the standard normal distribution.

5.7 Consider a sequence of independent Bernoulli trials and assume that the odds of failure at the i the trial is given by

$$\lambda_i = \frac{q}{1-q}(1-\theta q^{i-1}), \quad i = 1, 2, \ldots,$$

where $0 < \theta < 1$ and $0 < q < 1$ or $1 < q < \infty$. Let Z_i be the number of successes at the ith trial, for $i = 1, 2, \ldots$. Show that the number $X_n = \sum_{i=1}^{n} Z_i$ of successes in n trials, suitably standardized, converges in distribution to the standard normal distribution.

5.8 Consider the q-deformed negative binomial distribution of the first kind, with probability function

$$f_{[U_n]_{q^{-1}}}([u]_{q^{-1}}) = \begin{bmatrix} n+u-1 \\ u \end{bmatrix}_q \frac{\theta^u q^{\binom{u}{2}}}{\prod_{i=1}^{n+u}(1+\theta q^{i-1})}, \quad u = 0, 1, \ldots,$$

for $0 < \theta < \infty$ and $0 < q < 1$. The mean and variance of the q-random variable $[U_n]_{q^{-1}}$ are given in Exercise 2.11 as

$$E([U_n]_{q^{-1}}) = [n]_q\theta, \quad V([U_n]_{q^{-1}}) = [n]_q\theta(1+\theta q^{-1}).$$

Assume that the parameter θ depends on n, θ_n, and that $\theta_n \to \infty$ and $\theta_n q^n \to 0$ as $n \to \infty$.

(a) Derive the following approximate values of the mean and variance of $[U_n]_{q^{-1}}$:

$$E([U_n]_{q^{-1}}) \simeq \theta_n(1-q)^{-1}, \quad V([U_n]_{q^{-1}}) \simeq (\theta_n(1-q)^{-1})^2(q^{-1}-1)$$

and from the corresponding approximation of the standardized q-random variable $[U_n]_{q^{-1}}$,

$$Y_n = \frac{[U_n]_{q^{-1}} - \theta_n(1-q)^{-1}}{\theta_n(1-q)^{-1}\sqrt{q^{-1}-1}},$$

deduce the expression

$$[U_n]_{q^{-1}} = \theta_n q(1-q)^{-1}W_n, \quad W_n = Y_n q^{-1}\sqrt{q^{-1}-1} + q^{-1}.$$

(b) Show that the limiting distribution of W_n, as $n \to \infty$, is the Stieltjes–Wigert, with density function (5.8).

5.9 Consider the q-deformed Heine distribution, with probability function

$$f_{[X_\lambda]_{q^{-1}}}([x]_{q^{-1}}) = e_q(-\lambda)\frac{q^{\binom{x}{2}}\lambda^x}{[x]_q!}, \quad x = 0, 1, \ldots ,$$

where $0 < \lambda < \infty$, $0 < q < 1$ and $e_q(t) = \prod_{i=1}^{\infty}(1 - t(1-q)q^{i-1})^{-1}$ is the q-exponential function (1.24). The mean and variance of the q-random variable $[X_\lambda]_{q^{-1}}$ are given in Exercise 2.14 as

$$E([X_\lambda]_{q^{-1}}) = \lambda, \quad V([X_\lambda]_{q^{-1}}) = \lambda + \lambda^2(q^{-1} - 1).$$

(a) Derive the following approximate values of the mean and variance of $[X_\lambda]_{q^{-1}}$, for large λ,

$$E([X_\lambda]_{q^{-1}}) = \lambda, \quad V([X_\lambda]_{q^{-1}}) \simeq \lambda^2(q^{-1} - 1)$$

and from the corresponding approximation of the standardized q-random variable $[X_\lambda]_{q^{-1}}$,

$$Y_\lambda = \frac{[X_\lambda]_{q^{-1}} - \lambda}{\lambda\sqrt{q^{-1} - 1}},$$

deduce the expression

$$[X_\lambda]_{q^{-1}} = \lambda q W_\lambda, \quad W_\lambda = Y_\lambda q^{-1}\sqrt{q^{-1} - 1} + q^{-1}.$$

(b) Show that the limiting distribution of W_λ, as $\lambda \to \infty$, is the Stieltjes–Wigert, with density function (5.8).

APPENDIX

HINTS AND ANSWERS TO EXERCISES

CHAPTER 1

1.1 (a) Multiply both members of the expression

$$[m]_{k,q} = [m]_q[m-1]_q \cdots [m-k+1]_q$$

by $[m-k]_q! = [1]_q[2]_q \cdots [m-k]_q$ to show that $[m]_{k,q}[m-k]_q! = [m]_q!$ and equivalently that $[m]_{k,q} = [m]_q!/[m-k]_q!$ Then, using the definition of the q-binomial coefficient,

$$\begin{bmatrix} m \\ k \end{bmatrix}_q = \frac{[m]_{k,q}}{[k]_q!},$$

deduce the expression

$$\begin{bmatrix} m \\ k \end{bmatrix}_q = \frac{[m]_q!}{[k]_q![m-k]_q!},$$

which readily implies the required relation.

(b) Use the last expression, together with the relations

$$[x]_{m,q} = [x]_{k,q}[x-k]_{m-k,q}, \quad [x]_{m,q} = [x]_{m-k,q}[x-m+k]_{k,q}$$

to obtain the first two relations. The last two relations can be similarly derived by using the relations

$$[x]_{m,q}[x-m]_{k,q} = [x]_{m+k,q}, \quad [x]_{m+k,q} = [x]_{k,q}[x-k]_{m,q}.$$

Discrete q-Distributions, First Edition. Charalambos A. Charalambides.

1.2 Use the triangular recurrence relation of the q-binomial coefficient

$$q^{r-k}\begin{bmatrix} r-1 \\ k-1 \end{bmatrix}_q = \begin{bmatrix} r \\ k \end{bmatrix}_q - \begin{bmatrix} r-1 \\ k \end{bmatrix}_q$$

and, alternatively, the equivalent triangular recurrence relation

$$\begin{bmatrix} r-1 \\ k-1 \end{bmatrix}_q = \begin{bmatrix} r \\ k \end{bmatrix}_q - q^k\begin{bmatrix} r-1 \\ k \end{bmatrix}_q$$

to split the left-hand side of the first and second sum, respectively, into a difference of two sums.

1.3 Use the recurrence relation

$$\begin{bmatrix} n+1 \\ r+1 \end{bmatrix}_q = \begin{bmatrix} n \\ r \end{bmatrix}_q + q^{r+1}\begin{bmatrix} n \\ r+1 \end{bmatrix}_q$$

to split the left-hand side sum into a difference of two sums.

1.4 Use the q-exponential function $e_q(u) = \prod_{i=1}^{\infty}(1-u(1-q)q^{i-1})^{-1}$ to show that the q-exponential generating function of the sequence of sums

$$s_n = \sum_{k=0}^{n}(-1)^k\begin{bmatrix} n \\ k \end{bmatrix}_q, \quad n = 0, 1, \ldots,$$

is given by

$$\sum_{n=0}^{\infty} s_n \frac{u^n}{[n]_q!} = \prod_{i=1}^{\infty}(1-u^2(1-q)^2q^{2(i-1)})^{-1} = \sum_{m=0}^{\infty}\left(\frac{1-q}{1+q}\right)^m \cdot \frac{u^{2m}}{[m]_{q^2}!}$$

and conclude the required formula.

1.5 Expand the q-factorial of y of order n,

$$[y]_{n,q} = (-1)^n q^{ny-\binom{n}{2}}[x+(-x-y-1+n)]_{n,q},$$

into q-factorials of x and $(-x-y-1+n)$, using Vandermonde's q-factorial convolution formula. Also, use the relation

$$[-x-y-1+n]_{n-k,q} = (-1)^{n-k}q^{-(n-k)(x-y-k)+\binom{n-k}{2}}[x+y-k]_{n-k,q}$$

and then divide both members by $[x+y]_{n,q}$.

1.6 Use the triangular recurrence relation

$$\begin{bmatrix} n \\ k \end{bmatrix}_q = \begin{bmatrix} n-1 \\ k \end{bmatrix}_q + q^{n-k}\begin{bmatrix} n-1 \\ k-1 \end{bmatrix}_q$$

to split the sum into a difference of two sums and derive the required recurrence relation.

1.7 Use the triangular recurrence relation

$$\begin{bmatrix} n \\ k \end{bmatrix}_q = q^k \begin{bmatrix} n-1 \\ k \end{bmatrix}_q + \begin{bmatrix} n-1 \\ k-1 \end{bmatrix}_q$$

to express $S_n(x)$ in terms of $S_{n-1}(x)$ and

$$T_{n-1}(x) = \sum_{r=0}^{n-1} (-1)^r q^{\binom{r+2}{2} - n(x+r+1)} \begin{bmatrix} n-1 \\ r \end{bmatrix}_q \frac{[x]_q}{[x+r+1]_q}, \qquad n = 1, 2, \ldots .$$

Also, use the relation

$$[x]_q = [x+k]_q - q^x [k]_q$$

to express $S_n(x)$ in terms of $T_{n-1}(x)$. Then, eliminating $T_{n-1}(x)$ from these two expressions, derive the first-order recurrence relation for $S_n(x)$.

1.8 Use the q-binomial formula to express the left-hand side sum as a double sum. Then, interchange the order of summation and use the relation

$$\begin{bmatrix} n \\ k \end{bmatrix}_q \begin{bmatrix} n-k \\ j \end{bmatrix}_q = \begin{bmatrix} n \\ n-j-k \end{bmatrix}_q \begin{bmatrix} k+j \\ j \end{bmatrix}_q .$$

Finally, interchange the order of summation in the last expression and use again the q-binomial formula to obtain the right-hand side sum of the required relation. Replace t by $-t$ and set $u = 1$ to deduce the required identity. Finally, set $w = t$ to conclude the last identity.

1.9 Use the relation

$$\frac{t^k q^{\binom{k}{2}}}{\prod_{i=1}^{k+1}(1+tq^{n+i-2})} = \frac{t^k q^{\binom{k}{2}}}{\prod_{i=1}^{k}(1+tq^{n+i-2})} - \frac{t^{k+1} q^{\binom{k}{2}+n+k-1}}{\prod_{i=1}^{k+1}(1+tq^{n+i-2})}$$

to express the left-hand side sum

$$s_n(t;q) = \sum_{k=0}^{\infty} \begin{bmatrix} n+k-1 \\ k \end{bmatrix}_q \frac{t^k q^{\binom{k}{2}}}{\prod_{i=1}^{k}(1+tq^{n+i-1})}, \qquad n = 1, 2 \ldots ,$$

as a difference of two sums. Then, using the recurrence relation

$$\begin{bmatrix} n+k-1 \\ k \end{bmatrix}_q = \begin{bmatrix} n+k-2 \\ k \end{bmatrix}_q + q^{n-1} \begin{bmatrix} n+k-2 \\ k-1 \end{bmatrix}_q ,$$

derive a first-order recurrence relation for $s_n(t;q)$, $n = 2, 3, \ldots$, which entails the required expression. The equivalent expression is deduced by replacing t and q by t^{-1} and q^{-1}, respectively.

1.10 (a) Use the relation

$$\prod_{j=1}^{k}(1-tq^{j-1})q^k = \frac{1}{t}\prod_{j=1}^{k}(1-tq^{j-1}) - \frac{1}{t}\prod_{j=1}^{k+1}(1-tq^{j-1}), \quad k=0,1,\ldots,$$

to express the right-hand side sum as a difference of two telescopic sums, which entail the required expression.

(b) Take the limit as $n \to \infty$ and use relation

$$\lim_{n\to\infty}\prod_{j=1}^{n}(1-tq^{j-1}) = \prod_{j=1}^{\infty}(1-tq^{j-1}) = E_q(-t/(1-q))$$

to deduce the limiting expression.

1.11 Expand the product $\prod_{i=1}^{n}(1+tq^{i-1}/[n]_q)$ into powers of $t/[n]_q$, by using the q-binomial formula. Then, take the limit of the expansion, as $n \to \infty$, and deduce the required limiting formula. Similarly, using the negative q-binomial formula, derive the other limiting formula.

1.12 (a) Interchange the order of summation and use the power series expansion of the q-exponential function $e_q(u) = \sum_{k=0}^{\infty} u^k/[k]_q!$ to get the required generating function.

(b) Derive the recurrence relation

$$\begin{bmatrix} n+1 \\ k \end{bmatrix}_q = \begin{bmatrix} n \\ k \end{bmatrix}_q + \begin{bmatrix} n \\ k-1 \end{bmatrix}_q - (1-q)[n]_q \begin{bmatrix} n-1 \\ k-1 \end{bmatrix}_q,$$

and use it to deduce the required recurrence relation for $H_n(t;q)$, $n = 0, 1, \ldots$.

1.13 Use the fact that the summation, for $n = 0, 1, \ldots$, of the multiple sum over all $k_i = 0, 1, \ldots, n$, $i = 1, 2, \ldots, r$, such that $k_1 + k_2 + \cdots + k_r = n$, is equivalent to the multiple sum over all $k_i = 0, 1, \ldots, i = 1, 2, \ldots, r$, without any restriction.

1.14 Expand the noncentral q-factorial $[t-r]_{n,q}$ into powers of

$$[t-r]_q = q^{-r}([t]_q - [r]_q),$$

by the aid of the q-Stirling numbers of the first kind, and then expand the powers of $[t]_q - [r]_q$ into powers of $[t]_q$. Comparing the resulting expansion with the definition of the noncentral q-Stirling numbers of the first kind, deduce the first expression.

Also, expand the noncentral q-factorial $[t-r]_{n,q}$ into q-factorials $[t]_{j,q}$, $j = 0, 1, \ldots, n$, using the q-Vandermonde's formula, and then expand these q-factorials into powers of $[t]_q$. Comparing the resulting expansion with the definition of the noncentral q-Stirling numbers of the first kind, deduce the second expression.

1.15 Expand the ascending q-factorial $[t+\theta+r+n-1]_{n,q^{-1}}$ into powers of

$$[t+\theta]_{q^{-1}} = [\theta]_{q^{-1}} + q^{-\theta}[t]_{q^{-1}},$$

by using the noncentral signless q-Stirling numbers of the first kind, and then expand these powers into powers of $[t]_{q^{-1}}$. Comparing the resulting expansion with the definition of the noncentral signless q-Stirling numbers of the first kind, deduce the required expression.

1.16 In the triangular recurrence relation of the q-Stirling numbers of the first kind, set $k=1$ and deduce the first-order recurrence relation

$$s_q(n,1) = -[n-1]_q s_q(n-1,1), \quad n=2,3,\ldots, \quad s_q(1,1)=s_q(0,0)=1.$$

Iterate it to get the first expression. Also, set $k=2$ and deduce the recurrence relation

$$s_q(n,2)+[n-1]_q s_q(n-1,2) = (-1)^{n-2}[n-2]_q!, \ n=3,4,\ldots, s_q(2,2)=1,$$

and solve it to get the required expression.

1.17 Expand the power $[t+r]_q^n$ into q-factorials $[t+r]_{j,q}$, $j=0,1,\ldots,n$, using the q-Stirling numbers of the second kind, and then expand the q-factorial $[t+r]_{j,q}$ into q-factorials $[t]_{k,q}$, $k=0,1,\ldots,j$, using the q-Vandermonde's formula. Comparing the resulting expansion with the definition of the noncentral q-Stirling numbers of the second kind, deduce the first expression.
Also, expand the power

$$[t+r]_q^n = (q^r[t]_q + [r]_q)^n$$

into powers $[t]_q^j, j=0,1,\ldots,n$, and then expand the power $[t]_q^j$ into q-factorials $[t]_{k,q}$, $k=0,1,\ldots,j$. Comparing the resulting expansion with the definition of the noncentral q-Stirling numbers of the second kind, deduce the second expression.

1.18 Multiply the expression defining the noncentral q-Stirling numbers of the first kind by $q^{\binom{n}{2}+rn}u^n/[n]_q!$, sum it for $n=0,1,\ldots$, and deduce, by virtue of the general q-binomial formula

$$\prod_{i=1}^{\infty}\frac{1+uq^{i-1}}{1+uq^{t+i-1}} = \sum_{k=0}^{\infty} q^{\binom{k}{2}}\begin{bmatrix} t-r \\ k \end{bmatrix}_q u^k,$$

the generating function

$$\sum_{n=0}^{\infty}\sum_{k=0}^{n} s_q(n,k;r)[t]_q^k\frac{u^n}{[n]_q!} = \prod_{i=1}^{\infty}\frac{1+uq^{r+i-1}}{1+uq^{t+i-1}},$$

which, on using the relation $q^t = 1-(1-q)[t]_q$ and then replacing $[t]_q$ by t, implies the first generating function. The second generating function may be derived by using the following explicit expression of the q-Stirling numbers of the second kind

$$S_q(n,k;r) = \frac{1}{[k]_q!}\sum_{j=0}^{k}(-1)^{k-j}q^{\binom{j+1}{2}-(r+j)k}\begin{bmatrix}k\\j\end{bmatrix}_q [r+j]_q^n.$$

1.19 Multiply both members of the explicit expression of the noncentral q-Stirling numbers of the first kind,

$$s_q(n,k;r) = \frac{1}{(1-q)^{n-k}}\sum_{i=k}^{n}(-1)^{i-k}q^{\binom{n-i}{2}+r(n-i)}\begin{bmatrix}n\\i\end{bmatrix}_q\binom{i}{k},$$

by $\binom{k}{j}(1-q)^{n-k}$ and sum the resulting expression for $k=j,j+1,\ldots,n$, to deduce the required relation. Also, multiply both members of the explicit expression of the noncentral q-Stirling numbers of the second kind,

$$S_q(n,k;r) = \frac{1}{(1-q)^{n-k}}\sum_{i=k}^{n}(-1)^{i-k}q^{r(i-k)}\binom{n}{i}\begin{bmatrix}i\\k\end{bmatrix}_q,$$

by

$$\begin{bmatrix}k\\j\end{bmatrix}_q q^{\binom{k-j}{2}+r(k-j)}(1-q)^{n-k}$$

and sum the resulting expression for $k=j,j+1,\ldots,n$ to obtain the required relation.

1.20 Multiply both members of the expression

$$S_q(n,k;r) = \frac{1}{[k]_q!}\sum_{i=0}^{k}(-1)^{k-i}q^{\binom{i+1}{2}-(r+i)k}\begin{bmatrix}k\\i\end{bmatrix}_q [r+i]_q^n$$

by $\binom{n+j}{j}(1-q)^{n-k}$ and sum the resulting expression for $n=0,1,\ldots$. Then, interchange the order of summation and use the q-binomial formula to deduce the required relation.

1.21 Expand the q-factorial $[st+r]_{n,q} = [s(t+r/s)]_{n,q}$ into q-factorials $[t+r/s]_{j,q^s}$, $j=0,1,\ldots,n$, using the generalized q-factorial coefficients, and then expand these q-factorials into q-factorials $[t]_{k,q^s}$, $k=0,1,\ldots,j$, using the q-Vandermonde's formula. Comparing the resulting expansion with the definition of $C_q(n,k;s,r)$, deduce the first expression.

Also, expand the q-factorial $[st+r]_{n,q}$ into q-factorials $[st]_{j,q}, j=0,1,\ldots,n$, using the q-Vandermonde's formula, and then expand these q-factorials into q-factorials $[t]_{k,q^s}$, $k=0,1,\ldots,j$. Comparing the resulting expansion with the definition of $C_q(n,k;s,r)$, conclude the second expression.

1.22 Expand the q-factorial $[t+n-r-1]_{n,q}$ into q-factorials $[t]_{k,q}$, $k=0,1,\ldots,n$, using the q-Vandermonde's formula, and comparing this expansion with the definition of the noncentral signless q-Lah numbers, conclude the required expression.

1.23 (a) Replace t by $-t$ in the definition of q-Eulerian numbers and use the relations $[-t]_q = -q^{-t}[t]_q$ and

$$\begin{bmatrix} -t+n-k \\ n \end{bmatrix}_q = (-1)^n q^{-nt-nk-\binom{n+1}{2}} \begin{bmatrix} t+k-1 \\ n \end{bmatrix}_q .$$

(b) Use the definition of the q-Eulerian numbers to get the relation

$$[k-r]_q^n = \sum_{j=0}^{k-r} q^{\binom{j}{2}} A_q(n,j) \begin{bmatrix} n+k-r-j \\ k-r-j \end{bmatrix}_q$$

$$= q^{\binom{k}{2}-\binom{r}{2}} \sum_{j=0}^{k-r} (-1)^{k-r-j} q^{(n-r+1)(k-r-j)-j(k-j)} A_q(n,j) \begin{bmatrix} -n-1 \\ k-j-r \end{bmatrix}_q .$$

Multiply it by $(-1)^r q^{\binom{r}{2}} \begin{bmatrix} n+1 \\ r \end{bmatrix}_q$, sum for $r=0,1,\ldots,k$ and conclude the required expression for $A_q(n,k)$, since

$$\sum_{r=0}^{k-j} q^{(n-r+1)(k-j-r)} \begin{bmatrix} n+1 \\ r \end{bmatrix}_q \begin{bmatrix} -n-1 \\ k-j-r \end{bmatrix}_q = \begin{bmatrix} 0 \\ k-j \end{bmatrix}_q = \delta_{k,j}.$$

1.24 Expand into q-binomial coefficients, using the q-Eulerian numbers, both members of the identity $[t]_q^{n+1} = [t]_q [t]_q^n$ and then use the relation

$$\begin{bmatrix} t+n-k \\ n \end{bmatrix}_q [t]_q = [k]_q \begin{bmatrix} t+n-k+1 \\ n+1 \end{bmatrix}_q + q^k [n-k+1]_q \begin{bmatrix} t+n-k \\ n+1 \end{bmatrix}_q .$$

Equate the coefficients of the q-binomials $\begin{bmatrix} t+n-k \\ n+1 \end{bmatrix}_q$ in both sides of the resulting identity to deduce the required recurrence relation.

1.25 (a) Execute the multiplications and additions in the product $\prod_{i=1}^{n}(t+a_i)$ and deduce the coefficient of t^k.

(b) Expand both members of the recurrence relation

$$\prod_{i=1}^{n}(t+a_i) = (t+a_n)\prod_{i=1}^{n-1}(t+a_i)$$

into powers of t and conclude the triangular recurrence relation.

(c) Set $a_i = 1/(\theta q^{i-1})$ and use the q-binomial theorem to find the first relation. Also, set $a_i = \theta[r+i-1]_q$ and use the definition of the noncentral signless q-Stirling numbers of the first kind to deduce the second relation.

1.26 (a) Expand both members of the recurrence relation $t^n = t \cdot t^{n-1}$ into polynomials $\prod_{i=1}^{k}(t - a_i)$, for $k = 1, 2, \ldots, n$, and use the relation

$$t\prod_{i=1}^{j}(t - a_i) = \prod_{i=1}^{j+1}(t - a_i) + a_{j+1}\prod_{i=1}^{j}(t - a_i)$$

to derive the triangular recurrence relation.

(b) Using the triangular recurrence relation obtain for the sequence of generating functions

$$\phi_k(u) = \sum_{n=k}^{\infty} S(n, k; \boldsymbol{a})u^n$$

a first-order recurrence relation, which entails $\phi_k(u) = u^k \prod_{i=1}^{k+1} (1 - a_i u)^{-1}$.

Also, expand each term of the product $\prod_{i=1}^{k} u(1 - a_i u)^{-1}$ into powers of u and equate the coefficients of u^n to find the required expression for the generalized Stirling numbers of the second kind.

(c) Set $a_i = \theta q^{i-1}$ in the generating function in (b) and use the negative q-binomial theorem to find the first relation. Also, set $a_i = \theta[r + i - 1]_q$ and use the expansion

$$\sum_{n=k}^{\infty} S_q(n - 1, k - 1; r)u^n = u^k \prod_{j=1}^{k} (1 - [r + j - 1]_q u)^{-1}$$

to deduce the second relation.

1.27 (a) Expand the polynomial $\prod_{i=1}^{n}(t - a_i)$ into powers of t, using the generalized Stirling numbers of the first kind, and in the resulting expression expand the powers of t into polynomials $\prod_{i=1}^{k}(t - a_i)$, using the generalized Stirling numbers of the second kind. The final expression implies the first of the orthogonality relations. The second orthogonality relation can be similarly deduced.

(b) Set $u = 1/t$ in the expansion of part (b) of Exercise 1.26 and find the required expansion. Furthermore, replace t by $-t$ in the last expression, then multiply it by

$$|s(n - 1, k - 1; \boldsymbol{a})| = (-1)^{n-k}s(n - 1, k - 1; \boldsymbol{a})$$

and sum the resulting expression for $n = k, k + 1, \ldots,$ to conclude the required expansion.

1.28 (a) Expand the polynomial $\prod_{i=1}^{n}(t - a_i)$ into powers of t, using the generalized Stirling numbers of the first kind, and in the resulting expression expand the powers of t into polynomials $\prod_{j=1}^{k}(t - b_j)$, using the generalized Stirling numbers of the second kind. Comparing the resulting expansion with the definition of the generalized Lah numbers, conclude the required expression.

(b) Expand both members of the recurrence relation

$$\prod_{i=1}^{n}(t-a_i) = (t-a_n)\prod_{i=1}^{n-1}(t-a_i)$$

into polynomials $\prod_{j=1}^{k}(t-b_j)$, for $k = 1, 2, \ldots, n$, and use the relation

$$(t-a_n)\prod_{j=1}^{m}(t-b_j) = \prod_{j=1}^{m+1}(t-b_j) + (b_{m+1}-a_n)\prod_{j=1}^{m}(t-b_j)$$

to derive the triangular recurrence relation.

(c) Multiply both sides of the triangular recurrence relation

$$\frac{1}{\prod_{i=1}^{n}(t-a_i)} = \frac{t}{\prod_{i=1}^{n+1}(t-a_i)} - \frac{a_{n+1}}{\prod_{i=1}^{n+1}(t-a_i)}$$

and sum the resulting expression for $n = k, k+1, \ldots$ to derive a first-order recurrence relation for

$$C_k(t) = \sum_{n=k}^{\infty} C(n, k; \boldsymbol{a}, \boldsymbol{b}) \frac{1}{\prod_{i=1}^{n+1}(t-a_i)}.$$

Iterate it to get the expression $C_k(t) = 1/\prod_{j=1}^{k+1}(t-b_j)$.

1.29 (a) Apply the definition of the q-derivative operator.

(b) Use the expansions of the q-exponential functions and the q-derivative of a power.

1.30 (a) Apply the expansion of a function $f(t)$ into a q-power series,

$$f(t) = \sum_{k=0}^{\infty}\left[\frac{1}{[k]_q!}D_q^k f(t)\right]_{t=0} \cdot [t]_q^k,$$

to the function $f(t) = [t-r]_{n,q}$.

(b) Show that the q-Leibnitz formula holds true for $n = 1$, using the definition of the q-derivative, and by induction show that it holds true for any positive integer n.

1.31 Use the q-derivative of the nth power of t. Expand the integrand in the expression

$$-l_q(1-x) = \int_0^x \frac{d_q u}{1-u}$$

into powers of u and integrate the resulting expression by interchanging the summation and integral signs.

1.32 (a) Use the relation $d_q E_q(-t) = -E_q(-qt)d_q t$ and then apply a q-integration by parts. Iterate successively the first-order recurrence relation to conclude that $I_{n,q} = [n]_q!$.

(b) Write the q-gamma function as

$$\Gamma_q(x+1) = \prod_{i=1}^{\infty} \frac{(1-q^i)(1-q^{i+1})^x}{(1-q^{i+x})(1-q^i)^x}, \quad |q| < 1,$$

and take its limit as $q \to 1^-$. Then, use Euler's product formula

$$\lim_{n\to\infty} \prod_{i=1}^{n} \left(1+\frac{x}{i}\right)^{-1}\left(1+\frac{1}{i}\right)^x = \lim_{n\to\infty} \left(\frac{n}{n+1}\right)^x \lim_{n\to\infty} n^x \Big/ \binom{x+n}{n},$$

with

$$\lim_{n\to\infty} n^x \Big/ \binom{x+n}{n} = \Gamma(x+1).$$

1.33 (a) Apply a q-integration by parts to get the relation

$$J_{n,q} = [n]_q \int_0^{\infty} t^{n-1} e_q(-qt) d_q t$$

and then make the transformation $u = qt$ to deduce the required first-order recurrence relation. Iterate it successively to conclude that $J_{n,q} = q^{-\binom{n+1}{2}} [n]_q!$.

(b) Write the q-gamma function as

$$\Gamma_q(x+1) \equiv \gamma_{q^{-1}}(x+1) = \frac{1}{q^{-\binom{x}{2}}} \prod_{i=1}^{\infty} \frac{(1-q^{-i})(1-q^{-(i+1)})^x}{(1-q^{-(i+x)})(1-q^{-i})^x}, \quad |q| > 1,$$

and take its limit as $q^{-1} \to 1^-$. Then, proceed as in Exercise 1.32.

1.34 (a) Use the expression

$$\int_0^1 f(t) d_q t = (1-q) \sum_{k=0}^{\infty} f(q^k) q^k$$

and the general q-binomial formula to derive the expression

$$B_q(x,y) = \frac{(1-q)\prod_{i=1}^{\infty}(1-q^i)}{\prod_{i=1}^{\infty}(1-q^{x+i-1})} \cdot \frac{\prod_{i=1}^{\infty}(1-q^{x+y+i-1})}{\prod_{i=1}^{\infty}(1-q^{y+i-1})}$$

and introduce the q-gamma function to get the required relation.

(b) Use the relations

$$\prod_{i=1}^{\infty} \frac{1-tq^i}{1-tq^{n-r+i-1}} = \prod_{i=1}^{n-r-1} (1-tq^{i-1}), \quad \Gamma_q(r+1) = [r]_q!.$$

1.35 (a) Show that $q^{\Theta} t^m = (qt)^m$, by expanding $q^{\Theta} = e^{\Theta \log q}$ into powers of Θ, and conclude that $q^{r\Theta} f(t) = f(q^r t)$, for $r = 1, 2, \ldots$. Furthermore, using the q-derivative operator, deduce for $\Theta_q = tD_q$ the expression $\Theta_q = [\Theta]_q$.

(b) Let

$$\Theta_q^n f(t) = \sum_{k=0}^{n} q^{\binom{k}{2}} c_{n,k} t^k \mathcal{D}_q^k f(t)$$

and determine the coefficients $c_{n,k}$, $k = 1, 2, \ldots, n$, by choosing the most convenient function $f(t) = t^u$, with u a real number. Finally, invert the last expression.

1.36 (a) Use the definition of the q-difference operator to show that it holds true for $k = 1, 2$ and by induction show that it holds true for any positive integer k.

(b) Apply the expansion of a polynomial $f_n(t)$ in q^t into a polynomial of q-factorials,

$$f_n(t) = \sum_{k=0}^{n} \left[\frac{1}{[k]_q!} \Delta_q^k f_n(t)\right]_{t=0} \cdot [t]_{k,q},$$

to the polynomial $f_n(t) = [t + r]_q^n$ to get the first expression. Also, apply this expansion, with q replaced by q^s, to the polynomial $f_n(t) = [st + r]_{n,q}$ to deduce the second expression.

1.37 For (a) take the mth derivative and for (b) take the mth q-derivative in both members of the power series expression of the probability generating function, $P_X(t) = \sum_{x=0}^{\infty} f(x) t^x$, with $|t| \leq 1$.

1.38 (a) Use the triangular recurrence relation of the q-binomial coefficients

$$q^{x-m} \begin{bmatrix} x \\ m \end{bmatrix}_q = \begin{bmatrix} x+1 \\ m+1 \end{bmatrix}_q - \begin{bmatrix} x \\ m+1 \end{bmatrix}_q$$

and get the expression

$$E([X_n]_{m,q}) = \frac{q^m [n-1]_{m,q}}{[m+1]_q}.$$

Deduce the q-expected value and the q-variance as

$$E([X_n]_q) = \frac{q[n-1]_q}{[2]_q}, \quad V([X_n]_q) = \frac{q[n-1]_q [n+1]_q}{[2]_q^2 [3]_q}.$$

(b) Use the formula of the sum of a finite number of terms of a geometric progression and express the probability generating function of X_n as

$$P_{X_n}(t) = [n]_{qt}/[n]_q.$$

CHAPTER 2

2.1 (a) Expand the probability generating function $E(t^{X_n})$ into powers of t, using the formula in Exercise 1.25 that defines the generalized signless Stirling numbers of the first kind.

(b) Expand the factorial moment generating function $E((1+t)^{X_n})$ into powers of t.

(c) Use the expression $P(T_k = n) = P(X_{n-1} = k-1)p_n$.

2.2 Apply the results of Exercise 2.1.

2.3 Use the triangular recurrence relation of the q-binomial coefficients $\begin{bmatrix} n \\ m \end{bmatrix}_q$ and get the first-order relation for the mean

$$E(X_n) - E(X_{n-1}) = P_{n-1}, \quad n = 2, 3, \ldots, \quad E(X_1) = \theta/(1+\theta),$$

where

$$P_{n-1} = \sum_{m=1}^{n} q^{n-m} \begin{bmatrix} n-1 \\ m-1 \end{bmatrix}_q [m-1]_q!(1-q)^{m-1} \frac{\theta^m q^{\binom{m}{2}}}{\prod_{i=1}^{m}(1+\theta q^{i-1})}.$$

Express the general term of the sum P_{n-1} as

$$\begin{aligned} q^{n-m} & \begin{bmatrix} n-1 \\ m-1 \end{bmatrix}_q [m-1]_q!(1-q)^{m-1} \frac{\theta^m q^{\binom{m}{2}}}{\prod_{i=1}^{m}(1+\theta q^{i-1})} \\ &= \begin{bmatrix} n-1 \\ m-1 \end{bmatrix}_q [m-1]_q!(1-q)^{m-1} \frac{\theta^m q^{\binom{m}{2}}}{\prod_{i=1}^{m}(1+\theta q^{i-1})} \\ &- \begin{bmatrix} n-1 \\ m \end{bmatrix}_q [m]_q!(1-q)^{m} \frac{\theta^{m+1} q^{\binom{m+1}{2}}}{\prod_{i=1}^{m+1}(1+\theta q^{i-1})} \\ &- \frac{1}{\theta q^m} \begin{bmatrix} n-1 \\ m \end{bmatrix}_q [m]_q!(1-q)^{m} \frac{\theta^{m+1} q^{\binom{m+1}{2}}}{\prod_{i=1}^{m+1}(1+\theta q^{i-1})} \end{aligned}$$

and derive the expression $P_{n-1} = \theta q^{n-1}/(1+\theta q^{n-1})$, which implies the required recurrence relation for the mean.

2.4 Use the triangular recurrence relation of the q-binomial coefficients $\begin{bmatrix} n \\ m \end{bmatrix}_q$ and get the first-order relation for the second factorial moment

$$\begin{aligned} E[(X_n)_2] - E[(X_{n-1})_2] &= Q_{n-1}, \quad n = 3, 4, \ldots, \\ E[(X_2)_2] &= 2\frac{\theta}{1+\theta} \cdot \frac{\theta q}{1+\theta q}, \end{aligned}$$

where

$$Q_{n-1} = 2\sum_{m=2}^{n} q^{n-m} \begin{bmatrix} n-1 \\ m-1 \end{bmatrix}_q [m-1]_q!(1-q)^{m-2} \zeta_{m-1,q} \frac{\theta^m q^{\binom{m}{2}}}{\prod_{i=1}^{m}(1+\theta q^{i-1})}.$$

Express the general term of the sum Q_{n-1} as

$$
\begin{aligned}
q^{n-m} & \begin{bmatrix} n-1 \\ m-1 \end{bmatrix}_q [m-1]_q!(1-q)^{m-2}\zeta_{m-1,q} \frac{\theta^m q^{\binom{m}{2}}}{\prod_{i=1}^{m}(1+\theta q^{i-1})} \\
&= \begin{bmatrix} n-1 \\ m-1 \end{bmatrix}_q [m-1]_q!(1-q)^{m-2}\zeta_{m-1,q} \frac{\theta^m q^{\binom{m}{2}}}{\prod_{i=1}^{m}(1+\theta q^{i-1})} \\
&- \begin{bmatrix} n-1 \\ m \end{bmatrix}_q [m]_q!(1-q)^{m-1}\zeta_{m,q} \frac{\theta^{m+1} q^{\binom{m+1}{2}}}{\prod_{i=1}^{m+1}(1+\theta q^{i-1})} \\
&- \frac{1}{\theta q^m}\begin{bmatrix} n-1 \\ m \end{bmatrix}_q [m]_q!(1-q)^{m-1}\zeta_{m,q} \frac{\theta^{m+1} q^{\binom{m+1}{2}}}{\prod_{i=1}^{m+1}(1+\theta q^{i-1})} \\
&- \begin{bmatrix} n-1 \\ m \end{bmatrix}_q [m-1]_q!(1-q)^{m-1} \frac{\theta^m q^{\binom{m}{2}}}{\prod_{i=1}^{m}(1+\theta q^{i-1})}
\end{aligned}
$$

and derive the expression $Q_{n-1} = 2E(X_{n-1})\theta q^{n-1}/(1+\theta q^{n-1})$, which implies the required recurrence relation for the second factorial moment.

2.5 Use the expression of the probability function of X_n in terms of its q-binomial moments,

$$
f_n(x) = \sum_{m=x}^{\infty} (-1)^{m-x} q^{\binom{m-x}{2}} \begin{bmatrix} m \\ x \end{bmatrix}_q E\left(\begin{bmatrix} X_n \\ m \end{bmatrix}_q\right), \quad x = 0, 1, \ldots,
$$

and the q-binomial formula (1.19).

2.6 (a) Use the expression $X_n = \sum_{i=1}^{n} Z_i$, where $Z_1, Z_2, \ldots, Z_n$ are independent zero-one Bernoulli random variables.

(b) Show by induction on n that

$$
\frac{d_q}{d_q t} \prod_{i=1}^{n}(1+\theta t q^{i-1}) = [n]_q \theta \prod_{i=1}^{n-1}(1+(\theta q)t q^{i-1}), \quad m = 1, 2, \ldots .
$$

Furthermore, by induction on m, show that

$$
\frac{d_q^m}{d_q t^m} \prod_{i=1}^{n}(1+\theta t q^{i-1}) = [n]_{m,q}\theta^m q^{\binom{m}{2}} \prod_{i=m+1}^{n}(1+\theta t q^{i-1}), \quad m = 1, 2, \ldots,
$$

and then use the expression

$$
E([X_n]_{m,q}) = \left[\frac{d_q^m}{d_q t^m} P_{X_n}(t)\right]_{t=1}, \quad m = 1, 2, \ldots,
$$

to obtain the mth q-factorial moment of X_n.

2.7 (a) Use the q-factorial moments

$$E([X_n]_{m,q}) = \frac{[n]_{m,q}\theta^m q^{\binom{m}{2}}}{\prod_{i=1}^{m}(1+\theta q^{i-1})}, \qquad m = 1, 2, \ldots,$$

and expression (1.59),

$$V([X_n]_q) = qE([X_n]_{2,q}) + E([X_n]_q) - [E([X_n]_q)]^2.$$

(b) Derive the mth q^{-1}-factorial moment $E([X_n]_{m,q^{-1}})$, by using the relation

$$[x]_{m,q^{-1}} = q^{-xm+\binom{m+1}{2}}[x]_{m,q}.$$

(c) Use the q^{-1}-factorial moments $E([X_n]_{m,q^{-1}})$, $m = 1, 2, \ldots,$ and expression (1.59),

$$V([X_n]_{q^{-1}}) = q^{-1}E([X_n]_{2,q^{-1}}) + E([X_n]_{q^{-1}}) - [E([X_n]_{q^{-1}})]^2.$$

2.8 Take the q-derivative of the product

$$P_{n,r}(t) = \prod_{i=1}^{n}(1+tq^{i-1})^{-1}\sum_{x=0}^{r}\begin{bmatrix} n \\ x \end{bmatrix}_q q^{\binom{x}{2}}t^x, \qquad r = 0, 1, \ldots, n,$$

by using the q-Leibnitz formula, and show the q-differential equation

$$\frac{d}{d_qt}P_{n,r}(t) = -\frac{[n]_q!q^{\binom{r+1}{2}}}{[r]_q![n-r-1]_q!}\cdot\frac{t^r}{\prod_{i=1}^{n+1}(1+tq^{i-1})}.$$

Finally, take its q-integral in the interval $[0, \theta]$ and deduce the required formula.

2.9 Use the expression of the probability function of W_n in terms of its q-binomial moments,

$$f_n(w) = \sum_{m=w}^{\infty}(-1)^{m-w}q^{\binom{m-w}{2}}\begin{bmatrix} m \\ w \end{bmatrix}_q E\left(\begin{bmatrix} W_n \\ m \end{bmatrix}_q\right), \qquad w = 0, 1, \ldots,$$

and the negative q-binomial formula.

2.10 (a) Use that expression

$$\begin{aligned} P(U_{n+1} = u|W_n = w) &= P(Z_{n+w+i} = 0, i = 1, 2, \ldots, u, Z_{n+w+u+1} = 1) \\ &= \prod_{i=1}^{u}P(Z_{n+w+i} = 0)P(Z_{n+w+u+1} = 1), \end{aligned}$$

where Z_i, $i = 1, 2, \ldots,$ is a sequence of independent zero-one Bernoulli random variables, with $P(Z_i = 1) = p_i$, $i = 1, 2, \ldots,$ to derive the required conditional probability function as

$$P(U_{n+1} = u|W_n = w) = \frac{\theta q^{n+w+u}}{\prod_{i=1}^{u+1}(1+\theta q^{n+w+i-1})}, \qquad u = 0, 1, \ldots .$$

(b) Derive the mth conditional q-factorial moments of U_{n+1}, given W_n, by using the alternative negative q-binomial formula in Exercise 1.9,

$$E([U_{n+1}]_{m,q}|W_n = w) = [m]_q!\theta^{-m}q^{-(n+w)-\binom{m}{2}}, \quad m = 0, 1, \dots .$$

(c) Use that expression of the usual conditional factorial moments in terms of the conditional q-factorial moments,

$$E[(U_{n+1})_j|\, W_n = w] = j!\sum_{m=j}^{\infty}(-1)^{m-j}(1-q)^{m-j}s_q(m,j)\frac{E([U_{n+1}]_{m,q}|W_n = w)}{[m]_q!}.$$

2.11 (a) Derive the mth q^{-1}-factorial moment $E([U_n]_{m,q^{-1}})$, by using the relation

$$[u]_{m,q^{-1}} = q^{-um+\binom{m+1}{2}}[u]_{m,q}$$

and the additional negative q-binomial formula in Exercise 1.9.

(b) Use the q^{-1}-factorial moments $E([U_n]_{m,q^{-1}})$, $m = 1, 2, \dots$, and expression (1.59),

$$V([U_n]_{q^{-1}}) = q^{-1}E([U_n]_{2,q^{-1}}) + E([U_n]_{q^{-1}}) - [E([U_n]_{q^{-1}})]^2.$$

2.12 (a) Use that expression

$$\begin{aligned}P(S_{n+1} = s|U_n = u) = & \; P(Z_{n+u+i} = 1, i = 1, 2, \dots, s, Z_{n+u+s+1} = 0)\\ = & \prod_{i=1}^{u} P(Z_{n+u+i} = 1)P(Z_{n+u+s+1} = 0),\end{aligned}$$

where Z_i, $i = 1, 2, \dots$, is a sequence of independent zero-one Bernoulli random variables, with $P(Z_i = 1) = p_i$, $i = 1, 2, \dots$, to derive the required conditional probability function as

$$P(S_{n+1} = s|U_n = u) = \frac{\theta^s q^{s(n+u)+\binom{s}{2}}}{\prod_{i=1}^{s+1}(1+\theta q^{n+u+i-1})}, \quad s = 0, 1, \dots .$$

(b) Derive the mth conditional q^{-1}-factorial moment S_{n+1}, given U_n, by using the relation

$$[s]_{m,q^{-1}} = q^{-sm+\binom{m+1}{2}}[s]_{m,q}$$

and the additional negative q-binomial formula in Exercise 1.9, as

$$E([S_{n+1}]_{m,q^{-1}}|U_n = u) = [m]_q!(\theta q^{n+u})^m, \quad m = 0, 1, \dots .$$

(c) Use that expression of the usual conditional factorial moments in terms of the conditional q^{-1}-factorial moments,

$$E[(S_{n+1})_j \mid U_n = u]$$
$$= j! \sum_{m=j}^{\infty} (-1)^{m-j}(1 - q^{-1})^{m-j} s_{q^{-1}}(m,j) \frac{E([S_{n+1}]_{m,q^{-1}} | U_n = u)}{[m]_{q^{-1}}!}.$$

2.13 (a) Use the power series expansion of the q-exponential function $E_q(t)$ to get the probability generating function of the Heine distribution and compare it with the distribution of the sum $X = \sum_{i=1}^{\infty} Z_i$, where Z_i, $i = 1, 2, \dots$, are zero–one Bernoulli random variables, with

$$P(Z_i = 1) = \frac{\lambda(1-q)q^{i-1}}{1 + \lambda(1-q)q^{i-1}}, \quad i = 1, 2, \dots .$$

(b) Use the expressions $E(X) = \sum_{i=1}^{\infty} E(Z_i)$ and $V(X) = \sum_{i=1}^{\infty} V(Z_i)$ to deduce the required formulae.

2.14 Work as in Exercise 2.7.

2.15 Consider each subinterval as a Bernoulli trial, with probability of success at the ith trial given by

$$p_{n,i}(t) = \frac{\lambda t q^{i-1}/[n]_q}{1 + \lambda t q^{i-1}/[n]_q}, \quad i = 1, 2, \dots, n.$$

Use Theorem 2.1 to conclude that the random variable $X_{t,n}$ obeys a q-binomial distribution of the first kind.

2.16 Use the relation satisfied by the stationary distribution $p_x = P(X = x)$, $x = 0, 1, \dots$,

$$p_x = \sum_{k=0}^{\infty} p_k p_{k,x}, \quad x = 0, 1, \dots,$$

and get the recurrence relation

$$p_x = \theta q^{x-1} p_{x-1} + q^x p_x, \quad x = 1, 2, \dots,$$

which implies the probability function of the Heine distribution.

2.17 Use the expression of the probability function of X_n in terms of its binomial moments,

$$f_n(x) = \sum_{j=x}^{\infty} (-1)^{j-x} \binom{j}{x} E\left[\binom{X_n}{j}\right], \quad x = 0, 1, \dots,$$

and the expression (1.39) of the q-Stirling number of the first kind.

2.18 (a) Multiply the probability function of the random variable X_n,

$$P(X_n = x) = \sum_{j=x}^{n} (-1)^{j-x} q^{\binom{j}{2}+rj} \begin{bmatrix} n \\ j \end{bmatrix}_q \binom{j}{x},$$

by $[x]_{m,q}$ and sum the resulting expression for $x = m, m+1, \ldots, n$. Interchange the order of summation and use expression (1.40) of the q-Stirling number of the second kind.

(b) Apply expression (1.61), of the jth factorial moment in terms of the q-factorial moments, and after interchanging the order of summation, use the orthogonality relation of the q-Stirling numbers (1.31) to deduce the required expression.

2.19 Find the probability generating function of the sum $X_n = \sum_{i=1}^{n} Z_i$, with Z_i, $i = 1, 2, \ldots, n$ independent zero–one Bernoulli random variables, and deduce the binomial moment generating function. Expand it to get the binomial moments of X_n and then use the expression of the probability function in terms of the binomial moments.

2.20 (a) Consider the selection, placement, and movement of a ball along the cells of the ith column as the ith Bernoulli trial and the event that a white ball comes to rest in the ith (diagonal) cell as success, for $i = 1, 2, \ldots, n$. Derive the probability of success at the ith trial as $\theta q^{i-1}/[n]_q$. Furthermore, consider the event A_i of success at the ith Bernoulli trial and deduce the probability

$$P(A_{i_1} A_{i_2} \cdots A_{i_j}) = \frac{\theta^j}{[n]_q^j} q^{i_1+i_2+\cdots+i_j-j}, \quad 1 \le i_1 < i_2 < \cdots < i_j \le n.$$

Then, use the inclusion and exclusion principle and expression (1.5),

$$\sum_{1 \le i_1 < i_2 < \cdots < i_j \le n} q^{i_1+i_2+\cdots+i_j} = q^{\binom{j+1}{2}} \begin{bmatrix} n \\ j \end{bmatrix}_q,$$

to obtain the required probability function.

(b) Use the limiting expression $\lim_{n\to\infty} [n]_{j,q}/[n]_q^j = 1$.

(c) Use the limiting expression $\lim_{q\to 1} [j]_q! = j!$.

2.21 (a) Use that expressions

$$\begin{aligned} &P(W_k = n, T_{k-1} = j - 1) \\ &\quad = P(T_k = j + n - 1 \mid T_{k-1} = j - 1) P(T_{k-1} = j - 1) \end{aligned}$$

and

$$\begin{aligned}
P(T_k = j+n-1 | T_{k-1} = j-1) \\
&= P(Z_{j+i} = 0, i = 0, 1, \dots, n-2, Z_{j+n-1} = 1) \\
&= \prod_{i=1}^{n-2} P(Z_{j+i} = 0) P(Z_{j+n-1} = 1),
\end{aligned}$$

where Z_i, $i = 1, 2, \dots$, is a sequence of independent zero–one Bernoulli variables, with $P(Z_i = 1) = p_i$, $i = 1, 2, \dots$, to derive the probability function of W_k. Also, using the relation

$$\frac{q^{i-1}}{[i+j-2]_q [i+j-1]_q} = \frac{q^{i-1}}{[i+j-2]_q} - \frac{q^i}{[i+j-1]_q},$$

find a single summation expression for the tail probability $P(W_k > n)$. Furthermore, use the geometric series expansion

$$\frac{1}{[n+j-1]_q} = \sum_{i=0}^{n} (-1)^i \frac{q^{\binom{i}{2}+(j-1)i} [n]_{i,q}}{[j+i-1]_{i+1,q}},$$

to express the probability $P(W_k > n)$ as a single finite sum and deduce the required formula.

(b) Use the expression $E(W_k) = \sum_{n=0}^{\infty} P(W_k > n)$ and the negative q-binomial formula.

2.22 Work as in Example 2.6.

2.23 Work as in Exercise 2.21.

2.24 (a) Show that $p_j = [\theta]_q / [\theta + r + j - 1]_q$ and derive the probability generating function $E(t^{X_n})$.

(b) Deduce the factorial moment generating function $E((1+t)^{X_n})$ and expand it into powers of t to get the factorial moments.

(c) Use the relation $P(T_k = n) = P(X_{n-1} = k-1) p_n$.

CHAPTER 3

3.1 (a) Expand the probability generating function $E(s^{T_k})$ into powers of s, using the formula in Exercise 1.26 that defines the generalized Stirling numbers of the second kind.

(b) Expand the ascending factorial moment generating function $E((1-s)^{-T_k})$ into powers of s.

(c) Use the expression $P(X_n = x) = P(T_{x+1} = n+1)/p_{x+1}$.

3.2 Divide each term of the recurrence relation by $\prod_{j=1}^{x} p_j = \prod_{j=1}^{x}(1-q_j)$ and deduce a recurrence relation for

$$S(n,x;\boldsymbol{q}) = P(X_n = x)\Big/\left(\prod_{j=1}^{x} p_j\right), \quad x = 0,1,\ldots,n.$$

Multiply it by u^n and sum the resulting expression for $n = x, x+1, \ldots,$ to deduce for the generating function

$$\phi_x(u;q) = \sum_{n=x}^{\infty} S(n,x;\boldsymbol{q})u^n, \quad x = 1,2,\ldots,$$

a first-order recurrence relation, which implies $\phi_x(u;q) = u^x \prod_{i=1}^{x+1}(1-q_i u)^{-1}$. Expand it in partial fractions to find the required expression.

3.3 Use the triangular recurrence relation

$$\begin{bmatrix} n+m-2 \\ m \end{bmatrix}_q = \begin{bmatrix} n+m-1 \\ m \end{bmatrix}_q - q^{n-1}\begin{bmatrix} n+m-2 \\ m-1 \end{bmatrix}_q,$$

together with the expression

$$\frac{1}{\prod_{i=1}^{m}(1-\theta q^{n+i-2})} = \frac{1}{\prod_{i=1}^{m}(1-\theta q^{n+i-1})} + q^{n-1}\frac{\theta(1-q)[m]_q}{\prod_{i=1}^{m+1}(1-\theta q^{n+i-2})}$$

to derive the required first-order recurrence relation for the mean $E(W_n)$.

3.4 Use successively the triangular recurrence relation

$$\begin{bmatrix} n+m-2 \\ m \end{bmatrix}_q = \begin{bmatrix} n+m-1 \\ m \end{bmatrix}_q - q^{n-1}\begin{bmatrix} n+m-2 \\ m-1 \end{bmatrix}_q$$

and the expressions

$$\frac{1}{\prod_{i=1}^{m}(1-\theta q^{n+i-2})} = \frac{1}{\prod_{i=1}^{m}(1-\theta q^{n+i-1})} + q^{n-1}\frac{\theta(1-q)[m]_q}{\prod_{i=1}^{m+1}(1-\theta q^{n+i-2})}$$

and

$$\zeta_{m,q} = \zeta_{m-1,q} + \frac{1}{[m]_q}$$

to derive the required first-order recurrence relation for the second factorial moment.

3.5 Use the expression of the probability function of W_n in terms of its q-binomial moments,

$$f_n(w) = \sum_{m=w}^{\infty} (-1)^{m-w} q^{\binom{m-w}{2}} \begin{bmatrix} m \\ w \end{bmatrix}_q E\left(\begin{bmatrix} W_n \\ m \end{bmatrix}_q\right), \quad w = 0,1,\ldots,$$

and the additional negative q-binomial formula in Exercise 1.9.

3.6 (a) Use the expression $W_n = \sum_{j=1}^{n} U_i$, where $U_1, U_2, \ldots, U_n$ are independent geometric random variables.

(b) Show inductively that

$$\frac{d_q^m}{d_q t^m} \prod_{j=1}^{n} (1-\theta t q^{j-1})^{-1} = [n+m-1]_{m,q} \theta^m \prod_{j=1}^{n+m} (1-\theta t q^{j-1})^{-1},$$

for $m = 1, 2, \ldots$, and use the expression

$$E([W_n]_{m,q}) = \left[\frac{d_q^m}{d_q t^m} P_{W_n}(t)\right]_{t=1}.$$

3.7 (a) Derive the mth q-factorial moment by using the negative q-binomial formula.

(b) Apply expression 1.61.

(c) Set $j = 1$ in the expression of the jth factorial moment and use $s_q(m, 1) = (-1)^{m-1}[m-1]_q!$ to find $E(W_n)$. Then, use the triangular recurrence relation

$$\begin{bmatrix} n+m-1 \\ m \end{bmatrix}_q = q^m \begin{bmatrix} n+m-2 \\ m \end{bmatrix}_q + \begin{bmatrix} n+m-2 \\ m-1 \end{bmatrix}_q$$

and deduce the first-order recurrence relation

$$E(W_n) - E(W_{n-1}) = q^{r-n+1} S_n, \quad n = 2, 3, \ldots, [r],$$

where

$$S_n = \frac{1}{1-q} \sum_{m=1}^{\infty} \begin{bmatrix} n+m-2 \\ m-1 \end{bmatrix}_q \frac{q^{(r-n+1)(m-1)}[m-1]_q!}{[r+m]_{m,q}}.$$

Also, use the triangular recurrence relation

$$\begin{bmatrix} n+m-2 \\ m \end{bmatrix}_q = \begin{bmatrix} n+m-1 \\ m \end{bmatrix}_q - q^{n-1} \begin{bmatrix} n+m-2 \\ m-1 \end{bmatrix}_q,$$

to express $E(W_{n-1})$ as a difference of two sums and use the expressions

$$q^{(r-n+2)m} = q^{(r-n+1)m} - q^{(r-n+1)m}(1-q)[m]_q$$

and

$$q^{(r-n+2)m+n-1} = q^{(r-n+1)(m-1)} - q^{(r-n+1)(m-1)}(1-q)[r+m]_q$$

in the first and second summands, respectively, to get the first-order recurrence relation

$$E(W_n) - E(W_{n-1}) = S_n - 1, \quad n = 2, 3, \ldots, [r].$$

Remove S_n from the two first-order recurrence relations for $E(W_n)$, and find the recurrence relation

$$E(W_n) - E(W_{n-1}) = \frac{q^{r-n+1}}{1-q^{r-n+1}}, \quad n = 2, 3, \ldots, [r],$$

which implies the required expression.

Set $j = 2$ in the expression of the jth factorial moment and use $s_q(m, 2) = (-1)^{m-2}[m-1]_q!\zeta_{m-1,q}$, with $\zeta_{m,q} = \sum_{j=1}^{m} 1/[j]_q$, to find $E[(W_n)_2]$. Then, proceed as in the case $E(W_n)$ and derive the recurrence relations

$$E[(W_n)_2] - E[(W_{n-1})_2] = q^{r-n+1}Q_n, \quad n = 2, 3, \ldots, [r],$$

and

$$E[(W_n)_2] - E[(W_{n-1})_2] = Q_n - 2E(W_n), \quad n = 2, 3, \ldots, [r],$$

with $E[(W_1)_2] = 2q^{2r}/(1-q^r)^2$, where

$$Q_n = \frac{2}{(1-q)^2} \sum_{m=2}^{\infty} \begin{bmatrix} n+m-2 \\ m-1 \end{bmatrix}_q \frac{q^{(r-n+1)(m-1)}[m-1]_q!\zeta_{m-1,q}}{[r+m]_{m,q}}.$$

Remove Q_n from the two recurrence relations and find for $E[(W_n)_2]$ the recurrence relation

$$E[(W_n)_2] - E[(W_{n-1})_2] = \frac{2q^{r-n+1}}{1-q^{r-n+1}} \cdot E(W_n), \quad n = 2, 3, \ldots, [r],$$

which implies the expression

$$E[(W_n)_2] = 2\sum_{j=1}^{n} \left(\frac{q^{r-j+1}}{1-q^{r-j+1}} \right)^2 + 2\sum_{j=1}^{n} \frac{q^{r-j+1}}{1-q^{r-j+1}} \sum_{i=1}^{j-1} \frac{q^{r-i+1}}{1-q^{r-i+1}}.$$

3.8 Consider the event

$$1^{j_1}0\ 1^{j_2}0\ \cdots\ 1^{j_{n-x}}0\ 1^{j_{n-x+1}},$$

that among the values $(z_1, z_2, \ldots, z_n)$, j_1 are nonzeroes followed by a zero, the next j_2 are nonzeroes followed by a zero and so on the last j_{n-x+1} are nonzeroes, with the total of nonzeroes being $j_1 + j_2 + \cdots + j_{n-x+1} = x$. Set

$$s_1 = j_1, \quad s_2 = j_1 + j_2, \quad \ldots, \quad s_{n-x} = j_1 + j_2 + \cdots + j_{n-x}$$

and, using the multiplication formula, express the probability of this event as

$$P(1^{j_1}0\ 1^{j_2}0\ \cdots\ 1^{j_{n-x}}0\ 1^{j_{n-x+1}}) = \prod_{j=1}^{x}(1-\theta q^{j-1})\theta^{n-x}q^{s_1+s_2+\cdots+s_{n-x}},$$

with $0 \leq s_1 \leq s_2 \leq \ldots \leq s_{n-x} \leq x$. Then, deduce the required expression of the probability function

$$P(Y_n = [x]_q) = \prod_{j=1}^{x}(1 - \theta q^{j-1})\theta^{n-x} \sum_{0 \leq s_1 \leq s_2 \leq \ldots \leq s_{n-x} \leq x} q^{s_1+s_2+\cdots+s_{n-x}}$$

by setting $r_i = s_i + 1$, for $i = 1, 2, \ldots, n - x$, and using the identity (1.3),

$$\sum_{1 \leq r_1 \leq r_2 \leq \ldots r_{n-x} \leq x+1} q^{r_1+r_2+\cdots+r_{n-x}} = q^{n-x}\begin{bmatrix} n \\ n-x \end{bmatrix}_q = q^{n-x}\begin{bmatrix} n \\ x \end{bmatrix}_q.$$

3.9 (a) Take the q-derivative of the sum

$$P_{n,r}(t) = \sum_{x=0}^{r}\begin{bmatrix} n \\ x \end{bmatrix}_q t^x \prod_{j=1}^{n-x}(1 - tq^{j-1}),$$

by using the q-Leibnitz formula, and show the q-differential equation

$$\frac{d}{d_q t}P_{n,r}(t) = -\frac{[n]_q! t^r}{[r]_q![n-r-1]_q!}\prod_{j=1}^{n-r-1}(1 - tq^j).$$

Finally, take its q-integral in the interval $[0, \theta]$ and deduce the required formula.

(b) Follow the same reasoning with the sum

$$Q_{n,r}(t) = \sum_{y=0}^{r}\begin{bmatrix} n \\ y \end{bmatrix}_q t^{n-y} \prod_{j=1}^{y}(1 - tq^{j-1})$$

instead of the sum $P_{n,r}(t)$.

3.10 (a) Multiply the probability function (3.14),

$$P(Y_n = y) = \begin{bmatrix} n \\ y \end{bmatrix}_q q^{(n-y)(r-y)}(1 - q)^y [r]_{y,q}, \quad y = 0, 1, \ldots, n,$$

by $[y]_{m,q}$, sum the resulting expression, for $y = m, m + 1, \ldots, n$, and use the q-binomial formula (1.20) to deduce the required expression for the mth q-factorial moment of Y_n.

(b) Apply expression (1.61).

3.11 (a) The conditional probability $P(S_{i+1} = j + 1 | S_i = j)$ may be expressed as

$$\begin{aligned} &P(\text{there exists a } u \in U_1 \text{ such that } (u, u_{i+1}) \in E \| U_1 | = j) \\ &\qquad = 1 - P(\text{for every } u \in U_1 \ (u, u_{i+1}) \notin E \| U_1 | = j) \end{aligned}$$

from which the required expression is readily deduced.

(b) The $n-1$ sequential additions of vertices to construct the random acyclic digraph $G_{n,q}$ from $G_{1,q}$ may be considered as a sequence of $n-1$ independent Bernoulli trials with success the addition of a vertex in the set U_1. Then, the probability of success at the ith trial, given that $j-1$ successes occur in the previous $i-1$ trials is given by

$$p_j = q^j, \quad j = 1, 2, \ldots, i, \quad i = 1, 2, \ldots, \quad 0 < q < 1.$$

Apply Theorem 3.2, with $X_{n-1} = n - S_n$ and $\theta = q$ to obtain the probability function of S_n as

$$P(S_n = k) = [n-1]_{k-1,q} q^{n-k}, \quad k = 1, 2, \ldots, n,$$

with $0 < q < 1$.

3.12 (a) Consider the event A_r that a person attempting to cross the minefield steps on rth mine (and is killed), for $r = 1, 2, \ldots, n$, and the event B_{j-1} that $j-1$ persons are killed, for $j = 1, 2, \ldots, n$. Then, deduce the conditional probability $p_j = P(A_{r_1} \cup A_{r_2} \cup \cdots \cup A_{r_{m-j+1}} | B_{j-1})$ that a person attempting to cross the minefield is killed, given that $j-1$ persons are killed as $p_j = 1 - q^{m-j+1}$, for $j = 1, 2, \ldots, m$.

(b) Consider an attempt to cross the minefield as a Bernoulli trial and the killing of a person as a success and conclude that the random variable X_n follows an absorption distribution with probability function given by (3.14),

$$P(X_n = x) = \begin{bmatrix} n \\ x \end{bmatrix}_q q^{(n-x)(m-x)} (1-q)^x [m]_{x,q}, \quad x = 0, 1, \ldots, n.$$

3.13 (a) Assuming that $Y = y$, the conditional probability p_j that a particle is absorbed, given that $j-1$ particles are absorbed, may be obtained as $p_j = 1 - q^{y-j+1}$, for $j = 1, 2, \ldots, y$, by following the reasoning in (a) of Exercise 3.12. Then, the conditional probability $P(X_n = x | Y = y)$ is readily deduced.

(b) Use the relation

$$P(X_n = x) = \sum_{y=x}^{\infty} P(Y = y) P(X_n = x | Y = y), \quad x = 0, 1, \ldots,$$

together with the q-exponential functions $E_q(t)$ and $e_q(t)$.

(c) Follow the reasoning of the derivation of (b).

3.14 (a) Use the relation

$$P(W_n = w \mid Y = y) = P(X_{n+w-1} = n - 1 \mid Y = y) p_{n+w}$$

and (a) of Exercise 3.13.

(b) Use the relation

$$P(W_n = w) = \sum_{y=n}^{\infty} P(Y = y)P(W_n = w|Y = y), \quad w = 0, 1, \ldots,$$

together with the q-exponential functions $E_q(t)$ and $e_q(t)$.

3.15 (a) Take the limit, as $n \to \infty$, of the expression

$$\begin{bmatrix} n \\ x \end{bmatrix}_q (1-q)^x \sum_{y=x}^{\infty} q^{(n-x)(y-x)}[y]_{x,q} P(Y = y) = P(X_n = x), \quad 0 < q < 1,$$

and deduce that $P(Y = x) = \lim_{n\to\infty} P(X_n = x)$.

(b) Use the fact that the limit, as $n \to \infty$, of a q-binomial distribution of the first or second kind is a Heine or an Euler distribution.

3.16 (a) Use the power series expansion of the q-exponential function $e_q(t)$ to get the probability generating function of the Euler distribution and compare it with the distribution of the sum $X = \sum_{j=1}^{\infty} U_j$, where $U_j, j = 1, 2, \ldots,$ are independent geometric random variables, with

$$P(U_j = u) = (1 - \lambda(1 - q)q^{j-1})(\lambda(1 - q)q^{j-1})^u,$$

for $u = 0, 1, \ldots,$ and $j = 1, 2, \ldots$.

(b) Use the expressions $E(X) = \sum_{j=1}^{\infty} E(U_j)$ and $V(X) = \sum_{j=1}^{\infty} V(U_j)$ to deduce the required formulae.

3.17 (a) Replace the parameter λ by a new parameter $\lambda = [\theta]_q$ and deduce for the probability function $f(x) = P(X = x), x = 0, 1, \ldots,$ of the q-Poisson distributions the recurrence relation

$$f(x) = \frac{[\theta]_q}{[x]_q} f(x-1), \quad x = 1, 2, \ldots, \quad f(0) = E_q(-[\theta]_q)$$

where $0 < \theta < \infty$ and $0 < q < 1$ or $1 < q < \infty$. This recurrence relation implies that, if θ is not an integer, then $f(x)$ assumes its maximum value at $x = [\theta]$, the integer part of θ, while if θ is an integer, then $f(x)$ assumes its maximum value at the points $x = \theta - 1$ and $x = \theta$.

(b) Compute the difference of the inverse failure rate at two successive points and conclude that

$$\frac{1}{r(x-1)} - \frac{1}{r(x)} > 0, \quad x = 1, 2, \ldots,$$

which implies the monotonicity of the failure rate $r(x)$, for $x = 0, 1, \ldots$.

3.18 Consider each subinterval as a Bernoulli trial, with conditional probability of success at the any trial, given that $j - 1$ successes occur in the previous trials, given by

$$p_{n,i}(t) = 1 - \lambda t q^{j-1}/[n]_q, \quad j = 1, 2, \ldots, n.$$

Use Theorem 3.2 to conclude that the random variable $X_{t,n}$ obeys a q-binomial distribution of the second kind.

3.19 Derive the probability function of the sum $Z = X + Y$ as

$$P(Z = z) = e_q(-\lambda_1)E_q(-\lambda_2)\prod_{i=1}^{z}(1 + (\lambda_1/\lambda_2)q^{i-1})\frac{\lambda_2^z}{[z]_q!}, \quad z = 0, 1, \ldots,$$

where $0 < \lambda_1 < \infty, 0 < \lambda_2 < 1/(1-q)$ and $0 < q < 1$ and then use the expression

$$P(X = x|X + Y = n) = \frac{P(X = x)P(Y = n - x)}{P(X + Y = n)}$$

or

$$P(n - X = y|X + Y = n) = \frac{P(X = n - y)P(Y = y)}{P(X + Y = n)}.$$

3.20 Show that the conditional probability $c(x, n)$ of X, given that $X + Y = n$, satisfies the condition

$$\frac{c(x + y, x + y)c(0, y)}{c(x, x + y)c(y, y)} = \frac{h(x + y)}{h(x)h(y)}, \quad x = 0, 1, \ldots, \quad y = 0, 1, \ldots,$$

with

$$h(x) = q^{\binom{x}{2}}/[x]_q!, \quad x = 0, 1, \ldots,$$

and use a well-known theorem of Patil and Seshadri (1964) to conclude that the random variables X and Y follow a Heine and an Euler distributions, respectively.

3.21 Verify that the q-factorial moments of a q-Poisson distribution are given by

$$E([X]_{m,q}) = \lambda^m, \quad m = 1, 2, \ldots,$$

with $0 < \lambda < 1/(1-q)$ and $0 < q < 1$ or $0 < \lambda < \infty$ and $1 < q < \infty$ and conclude the relation $E([X]_{2,q}) = [E([X]_q)]^2$.

Inversely, assume that the last relation holds true and express it in terms of the power series probability function of X. Equate the coefficients of λ^x in both sides of the power series identity to get the relation

$$\sum_{k=0}^{x}[k + 1]_q[k + 2]_q a_{k+2}a_{x-k} = \sum_{k=0}^{x}[k + 1]_q[x - k + 1]_q a_{x+1}a_{x-k},$$

for $x = 0, 1, \ldots$. Show successively that $a_k = a^k/[k]_q!$, for $k = 2, 3$, with $a = a_1 > 0$, and by mathematical induction conclude that

$$a_x = a^x/[x]_q!, \qquad x = 0, 1, \ldots,$$

which implies that X follows a q-Poisson distribution.

3.22 (a) Use the expression $T_k = \sum_{j=1}^{k} U_j$, where $U_1, U_2, \ldots, U_k$ are independent geometric random variables.

(b) Use the relation connecting the ascending binomial moment generating function,

$$B_{T_k}(s) = \sum_{j=0}^{\infty} E\left[\binom{T_k + j - 1}{j}\right] s^j,$$

with the probability generating function:

$$B_{T_k}(s) = P_{T_k}((1-s)^{-1}) = E((1-s)^{-T_k})$$

and expand it into powers of s^j, using the negative q-binomial formula.

3.23 (a) Consider the selection, placement, and movement of a ball along the cells of a column as a Bernoulli trial and the event that a white ball comes to rest in the diagonal cell as success. Then, the sequential placement and step-by-step movement of balls in a column constitutes a geometric sequence of trials. Derive the probability of success at the jth geometric sequence of trials, which is the probability that a white ball comes to rest in the jth cell of the jth column, as $\theta q^{j-1}/[n]_q$. Then, use Theorem 3.1 to deduce the required probability function.

(b) Use the limits

$$\lim_{n\to\infty} \begin{bmatrix} n+w-1 \\ w \end{bmatrix}_q = \frac{1}{(1-q)^w [w]_q!}, \qquad \lim_{n\to\infty} [n]_q = \frac{1}{1-q},$$

and

$$\lim_{n\to\infty} \prod_{j=1}^{n} \left(1 - \frac{\theta q^{j-1}}{[n]_q}\right) = E_q(-\theta).$$

3.24 The probability function $f(x) = P(X = x)$, $x = 0, 1, \ldots$, when it exists satisfies the relations

$$f(x) = \sum_{k=x-1}^{\infty} f(k) p_{k,x}, \quad x = 1, 2, \ldots, \quad f(0) = \sum_{k=0}^{\infty} p_{k,0}.$$

Set $p_{k,x} = q^x/[k+2]_q$, $x = 0, 1, \ldots, k+1$, and deduce the recurrence relation

$$f(x) = qf(x-1) - \frac{q^x}{[x]_q} f(x-2), \quad x = 2, 3, \ldots, \quad f(1) = qf(0),$$

where $0 < q < 1$ or $1 < q < 2$. Use it to show that

$$f(k) = f(0)\frac{q^{\binom{k}{2}}q^k}{[k]_q!}, \quad k = 2, 3,$$

and by mathematical induction conclude that

$$f(x) = f(0)\frac{q^{\binom{x}{2}}q^x}{[x]_q!}, \quad x = 0, 1, \ldots,$$

which implies that X follows a Heine distribution for $0 < q < 1$ and an Euler distribution for $1 < q < 2$.

CHAPTER 4

4.1 Use the conditional probability $p_{i,j} = (1 - b_j)/(1 + a_i)$ to derive a recurrence relation for the probability function $P(X_n = k)$ and deduce for the double sequence

$$c_{n,k} = \frac{\prod_{i=1}^{n}(1 + a_i)}{\prod_{j=1}^{k}(1 - b_j)}P(X_n = k), \quad k = 0, 1, \ldots, n, \quad n = 0, 1, \ldots,$$

a triangular recurrence relation. Then, multiply both members of the recurrence relation by $\prod_{j=1}^{k}(t - b_j)$ and sum it for $x = 1, 2, \ldots, n$ to deduce for

$$c_n(t) = \sum_{k=1}^{n} c_{n,k} \prod_{j=1}^{k}(t - b_j), \quad n = 1, 2, \ldots,$$

a first-order recurrence relation, which implies $c_n(t) = \prod_{i=1}^{n}(t + a_i)$. Finally, use Exercise 1.28 to get the required expression.

4.2 Use the expression

$$C(n, k; -\boldsymbol{a}, \boldsymbol{b}) = \sum_{m=k}^{n} |s(n, m; \boldsymbol{a})| S(m, k; \boldsymbol{b}).$$

4.3 Use the expression

$$C(n-1, k-1; -\boldsymbol{a}, \boldsymbol{b}) = \sum_{m=k}^{n} |s(n-1, m-1; \boldsymbol{a})| S(m-1, k-1; \boldsymbol{b}).$$

4.4 Consider the event

$$1^{j_1}0\ 1^{j_2}0\ \cdots\ 1^{j_{n-x}}0\ 1^{j_{n-x+1}},$$

that among the values $(z_1, z_2, \ldots, z_n)$, j_1 are nonzeroes followed by a zero, the next j_2 are nonzeroes followed by a zero and so on the last j_{n-x+1} are nonzeroes, with the total of nonzeroes being $j_1 + j_2 + \cdots + j_{n-x+1} = x$. Set

$$s_1 = j_1, \quad s_2 = j_1 + j_2, \quad \ldots, \quad s_{n-x} = j_1 + j_2 + \cdots + j_{n-x}$$

and, using the multiplication formula, express the probability of this event as

$$P(1^{j_1}0\ 1^{j_2}0\ \cdots\ 1^{j_{n-x}}0\ 1^{j_{n-x+1}}) = \frac{[\alpha]_{x,q^{-k}}[\beta]_{n-x,q^{-k}}}{[\alpha+\beta]_{n,q^{-k}}} q^{-k\alpha(n-x)} \times q^{k(s_1+s_2+\cdots+s_{n-x})},$$

with $0 \leq s_1 \leq s_2 \leq \ldots \leq s_{n-x} \leq x$. Then, deduce the required expression of the probability function

$$P(Y_n = [x]_q) = \frac{[\alpha]_{x,q^{-k}}[\beta]_{n-x,q^{-k}}}{[\alpha+\beta]_{n,q^{-k}}} q^{-k\alpha(n-x)} \times \sum_{0 \leq s_1 \leq s_2 \leq \ldots \leq s_{n-x} \leq x} q^{k(s_1+s_2+\cdots+s_{n-x})}$$

by setting $r_i = s_i + 1$, for $i = 1, 2, \ldots, n-x$, and using the identity (1.3),

$$\sum_{1 \leq r_1 \leq r_2 \leq \ldots \leq r_{n-x} \leq x+1} q^{k(r_1+r_2+\cdots+r_{n-x})} = q^{k(n-x)} \begin{bmatrix} n \\ n-x \end{bmatrix}_{q^k}.$$

4.5 (a) Use the conditional probability $p_{i,j} = [j]_q/[i+1]_q$ to derive a recurrence relation for the probability function $P(Y_n = y)$, $y = 0, 1, \ldots, n$, and deduce for the double sequence

$$c_{n,y} = q^{-y}[n+1]_{y+1,q} P(Y_n = y), \quad y = 0, 1, \ldots, n, \quad n = 0, 1, \ldots,$$

a triangular recurrence relation, which implies $c_{n,y} = [n]_{y,q}$.

(b) Use the triangular recurrence relation

$$\begin{bmatrix} y+1 \\ m+1 \end{bmatrix}_q = \begin{bmatrix} y \\ m+1 \end{bmatrix}_q + q^{y-m} \begin{bmatrix} y \\ m \end{bmatrix}_q$$

to obtain the mth q-factorial moment of Y_n as

$$E([Y_n]_{m,q}) = \frac{q^m [n]_{m,q}}{[m+1]_q}.$$

In particular, deduce the q-mean as $E([Y_n]_q) = q[n]_q/[2]_q$ and use (1.59) to get the q-variance as

$$V([Y_n]_q) = \frac{q[n]_q(q^2[n]_q + [2]_q)}{[2]_q^2[3]_q}.$$

(c) Apply expression (1.61) to obtain the jth factorial moment of Y_n as

$$E[(Y_n)_j] = j! \sum_{m=j}^{n} (-1)^{m-j} \begin{bmatrix} n \\ m \end{bmatrix}_q \frac{(1-q)^{m-j} q^m s_q(m,j)}{[m+1]_q}.$$

4.6 Note that the sum $S_n = \sum_{i=1}^{n} Z_i$, if $Y_n = y$, takes the value $S_n = [n-y]_q$. Furthermore, consider the event

$$0^{k_1} q^0 \, 0^{k_2} q^1 \cdots 0^{k_{n-y}} q^{n-y+1} \, 0^{k_{n-y+1}},$$

that among the values $(z_1, z_2, \dots, z_n)$, k_1 are zeroes followed by $q^0 = 1$, the next k_2 are zeroes followed by q^1 and so on the last k_{n-y+1} are zeroes, with the total number of zeroes being $k_1 + k_2 + \cdots + k_{n-y+1} = y$. Set

$$s_1 = k_1, \quad s_2 = k_1 + k_2, \dots, s_{n-y} = k_1 + k_2 + \cdots + k_{n-y}$$

and, using the multiplication formula, express the probability of this event as

$$P(0^{k_1} q^0 \, 0^{k_2} q^1 \cdots 0^{k_{n-y}} q^{n-y+1} \, 0^{k_{n-y+1}}) = \frac{[y]_q! [n-y]_q!}{[n+1]_q!} q^y q^{s_1+s_2+\cdots+s_{n-y}},$$

with $0 \le s_1 \le s_2 \le \dots \le s_{n-y} \le y$. Then, deduce the required expression of the probability function of Y_n

$$P(S_n = [n-y]_q) = \frac{[y]_q! [n-y]_q!}{[n+1]_q!} q^y \sum_{0 \le s_1 \le s_2 \le \dots \le s_{n-y} \le y} q^{s_1+s_2+\cdots+s_{n-y}}$$

by setting $r_i = s_i + 1$, for $i = 1, 2, \dots, n-y$, and using the identity (1.3),

$$\sum_{1 \le r_1 \le r_2 \le \dots \le r_{n-y} \le y+1} q^{r_1+r_2+\cdots+r_{n-y}} = q^{n-y} \begin{bmatrix} n \\ n-y \end{bmatrix}_q.$$

4.7 Use the probability functions

$$P(W = w) = (1-\theta)\theta^w, \quad w = 0, 1, \dots,$$

and

$$P(U = u) = (1-\theta q)(\theta q)^u, \quad u = 0, 1, \dots .$$

4.8 (a) Use the expression

$$P(Y_{m,n} \le y, Z_{m,n} \le z) = P(Z_{m,n} \le z) - P(Y_{m,n} > y, Z_{m,n} \le z)$$

together with the expressions

$$P(Z_{m,n} \le z) = [P(X_{i,n} \le z)]^m$$

and

$$P(Y_{m,n} > y, Z_{m,n} \le z) = [P(y < X_{i,n} \le z)]^m$$

and derive the joint distribution function of $Y_{m,n}$ and $Z_{m,n}$ as

$$F_{m,n}(y,z) = \frac{1}{[n+1]_q^m}\{[z+1]_q^m - ([z+1]_q - [y+1]_q)^m\},$$

for $y = 0, 1, \ldots, z$ and $z = 0, 1, \ldots, n$. Deduce the joint probability function $f_{m,n}(y,z) = P(Y_{m,n} = y, Z_{m,n} = z)$, by using its relation with the joint distribution function $F_{m,n}(y,z) = P(Y_{m,n} \leq y, Z_{m,n} \leq z)$, in the form

$$f_{m,n}(y,z) = \frac{1}{[n+1]_q^m}\{ ([z+1]_q - [y]_q)^m - ([z+1]_q - [y+1]_q)^m + ([z]_q - [y+1]_q)^m - ([z]_q - [y]_q)^m \},$$

which can be reduced to

$$f_{m,n}(y,z) = \frac{q^{my}}{[n+1]_q^m}\{[z-y+1]_q^m - [2]_{q^m}[z-y]_q^m + q^m[z-y-1]_q^m\},$$

for $y = 0, 1, \ldots, z$ and $z = 0, 1, \ldots, n$. Furthermore, use the expressions

$$P(Y_{m,n} > y) = [P(X_{i,n} > y)]^m, \quad P(Z_{m,n} \leq y) = [P(X_{i,n} \leq y)]^m$$

and derive the marginal distribution functions of $Y_{m,n}$ and $Z_{m,n}$ as

$$G_{m,n}(y) = P(Y_{m,n} \leq y) = 1 - \frac{q^{m(y+1)}[n-y]_q^m}{[n+1]_q^m},$$

for $y = 0, 1, \ldots, n$, and

$$H_{m,n}(z) = P(Z_{m,n} \leq z) = \frac{[z+1]_q^m}{[n+1]_q^m},$$

for $z = 0, 1, \ldots, n$, respectively. Deduce the marginal probability functions

$$g_{m,n}(y) = P(Y_{m,n} = y) = \frac{q^{my}[n-y+1]_q^m - q^{m(y+1)}[n-y]_q^m}{[n+1]_q^m},$$

for $y = 0, 1, \ldots, n$ and

$$h_{m,n}(z) = P(Z_{m,n} = z) = \frac{[z+1]_q^m - [z]_q^m}{[n+1]_q^m}, \quad z = 0, 1, \ldots, n,$$

respectively.

(b) The probability function of the range $R_{m,n}$,

$$f_{R_{m,n}}(r) = \sum_{z=r}^{n} f_{m,n}(z-r, z),$$

is readily obtained as

$$f_{R_{m.n}}(r) = \frac{[n-r+1]_{q^m}}{[n+1]_q^m}\{[r+1]_q^m - [2]_{q^m}[r]_q^m + q^m[r-1]_q^m\},$$

for $r = 0, 1, \ldots, n$.

4.9 Note that player a_x wins the game if the number X of coin tosses until the first occurrence of heads assumes any value of the set

$$C_x = \{(x-1) + kn : \ k = 0, 1, \ldots \}, \quad x = 1, 2, \ldots, n\,.$$

Use the relation $P(X_n = x) = P(X \in C_x)$ to get the required expression.

4.10 Show the probability formula for σ_n by induction on n. More precisely, verify it for $n = 2$, by using the result of Exercise 4.9. Furthermore, assume that it holds true for all permutations σ_{n-1} of $n-1$ players. Then, in the permutation $\sigma_n = (i_1, i_2, \ldots, i_n)$ suppose that $i_1 = j$ and consider the permutation $\sigma_{n-1} = (i_2, i_3, \ldots, i_n)$ of the $n-1$ players $\{1, 2, \ldots, j-1, j+1, \ldots, n\}$. Finally, use the relation $P(\sigma_n) = P(\{j\})P(\sigma_{n-1} | \{j\})$, together with the result of Exercise 4.9 and the induction hypothesis, to conclude that the probability formula holds true for σ_n.

4.11 (a) Sum the probabilities $P(\sigma_n)$, $\sigma_n \in S_{n,k}$, given in Exercise 4.10, and use the relation $\sum_{\sigma_n \in S_{n,k}} 1 = a(n,k)$.

(b) Use the summation formula $\sum_{k=0}^{m_n} a(n,k)q^k = [n]_q!$.

(c) Establish the relation $S_{n,k} = \cup_{j=1}^{n} S_{n-1,k-j+1}$ and take its cardinality to deduce the recurrence relation.

4.12 Let X_m be the number of white balls drawn from the first urn. The distribution of X_m and the conditional distribution of Y_n, given $X_m = x$, are q-hypergeometric with parameters m, r, s, q, and n, x, $m-x$, q, respectively. Use the probability functions of these distributions and the expression

$$P(Y_n = y) = \sum_{x=y}^{m} P(X_m = x)P(Y_n = y \mid X_m = x)$$

to derive the probability function of Y_n as

$$P(Y_n = y) = q^{(n-y)(r-y)} \begin{bmatrix} r \\ y \end{bmatrix}_q \begin{bmatrix} s \\ n-y \end{bmatrix}_q \Big/ \begin{bmatrix} r+s \\ n \end{bmatrix}_q, \quad y = 0, 1, \ldots, n.$$

4.13 (a) Consider the number X_u of white balls drawn in u random q-drawings and use the relation

$$P(U = u) = P(X_u = u)(1 - p_{u+1,u+1}),$$

where $p_{u+1,u+1}$ is the conditional probability of drawing a white ball at the $(u+1)$st q-drawing, given that u white balls are drawn in the previous u q-drawings.

(b) Use the relation

$$\begin{bmatrix} x \\ k \end{bmatrix}_q = q^{k(x-k)} \begin{bmatrix} x \\ k \end{bmatrix}_{q^{-1}}$$

and the identity derived in Example 1.6.

(c) Use expression (1.61).

4.14 (a) Consider the number X_n of white balls drawn in n random q-drawings and use the relation

$$P(U_n = u) = P(X_{n+u-1} = u)(1 - p_{n+u,u+1}),$$

where $p_{i,j}$ is the conditional probability of drawing a white ball at the ith q-drawing, given that $j-1$ white balls are drawn in the previous $i-1$ q-drawings.

(b) Rewrite the probability generating function of U_n in the form

$$P(U_n = u) = \begin{bmatrix} n+u-1 \\ u \end{bmatrix}_{q^k} q^{uk(\beta-n+1)} \frac{[\alpha]_{u,q^k}[\beta]_{n,q^k}}{[\alpha+\beta]_{n+u,q^k}},$$

and obtain the ith q-factorial moment $E([U_n]_{i,q^k})$, by using the negative q-Vandermonde formula (1.12).

(c) Use expression (1.61).

4.15 (a) Consider the event A_i of drawing a black ball at the ith q-drawing, for $i = 1, 2, \ldots, n$, and derive the probability

$$\begin{aligned} &P(A_{i_1}A_{i_2}\cdots A_{i_k}A'_{i_{k+1}}\cdots A'_{i_n}) \\ &= \frac{[s]_q^k q^{s(n-k)}}{[r+s+n-1]_{n,q}}[r+i_{k+1}-1]_q[r+i_{k+2}-1]_q\cdots[r+i_n-1]_q. \end{aligned}$$

The probability

$$\begin{aligned} P(X_n = k) = {} & \frac{[s]_q^k q^{s(n-k)}}{[r+s+n-1]_{n,q}} \\ & \times \sum [r+i_{k+1}-1]_q[r+i_{k+2}-1]_q\cdots[r+i_n-1]_q, \end{aligned}$$

where the summation is extended over all $(n-k)$-combinations $\{i_{k+1}, i_{k+2}, \ldots, i_n\}$ of the n positive integers $\{1, 2, \ldots, n\}$, may be obtained, by putting $j_m = i_{k+m} - 1$, for $m = 1, 2, \ldots, n-k$, and using (1.37) together with (1.30), as

$$P(X_n = k) = \frac{[s]_q^k q^{(s-1)(n-k)}|s_q(n,k;r)|}{[r+s+n-1]_{n,q}}, \quad k = 0, 1, \ldots, n.$$

(b) Use the relation

$$P(T_k = n) = P(X_{n-1} = k - 1)p_n,$$

where $p_n = P(A_n) = [s]_q/[r+s+n-1]_q$ to obtain the probability function

$$P(T_k = n) = \frac{[s]_q^k q^{(s-1)(n-k)}|s_q(n-1, k-1; r)|}{[r+s+n-1]_{n,q}}, \quad n = k, k+1, \dots .$$

4.16 (a) The probabilities of drawing a black or a white ball at any q-drawing, given that j black balls are drawn in the previous q-drawings, are given by

$$p_j = \frac{q^{r+j}[s-j]_q}{[r+s]_q} \quad \text{and} \quad q_j = 1 - p_j = \frac{[r+j]_q}{[r+s]_q},$$

for $j = 0, 1, \dots, s$, respectively. Use these probabilities to get a recurrence relation for the probability function $P(X_n = k)$, $k = 0, 1, \dots, n$. Then, deduce for the double sequence

$$c_{n,k} = \frac{[r+s]_q^n}{[s]_{k,q}} q^{-\binom{k}{2}-rk} P(X_n = k), \quad k = 0, 1, \dots, n, \quad n = 0, 1, \dots,$$

a triangular recurrence relation, which implies $c_{n,k} = S_q(n, k; r)$ and so

$$P(X_n = k) = \frac{q^{\binom{k}{2}+rk} S_q(n, k; r)[s]_{k,q}}{[r+s]_q^n}, \quad k = 0, 1, \dots, n.$$

(b) Use the relation

$$P(T_k = n) = P(X_{n-1} = k - 1)p_{k-1},$$

where $p_{k-1} = q^{r+k-1}[s-k+1]_q/[r+s]_q$, to obtain the probability function

$$P(T_k = n) = \frac{q^{\binom{k}{2}+rk} S_q(n-1, k-1; r)[s]_{k,q}}{[r+s]_q^n}, \quad n = k, k+1, \dots .$$

4.17 The first-step transition probabilities are given by

$$p_{k,k-1} = \frac{[k]_q}{[m]_q}, \quad k = 1, 2, \dots, m, \quad p_{k,k+1} = \frac{q^k[m-k]_q}{[m]_q}, \quad k = 0, 1, \dots, m-1,$$

with $0 < q < 1$ or $1 < q < \infty$. Use these probabilities to get for the stationary probability function $f(x) = P(X = x) = \lim_{n\to\infty} P(X_n = x)$, $x = 0, 1, \dots, m$, the system of equations

$$f(0) = \frac{[1]_q}{[m]_q} f(1),$$

$$f(x) = \frac{q^{x-1}[m - x+1]_q}{[m]_q} f(x-1) + \frac{[x+1]_q}{[m]_q} f(x+1), \quad x = 1, 2, \ldots, m - 1,$$

$$f(m) = \frac{q^{m-1}}{[m]_q} f(m-1).$$

Multiply both members of these equations by $[m]_q$ and use the relation $[m]_q = [x]_q + q^x[m-x]_q$ to get the recurrence relation

$$[x+1]_q f(x+1) - q^x[m-x]_q f(x) = [x]_q f(x) - q^{x-1}[m-x+1]_q f(x-1),$$

for $x = 1, 2, \ldots, m-1$, with initial condition $[1]_q f(1) - [m]_q f(0) = 0$. Apply it repeatedly and use the q-binomial formula to derive the expression

$$f(x) = \begin{bmatrix} m \\ x \end{bmatrix}_q \frac{q^{\binom{x}{2}}}{\prod_{i=1}^{m}(1+q^{i-1})}, \quad x = 0, 1, \ldots, m.$$

4.18 The first-step transition probabilities are given by

$$p_{k,k-1} = \frac{[k]_q}{[m]_q} \frac{1}{1+\theta}, \quad k = 1, 2, \ldots, m,$$

$$p_{k,k} = \frac{[k]_q}{[m]_q} \frac{\theta}{1+\theta} + \frac{q^k[m-k]_q}{[m]_q} \frac{1}{1+\theta}, \quad k = 0, 1, \ldots, m,$$

and

$$p_{k,k+1} = \frac{q^k[m-k]_q}{[m]_q} \frac{\theta}{1+\theta}, \quad k = 0, 1, \ldots, m-1,$$

for $0 < \theta < \infty$ and $0 < q < 1$ or $1 < q < \infty$. Use these probabilities to get for the stationary probability function $f(x) = P(X = x) = \lim_{n\to\infty} P(X_n = x)$, $x = 0, 1, \ldots, m$, the system of equations

$$f(0) = \frac{1}{1+\theta} f(0) + \frac{[1]_q}{[m]_q} \frac{1}{1+\theta} f(1),$$

$$f(x) = \frac{[m-x+1]_q}{[m]_q} \frac{\theta q^{x-1}}{1+\theta} f(x-1) + \left(\frac{[x]_q}{[m]_q} \frac{\theta}{1+\theta} + \frac{[m - x]_q}{[m]_q} \frac{q^x}{1+\theta} \right) f(x) + \frac{[x+1]_q}{[m]_q} \frac{1}{1+\theta} f(x+1),$$

for $x = 1, 2, \ldots, m-1$, and

$$f(m) = \frac{[1]_q}{[m]_q} \cdot \frac{\theta q^{m-1}}{1+\theta} f(m-1) + \frac{\theta}{1+\theta} f(m).$$

Multiply both members of these equations by $[m]_q$ and use the relation $[m]_q = [x]_q + q^x[m-x]_q$ to get the recurrence relation

$$[x+1]_q f(x+1) - [m-x]_q \theta q^x f(x) = [x]_q f(x) - [m-x+1]_q \theta q^{x-1} f(x-1),$$

for $x = 1, 2, \ldots, m-1$, with initial conditions $[1]_q f(1) - [m]_q \theta f(0) = 0$ and $\theta q^{m-1} f(m-1) - [m]_q f(m) = 0$. Apply it repeatedly and use the q-binomial formula to derive the expression

$$f(x) = \begin{bmatrix} m \\ x \end{bmatrix}_q \frac{\theta^x q^{\binom{x}{2}}}{\prod_{i=1}^{m}(1+\theta q^{i-1})}, \qquad x = 0, 1, \ldots, m.$$

4.19 (a) Set $u = n - x$ and $v = m - n - 1$ in the q-Cauchy's formula,

$$\begin{bmatrix} u+v \\ m \end{bmatrix}_{q^{-1}} = \sum_{k=0}^{m} q^{-k(v-m+k)} \begin{bmatrix} v \\ m-k \end{bmatrix}_{q^{-1}} \begin{bmatrix} u \\ k \end{bmatrix}_{q^{-1}},$$

to find the required combinatorial identity.

(b) Use the relation

$$E\left(\begin{bmatrix} n - X_n \\ k \end{bmatrix}_{q^{-b}} \right) = E\left(\begin{bmatrix} Y_n \\ k \end{bmatrix}_{q^{-b}} \right),$$

where Y_n is the number of failures in a sequence of independent Bernoulli trials, with probability of failure at the ith trial, given that $j-1$ failures occur in the $i-1$ previous trials, given by

$$q_{i,j} = 1 - p_{i,i-j} = q^{-(a-b)(i-1)+b(j-1)+c}, \quad j = 1, 2, \ldots, i, \quad i = 1, 2, \ldots,$$

to derive the expression

$$E\left(\begin{bmatrix} n - X_n \\ k \end{bmatrix}_{q^{-b}} \right) = q^{(2b-a)\binom{k}{2}+ck} \begin{bmatrix} n \\ k \end{bmatrix}_{q^{b-a}}.$$

Then, use it together with the combinatorial identity to obtain the required expression of the q-binomial moments of X_n.

CHAPTER 5

5.1 Derive the mean and variance of $\overline{X}_n$ as

$$E(\overline{X}_n) = \overline{\mu}_n, \quad V(\overline{X}_n) = \frac{1}{n^2}\sum_{i=1}^{n} V(X_i) + \frac{2}{n^2}\sum_{j=2}^{n}\sum_{i=1}^{j-1} C(X_i, X_j)$$

and use the assumptions $V(X_i) = \sigma^2 \leq c < \infty$ and $C(X_i, X_j) < 0$, to deduce the limiting expression $\lim_{n\to\infty} V(\overline{X}_n) = 0$. Then, apply Chebyshev's inequality to conclude the convergence of sequence $\overline{X}_n - \overline{\mu}_n$, $n = 1, 2, \ldots$, to zero.

5.2 Work as in Exercise 5.1 and use the expression

$$V(\overline{X}_n) = \frac{1}{n^2}\sum_{i=1}^{n} V(X_i) + \frac{2}{n^2}\sum_{j=2}^{n}\sum_{i=1}^{j-1} C(X_i, X_j)$$

together with

$$\sum_{r=1}^{n-1} \log r \leq n(\log n - 1) + 2(\log 2 - 1),$$

to show that $\lim_{n\to\infty} V(\overline{X}_n) = 0$. Then, apply Chebyshev's inequality to conclude the convergence of the sequence $\overline{X}_n - \overline{\mu}_n$, $n = 1, 2, \ldots,$ to zero.

5.3 Work as in Exercise 5.1 and use the expression

$$V(\overline{X}_n) = \frac{1}{n^2}\sum_{i=1}^{n} V(X_i) + \frac{2}{n^2}\sum_{j=2}^{n}\sum_{i=1}^{j-1} C(X_i, X_j)$$

together with

$$C(X_i, X_j) = 0, \quad i = 1, 2, \ldots, j-2, \quad j = 3, 4, \ldots, n,$$

and

$$C(X_i, X_j) =\leq \frac{1}{2}[V(X_{j-1}) + V(X_j)] \leq c,$$

to show that $\lim_{n\to\infty} V(\overline{X}_n) = 0$. Then, apply Chebyshev's inequality to conclude the convergence of the sequence $\overline{X}_n - \overline{\mu}_n$, $n = 1, 2, \ldots,$ to zero.

5.4 Consider the number U_j of failures between the $(j-1)$th and the jth success, for $j = 1, 2, \ldots, n$ and derive the expected value of $W_n = \sum_{j=1}^{n} U_j$ as

$$E(W_n) = \sum_{j=1}^{n} E(U_j) = \frac{1}{1-q}\sum_{j=1}^{n} \frac{q^{r+j-1}}{[r+j-1]_q} = \frac{h_{n+r-1,q} - h_{r-1,q}}{1-q},$$

where $h_{m,q} = \sum_{i=1}^{m} q^i/[i]_q$ is the incomplete q-harmonic series, for which $\lim_{m\to\infty} h_{m,q} = -l_q(1-q)$, the q-logarithmic function. Then, apply the law of large numbers.

5.5 (a) Consider the zero–one Bernoulli random variable $X_i = 1$, if the ball drawn from urn u_i is white and $X_i = 0$ if it is black, for $i = 1, 2, \ldots, n$, and derive the expected value of $S_n = \sum_{i=1}^{n} X_i$ as

$$E(S_n) = (1-q)n + h_{n,q}, \quad h_{n,q} = \sum_{i=1}^{n} \frac{q^i}{[i]_q}.$$

Then, apply the law of large numbers.

(b) Derive the variance of $S_n = \sum_{i=1}^{n} X_i$ as

$$V(S_n) = q(1-q)n + qh_{n,q} - h_{n,q}(2), \quad h_{n,q}(2) = \sum_{i=1}^{n} \frac{q^i}{[i]_q^2}$$

and apply the central limit theorem.

5.6 Derive the expected value and variance of $X_n = \sum_{i=1}^{n} Z_i$ as

$$E(X_n) = (1-q^s)n + (1-q)\{h_{n+r+s-1,q}(1) - h_{r+s-1,q}(1)\}$$

and

$$V(X_n) = q^s(1-q^s)n + (1-q)q^s\{h_{n+r+s-1,q}(1) - h_{r+s-1,q}(1)\} \\ - [s]_q^2\{h_{n+r+s-1,q}(2) - h_{r+s-1,q}(2)\},$$

where

$$h_{m,q}(k) = \sum_{j=1}^{m} \frac{q^j}{[j]_q^k}, \quad 0 < q < 1, \quad k \geq 1, \quad m = 1, 2, \ldots,$$

is the incomplete q-zeta function, which for $m \to \infty$ converges to the q-zeta function

$$h_q(k) = \sum_{j=1}^{\infty} \frac{q^j}{[j]_q^k}, \quad 0 < q < 1, \quad k \geq 1\,.$$

In particular

$$h_q(1) = \sum_{j=1}^{\infty} \frac{q^j}{[j]_q} = -l_q(1-q), \quad 0 < q < 1,$$

with $l_q(1-t)$ the q-logarithmic function. Note that $\lim_{n\to\infty} V(X_n) = \infty$ and conclude the required limiting relation.

5.7 Derive the expected value and variance of $X_n = \sum_{i=1}^{n} Z_i$ as

$$E(X_n) = (1-q)n + \theta \sum_{i=1}^{n} \frac{q^i}{(1-\theta q^i)/(1-q)}$$

and

$$V(X_n) = q(1-q)n + \theta q \sum_{i=1}^{n} \frac{q^i}{(1-\theta q^i)/(1-q)} - \theta \sum_{i=1}^{n} \frac{q^i}{(1-\theta q^i)^2/(1-q)^2}.$$

Use the inequalities

$$\sum_{i=1}^{n} \frac{q^i}{(1-\theta q^i)/(1-q)} \leq \sum_{j=1}^{n} \frac{q^j}{[j]_q}, \quad \sum_{i=1}^{n} \frac{q^i}{(1-\theta q^i)^2/(1-q)^2} \leq \sum_{j=1}^{n} \frac{q^j}{[j]_q^2}$$

to show that

$$\lim_{n\to\infty} \{E(X_n) - (1-q)n\} \simeq -l_q(1-q),$$

and

$$\lim_{n\to\infty} \{V(X_n) - q(1-q)n\} \simeq -\theta q l_q(1-q) - \theta h_q(2),$$

with $h_q(2) = \sum_{i=1}^{\infty} q^i/[i]_q^2 < \infty$ and $\lim_{n\to\infty} V(X_n) = \infty$, and conclude that

$$\lim_{n\to\infty} P\left(\frac{X_n - (1-q)n}{\sqrt{q(1-q)n}} \leq z\right) = \Phi(z),$$

where $\Phi(z)$ is the distribution function of the standard normal distribution.

5.8 (a) Work as in Remark 5.1.

(b) Work as in the proof of Theorem 5.6.

5.9 (a) Work as in Remark 5.1.

(b) Work as in the proof of Theorem 5.6.

REFERENCES

Andrews, G. E. (1976) *The Theory of Partitions*. Addison-Wesley, Reading, MA.

Andrews, G. E., Askey, R. and Roy, R. (1999) *Special Functions*. Cambridge University Press, Cambridge, NY.

Andrews, G. E., Crippa, D. and Simon, K. (1997) q-Series arising from the study of random graphs. *SIAM J. Discrete Math.* **10**, 41–56.

Balakrishnan, N. and Nevzorov, V. B. (1997) Stirling numbers and records. In: Balakrishnan, N. (Ed.), *Advances in Combinatorial Methods and Applications to Probability and Statistics* Birkhäuser, Boston, MA, pp. 189–200.

Barakat, R. (1985) Probabilistic aspects of particles transiting a trapping field: an exact combinatorial solution in terms of Gauss polynomials. *J. Appl. Math. Phys.* **36**, 422–432.

Benkherouf, L. and Alzaid, A. A. (1993) On the generalized Euler distribution. *Statist. Probab. Lett.* **18**, 323–326.

Benkherouf, L. and Bather, J. A. (1988) Oil exploration: sequential decisions in the face of uncertainty. *J. Appl. Probab.* **25**, 529–543.

Bickel, T., Galli, N. and Simon, K. (2001) Birth processes and symmetric polynomials. *Ann. Comb.* **5**, 123–139.

Blomqvist, N. (1952) On an exhaustion process. *Skand. Akt.* **35**, 201–210.

Borenius, G. (1953) On the statistical distribution of mine explosions. *Skand. Akt.* **36**, 151–157.

Carlitz, L. (1933) On Abelian fields. *Trans. Am. Math. Soc.* **35**, 112–136.

Carlitz, L. (1948) q-Bernoulli numbers and polynomials. *Duke Math. J.* **15**, 987–1000.

Discrete q-Distributions, First Edition. Charalambos A. Charalambides.

Cauchy, A.-L. (1843) Mémoire sur les fonctions dont plusieurs valuers sont liées entre elles par une équation linéaire, et sur deverses transformations de produits composés d' un nombre indéfini de facteurs, *Comptes Rendus de l'Académie des Sciences Paris* vol. XVII, p. 523, *Oeuvres de Cauchy*, 1re, VIII, Gauthier-Villars, Paris, 1893, 42–50.

Charalambides, Ch. A. (1996) On the q-differences of the generalized q-factorials. *J. Statist. Plann. Inference* **54**, 31–43.

Charalambides, Ch. A. (2002) *Enumerative Combinatorics*. Chapman & Hall/CRC, Boca Raton, FL.

Charalambides, Ch. A. (2004) Non-central generalized q-factorial coefficients and q-Stirling numbers. *Discrete Math.* **275**, 67–85.

Charalambides, Ch. A. (2005a) Moments of a class of discrete q-distributions. *J. Statist. Plann. Inference* **135**, 64–76.

Charalambides, Ch. A. (2005b) *Combinatorial Methods in Discrete Distributions*. John Wiley & Sons, Inc., Hoboken, NJ.

Charalambides, Ch. A. (2007) Distributions of record statistics in a geometrically increasing population. *J. Statist. Plann. Inference* **137**, 2214–2225.

Charalambides, Ch. A. (2009) Distributions of record statistics in a q-factorially increasing population. *Commun. Statist. Theory Methods* **38**, 1–14.

Charalambides, Ch. A. (2010a) The q-Bernstein basis as a q-binomial distribution. *J. Statist. Plann. Inference* **140**, 2184–2190.

Charalambides, Ch. A. (2010b) Discrete q-distributions on Bernoulli trials with geometrically varying success probability. *J. Statist. Plann. Inference* **140**, 2355–2383.

Charalambides, Ch. A. (2012a) A q-Pólya urn model and the q-Pólya and inverse q-Pólya distributions. *J. Statist. Plann. Inference* **142**, 276–288.

Charalambides, Ch. A. (2012b) On the distributions of absorbed particles in crossing a field containing absorption points. *Fund. Inform.* **117**, 147–154.

Charalambides, Ch. A. and Papadatos, N. (2005) The q-factorial moments of discrete q-distributions and a characterization of the Euler distribution. In: Balakrishnan, N., Bairamov, I. G. and Gebizlioglu, O. L. (Eds), *Advances on Models, Characterizations and Applications*, Chapman & Hall/CRC Press, Boca Raton, FL, pp. 57–71.

Chaudry, T. W. (1962) Frank Hilton Jackson. *J. Lond. Math. Soc.* **37**, 126–128.

Christiansen, J. S. (2003) The moment problem associated with the Stieltjes-Wigert polynomials. *J. Math. Anal. Appl.* **277**, 218–245.

Comtet, L. (1972) Nombres de Stirling généraux et fonctions symmetriques. *C. R. Acad. Sci., Paris*, **275**, 747–750.

Crippa, D. and Simon, K. (1997) q-Distributions and Markov processes. *Discrete Math.* **170**, 81–98.

Crippa, D., Simon, K. and Trunz, P. (1997) Markov processes involving q-Stirling numbers. *Combin. Probab. Comput.* **6**, 165–178.

Dubman, M. and Sherman, B. (1969) Estimation of parameters in a transient Markov chain arising in a reliability growth model. *Ann. Math. Statist.* **40**, 1542–1556.

Dunkl, C. F. (1981) The absorption distribution and the q-binomial theorem. *Commun. Statist. Theory Methods A* **10**, 1915–1920.

Euler, L. (1748) *Introductio in Analysin Infinitorum*. Marcum-Michaelem Bousquet, Lausannae.

Flajolet, P. (1985) Approximate counting: a detailed analysis. *BIT* **25**, 113–134.

Garsia, A. M. and Remmel, J. B. (1980) A combinatorial interpretation of the q-derangement and q-Laguerre numbers. *European J. Combin.* **1**, 47–59.

Gasper, G. and Rahman, M. (2004) *Basic Hypergeometric Series*. Second Edition, Cambridge University Press, Cambridge.

Gauss, C. F. (1863) *Werke*, Vol. **2**. Königliche Gesellschaft de Wissenschaften, Göttingen.

Goldman, J. and Rota, G.-C. (1970) On the foundations of conbinatorial theory IV: Finite vector spaces and Eulerian generating functions. *Stud. Appl. Math.* **49**, 239–258.

Goodman, T. M. T., Oruc, H. and Phillips, G. M. (1999) Convexity and generalized Bernstein polynomials. *Proc. Edinb. Math. Soc.* **42**, 179–190.

Gould, H. W. (1961) The q-Stirling numbers of the first and second kinds. *Duke Math. J.* **28**, 281–289.

Hahn, W. (1949) Über orthogonalpolynome, die q-Differenzengleichungen genügen. *Math. Nachr.* **2**, 4–34.

Hardy, G. H. (1940) *Ramanujan: Twelve Lectures on His Life and Works*, Cambridge University Press, Cambridge; reprinted by Chelsea, New York, 1978.

Heine, E. (1847) Untersuchungen über die Reihe $\cdots$. *J. Reine Angew. Math.* **34**, 285–328.

Heine, E. (1878) *Handbuch der Kugelfunctionen, Theorie und Anwendungen*, Vol. **1**, Reimer, Berlin.

Il'inskii, A. (2004) A probabilistic approach to q-polynomial coefficients, Euler and Stirling numbers I. *Matematicheskaya Fisika, Analiz. Geometriya* **11**, 434–448.

Il'inskii, A. and Ostrovska, S. (2002) Convergence of generalized Bernstein polynomials. *J. Approx. Theory* **116**, 100–112.

Jackson, F. H. (1910a) q-Difference equations. *Amer. J. Math.* **32**, 305–314.

Jackson, F. H. (1910b) On q-definite integrals. *Quart. J. Pure Appl. Math.* **41**, 193–203.

Jackson, F. H. (1951) Basic integration. *Q. J. Math.* **2**, 1–16.

Jacobi, C. G. J. (1846) Über einige der Binomialreihe analoge Reihen. *J. Reine Angew. Math.* **32**, 197–204.

Jing, S. C. (1994) The q-deformed binomial distribution and its behaviour. *J. Phys. A: Math. Gen.* **27**, 493–499.

Jing, S. C. and Fan, H. Y. (1993) q-deformed binomial state. *Phys. Rev. A* **49**, 2277–2279.

Kemp, A. (1987) A Poissonian binomial model with constrained parameters. *Naval Res. Logist.* **34**, 853–858.

Kemp, A. (1992a) Heine-Euler extensions of the Poisson distribution. *Commun. Statist. Theory Methods* **21**, 791–798.

Kemp, A. (1992b) Steady-state Markov chain models for the Heine and Euler distributions. *J. Appl. Probab.* **29**, 869–876.

Kemp, A. (1997) On modified q-Bessel functions and their statistical applications. In: Balakrishnan, N. (Ed.), *Advances in Combinatorial Methods and Applications to Probability and Statistics*, Birkhäuser, Boston, MA, pp. 451–463.

Kemp, A. (1998) Absorption sampling and the absorption distribution. *J. Appl. Probab.* **35**, 489–494.

Kemp, A. (2001) A characterization of a distribution arising from absorption sampling. In: Charalambides, Ch. A., Koutras, M. V. and Balakrishnan, N. (Eds.) *Probability and Statistical Models with Applications*, Chapman & Hall/CRC Press, Boca Raton, FL, pp. 239–246.

Kemp, A. (2002a) Existence conditions and properties for the generalized Euler family of distributions. *J. Statist. Plann. Inference* **101**, 169–178.

Kemp, A. (2002b) Certain q-analogues of the binomial distribution. *Sankhyā A* **64**, 293–305.

Kemp, A. (2003) Characterizations involving $U|(U + V = m)$ for certain discrete distributions. *J. Statist. Plann. Inference* **109**, 31–41.

Kemp, A. (2005) Steady-state Markov chain models for certain q-confluent hypergeometric distributions. *J. Statist. Plann. Inference* **135**, 107–120.

Kemp, A. and Kemp, C. D. (1991) Weldon's dice data revisited. *Amer. Statist.* **45**, 216–222.

Kemp, A. and Kemp, C. D. (2009) The q-cluster distribution. *J. Statist. Plann. Inference* **139**, 1856–1866.

Kemp, A. and Newton, J. (1990) Certain state-dependent processes for dichotomised parasite populations. *J. Appl. Probab.* **27**, 251–258.

Kemp, C. D. (1997) A q-logarithmic distribution. In: Balakrishnan, N. (Ed.), *Advances in Combinatorial Methods and Applications to Probability and Statistics*, Birkhäuser, Boston, MA, pp. 465–570.

Kupershmidt, B. A. (2000) q-Probability: 1. Basic discrete distributions. *J. Nonlinear Math. Phys.* **7**, 73–93.

Kyriakoussis, A. and Vamvakari, M. (2013) A q-analogue of the Stirling formula and a continuous limiting behaviour of the q-Binomial distribution-Numerical calculations. *Methodol. Comput. Appl. Probab.* **15**, 187–213.

Kyriakoussis, A. and Vamvakari, M. (2015) Heine process as a q-analogue of the Poisson process-Waiting and interarrival times. *Commun. Statist. Theory Methods.*

Louchard, G. and Prodinger, H. (2008) Generalized approximate counting revisited. *Theor. Comput. Sci.* **392**, 109–125.

Moritz, R. H. and Williams, R. C. (1988) A coin-tossing problem and some related combinatorics. *Math. Mag.* **61**, 24–29.

Nevzorov, V. B. (1984) Record times in the case of non-identically distributed random variables. *Theory Probab. Appl.* **29**, 845–846.

Newby, M. (1999) Moments and generating functions for the absorption distribution and its negative binomial analogue. *Commun. Statist. Theory Methods* **28**, 2935–2945.

Ostrovska, S. (2003) q-Bernstein polynomials and their iterates. *J. Approx. Theory* **123**, 232–255.

Ostrovska, S. (2006) On the Lupas q-analogue of the Bernstein operator. *Rocky Mountain J. Math.* **36**, 1615–1629.

Patil, G. P. and Seshadri, V. (1964) Characterization theorems for some univariate probability distributions. *J. R. Statist. Soc. Ser. B Stat. Methodol.* **26**, 286–292.

Phillips, G. M. (1997) Bernstein polynomials based on q-integers. *Ann. Numer. Anal.* **4**, 511–518.

Platonov, M. L. (1976) Elementary applications of combinatorial numbers in probability theory. *Theory Probab. Math. Statist.* **11**, 129–137.

Poisson, S. D. (1837) *Recherchés sur la Probabilité des Jugements en Matière Criminelle te en Matière Civile, Prècèdèes des Regles Générales du Calcul des Probabilités*, Bachelier, Imprimeur-Libraine pour des Mathematiques, la Physique, etc, Paris.

Rawlings, D. (1994a) Bernoulli trials and number theory. *Amer. Math. Monthly* **101**, 948–952.

Rawlings, D. (1994b) Limit formulas for q-exponetial functions. *Discrete Math.* **126**, 379–383.

Rawlings, D. (1997) Absorption processes: models for q-identities. *Adv. Appl. Math.* **18**, 133–148.

Rawlings, D. (1998) A probabilistic approach to some of Euler's number theoretic identities. *Trans. Amer. Math. Soc.* **350**, 2939–2951.

Sen A. and Balakrishnan, N. (1999) Convolution of geometrics and a reliability problem. *Statist. Probab. Lett.* **43**, 421–426.

Simon, K. (1988) Improved algorithm for transitive closure on acyclic digraphs. *Theor. Comput. Sci.* **58**, 325–346.

Simon, K., Crippa, D. and Collenberg, F. (1993) On the distribution of the transitive closure in a random acyclic digraph. In: Lengauer, T. (Ed.) *Lecture Notes in Computer Science*, Vol. **726**, 345–356.

Sylvester, J. J. (1882) A constructive theory of partitions in three acts, an interact and an exodion. *Amer. J. Math.* **5**, 251–330 (and ibid **6** (1884), 334-336); reprinted in *Collected Mathematical Papers of J. J. Sylvester*, Vol. 4, pp. 1–83. Cambridge University Press, London, 1912; reprinted by Chelsea, New York, 1974.

Tauber, S. (1962) On quasi-orthogonal numbers. *Amer. Math. Monthly*, **69**, 365–372.

Tauber, S. (1965) On generalized Lah-numbers. *Proc. Edinb. Math. Soc.*, **14**, 229–232.

Treadway, J. and Rawlings, D. (1994) Bernoulli trials and Mahonian statistics: a tale of two q's. *Math. Mag.*, **67**, 345–354.

Wigert, S. (1923) Sur les polynômes orthogonaux et l' approximation des fonctions continues. *Ark. Mat. Astron. Fys.*, **17**, 1–15.

Woodbury, M. A. (1949) On a probability distribution. *Ann. Math. Statist.* **20**, 311–313.

Yang, M. C. K. (1975) On the distributions of the inter-record times in an increasing population. *J. Appl. Probab.* **12**, 148–154.

Zacks, S. and Goldfard, D. (1966) Survival probabilities in crossing a field containing absorption points. *Naval. Res. Logist.* **13**, 35–48.

INDEX

Discrete q-Distributions, First Edition. Charalambos A. Charalambides.